*Ecology and Conservation of the
Diamond-backed Terrapin*

ECOLOGY and CONSERVATION of the DIAMOND-BACKED TERRAPIN

EDITED BY

Willem M. Roosenburg and Victor S. Kennedy

JOHNS HOPKINS
UNIVERSITY PRESS | BALTIMORE

© 2018 Johns Hopkins University Press
All rights reserved. Published 2018
Printed in the United States of America on acid-free paper
9 8 7 6 5 4 3 2 1

Johns Hopkins University Press
2715 North Charles Street
Baltimore, Maryland 21218-4363
www.press.jhu.edu

Library of Congress Cataloging-in-Publication Data

Names: Roosenburg, Willem M., editor. | Kennedy, Victor S., editor.
Title: Ecology and conservation of the diamond-backed terrapin /
    Willem M. Roosenburg and Victor S. Kennedy, eds.
Description: Baltimore : Johns Hopkins University Press, 2018. |
    Includes bibliographical references and index.
Identifiers: LCCN 2018000279 | ISBN 9781421426266
    (hardcover : alk. paper) | ISBN 1421426269
    (hardcover : alk. paper) | ISBN 9781421426273
    (electronic) | ISBN 1421426277 (electronic)
Subjects: LCSH: Diamondback terrapin—Ecology. | Diamondback
    terrapin—Conservation.
Classification: LCC QL666.C547 E26 2018 | DDC 597.92/59—dc23
LC record available at https://lccn.loc.gov/2018000279

A catalog record for this book is available from the British Library.

*Special discounts are available for bulk purchases of this book. For
more information, please contact Special Sales at 410-516-6936 or
specialsales@press.jhu.edu.*

Johns Hopkins University Press uses environmentally friendly
book materials, including recycled text paper that is composed of
at least 30 percent post-consumer waste, whenever possible.

To Bill and El, who showed us that science and friends make good partners, and to Kate and Deb for unstinting support

Contents

Preface

THE DIAMOND-BACKED TERRAPIN HAS HAD AN IMPORTANT ROLE in North American history as a human food source that developed great economic value but quickly waned due to overexploitation. Watermen and entrepreneurs, first in Chesapeake Bay and then throughout the species' range, made the terrapin an epicurean delicacy in the late nineteenth and early twentieth centuries. The resulting harvest pressure on a species with a life history of slow growth, delayed maturity, and low fecundity resulted in range-wide population declines, starting in Chesapeake Bay and expanding to other coastal regions, as demand outpaced supply. By the end of the nineteenth century, stocks were depleted greatly, and Prohibition and the elimination of sherry as the key ingredient in terrapin soup helped reduce the fishery. In the modern day, terrapins encounter numerous anthropogenic threats ranging from toxins to capture in commercial crab pots and habitat loss due to development of coastal regions. However, there is some progress, as most states have recognized the unsustainability of terrapin harvesting and have banned commercial take. Thankfully, through research and outreach, terrapin conservation is making slow and steady headway that generates optimism among those who study these turtles that they will be here for future generations. This book reviews the rapidly growing body of literature that contributes to our understanding of terrapins throughout their range. We hope it will serve as a reference for terrapin enthusiasts and researchers, but also for those interested in turtles in general.

The book also represents a shared life-long interest in Chesapeake Bay, its denizens, and their natural history. The terrapin was a species much celebrated but little known in the Bay until Willem Roosenburg began graduate school at the University of Pennsylvania in 1986. But the beginnings of this book precede his start in graduate school by 20 years, when Vic Kennedy was himself a graduate student working at the Hallowing Point Field Station of the University of Maryland Chesapeake Biological Laboratory. Willem Roosenburg, Sr., collaborated with Vic on several studies of the effects of thermal addition to the Patuxent estuary, and Willem Jr. accompanied his dad to work during summer vacations to fish, crab, and participate in the science underway at the field station. Willem and Vic remained in touch over the years as their careers unfolded, eventually agreeing to unite Willem's expertise on terrapins with Vic's experience serving as a volume editor to produce this book.

Thirty years ago, in the summer of 1987, Willem began one of the few long-term studies of terrapins and the first of its kind in Chesapeake Bay. Although in most other areas of their range terrapins are cryptic and prove challenging to catch, in Chesapeake Bay they were common in some areas and proved to be relatively easy to catch with the right gear. Over the years he has worked to develop the terrapin as a model species for the study of turtle life history, temperature-dependent sex determination, chelonian conservation, and more.

Many individuals have been instrumental in the development and completion of this book. Vincent Burke, formerly of Johns Hopkins University Press, came to Willem with the idea and the desire to publish a book on terrapins. His guidance was instrumental in initiating this process, which Tiffany Gasbarrini saw through to completion. We are grateful to the authors who wrote chapters and were patient with our comments and editorial changes. We also are grateful to the many reviewers who provided detailed comments for individual chapters, including Kimberly Andrews, Aaron Baxter, Donna Marie Bilcovic, Ruth Boettcher, Bruce Bury, John Byrd, Kristen Cecala, Andrew Coleman, Justin Congdon, Paul Converse, Jon Costanzo, John Davenport, Peter Paul van Dijk, Michael Dorcas, Whit Gibbons, George Heinrich, John Iverson, Dale Jackson, Peter King, Chris Manis, Suzanne McGaugh, Brian Mealey, Timothy Muir, James Parham, Greg Pauly, Turk Rhen, Christopher Rowe, Rich Seigel, Kyle Selcer, Will Selman, Arun Sethuraman, Brad Shaffer, James Spotila, David Steen, Tony Tucker, Gordon Ultsch, and Richard Vogt. Copy editor Linda Strange did an excellent job of ensuring that issues of style and format were attended to, as well as ferreting out errors and inconsistencies in the narratives. We thank Elizabeth Freedlander, of Horn Point Laboratory, who helped raise funds to support publication of this book and the private donors that contributed to helping defer publishing costs. Finally, we are grateful to our families—Deb, Kate, Dirk, and Selinde—who have been patient and supportive as we usurped time from other activities to complete the text.

*Ecology and Conservation of the
Diamond-backed Terrapin*

1

Introduction and History

J. WHITFIELD GIBBONS

Many researchers proclaim that their study species has qualities that set it apart from other species and make it the "ideal model" for research pursuits. Some even define their study organism as "unique," which becomes meaningless when considering that every species on earth is unique when we take into account its evolutionary history and current genetic makeup coupled with its geographic range, ecology, and life history traits. Nonetheless, a primary feature that sets the diamond-backed terrapin apart from all other turtles is that it resides permanently in brackish water. No other turtle can claim more than a passing association with the relatively thin interface between the full saltwater of the ocean and the freshwater lakes and tributaries associated with the mainland on every turtle-inhabited island, isthmus, and continent. Other turtles also are called "terrapins," but of the 335 recognized extant species, all but the diamond-backed terrapin live mostly on one side or the other of the brackish ecotone. Thus, the diamond-backed terrapin truly is unique among turtles in making this habitat its home.

The history of the species can be viewed from many perspectives. Its evolutionary history began sometime in the Miocene, a half-dozen to dozen million years ago, when some freshwater ancestor began to thrive in an unexploited niche that excluded competitors. Its greatest biological competitors resided in fresh or marine waters but not in between. The historical importance of the species to humans began a few thousand years ago, when Native Americans living along the mid-Atlantic coastal regions of North America first began using terrapins as a food source, along with freshwater turtles on the mainland. When European settlers reached the New World, having never experienced an estuarine turtle, they simply called them "torope," the moniker bestowed by the Algonquian Indians of the region.

The scientific history of the species began in 1793, the year that George Washington held his first cabinet meeting as president of the United States and Johann David Schoepff described the species and gave it its initial binomial, *Testudo terrapin*. In 1844, the year James K. Polk defeated Henry Clay for the presidency, John Edward Gray placed the terrapin in the correct genus, *Malaclemys*. But turtle taxonomy of the times was muddled and convoluted, and Gray's contribution distinguished terrapins from *Graptemys* but did not recognize *Malaclemys* as a monotypic genus. The current scientific name, *Malaclemys terrapin*, was not used until Outram Bangs put the two together in 1896, the last year of Grover Cleveland's second term in office. How many of these political figures living in Maryland and Virginia dined on terrapins is unknown, but Cleveland's first term was marked by the publication of *The White House Cookbook*. Among the featured recipes was "Stewed Terrapin with Cream," a foreboding of the conservation problems that lay ahead for the diamond-backed terrapin. Terrapins are no longer thrown alive into a pot of boiling water, as recommended in the recipe, but drowning in a crab pot could be viewed as an equally gruesome fate. Have we solved one problem of terrapin conservation while being unable to curtail another?

Our current knowledge of the ecology and conservation of terrapins began in the early 1900s with classic papers by Robert Coker and Samuel Hildebrand that addressed both the natural history of terrapins and their commercial use. Since the pre-World War II work by Coker, Hildebrand, and others, most research on terrapins has been published in the two decades since the mid-1990s. The longest continual population studies were initiated on Kiawah Island, South Carolina, in 1983. Other studies soon began to encompass a broader portion of the geographic range. The breadth of questions on ecology, physiology, and behavior expanded, and the work is now escalating at an impressive rate. Conservation remains a central theme in many of these efforts—as it should for such elegant animals that face so many human-caused

threats. This book adds considerably to our knowledge of the diamond-backed terrapin and discloses how many capable researchers are now involved in acquiring the critical facts needed for its conservation.

One of the first practical questions someone doing ecological research might ask is, how do you catch a terrapin and what do you do once you have it in hand? The history of techniques for catching turtles extends into the distant past to the Old World, before the earliest settlers arrived on the banks of the Atlantic and Gulf of Mexico where terrapins thrived. No single capture technique can be deemed "the best," for it depends on season and circumstances of when turtles will be where. The collection of thousands of adults during the Kiawah Island studies speaks to the advantage of basic seining and setting trammel nets across tidal creeks. But many other approaches have been effectively employed, before and after. The cumulative number of techniques for catching turtles that have been published in books, papers, and scientific notes, when added to those that have been used and not published, is enormous. After capturing a turtle, the researcher must of course focus on appropriate measurements, which may include the use of digital photography and collection of molecular data. Virtually all standardized techniques that a researcher needs to know for studying terrapins are referred to in this book.

How did terrapins come to be one of the few tetrapods in the world to become permanent residents in a singular habitat that has existed on earth since elevated land masses first appeared? How does an organism live in the brackish interface between freshwater runoff and the sea? An obvious consideration for organisms living in habitats with high salinity is the necessity for physiological adjustments to allow salt removal. Osmoregulation by terrapins has thus been of some research interest. Yet, approximately half of the published research focusing on salt elimination and electrolyte balance expressly in *M. terrapin* was completed before the mid-1970s. Chapter 9, on osmoregulation, draws on these earlier works and more recent publications to place terrapins in the context of other reptiles for which salt elimination is a critical physiological issue. The consolidation of what was previously known, complemented by current research findings, will serve as a baseline for future research into the behavioral and physiological strategies required for *M. terrapin* to survive in an estuarine habitat.

Into how many genetically identifiable units can *M. terrapin* be partitioned? Many North American turtles that systematists accepted as single species for decades, if not centuries, have now been separated into two or more species by warriors armed with genetic tools. *Malaclemys terrapin* has the opposite situation: no one is proposing that more than one species is represented. In fact, separation into the seven putative subspecies traditionally reported is now challenged—a refreshing move toward lumping rather than splitting. Discussions of genetic variation, phylogeography, paleontology, and taxonomy and systematics in this book all contribute to current and future understanding of the evolutionary history of this linearly distributed species that has no natural barriers between present-day populations. Collectively, all topics pertinent to the phylogeny, ecology, and conservation of the diamond-backed terrapin are addressed herein.

Geographic variation is evident among terrapins—hence the early descriptions of subspecies—but patterns of consistency and regional conformity have been elusive, even for the few genetic profiles that have been accomplished. One of the enigmas surrounding the distribution pattern of terrapins that continue to intrigue investigators is, how did terrapins get to Bermuda? This issue is addressed and opinions, supported by data, are offered in this book. In taking a position that terrapins reached Bermuda by natural dispersal, should we also ask why terrapins are not found in Cuba's much closer tidal brackish habitats, to which US salt marsh snakes (*Nerodia clarkii*) successfully emigrated? The Bermuda terrapin case is not closed, and hints of titillating debates are on the horizon.

Like the life cycles of freshwater turtles in temperate regions everywhere, the basic life cycle of *M. terrapin* is one in which eggs are laid in early spring in a terrestrial nest dug by the female, with the young hatching in less than three months. Hatchlings usually emerge in autumn and sometimes remain in the nest cavity or in the surrounding habitat for the duration of fall and winter. In either case, after entering the aquatic habitat they begin feeding, approximately one year after being deposited as eggs; whether this is true for the southernmost populations in Florida remains equivocal. Growth rates are presumably influential in the attainment of maturity, and like many emydids and trionychids, the males in a region tend to reach maturity at a smaller size than females, which remain measurably larger as adults. Sexual dimorphism is unequivocal, but are growth rates among individuals within some populations faster than those in others, as observed in some turtles? If so, do differential growth rates affect age at maturity and ultimate adult size? Age-size relationships to maturity in the sexes are highly complex in turtles. Many ecological questions are expertly addressed in the chapters on ecology, including nesting and overwintering of hatchlings, foraging behavior and diet, environmental sex determination, and basic life history. Refining our understanding of these ecological factors for terrapins will be an exciting challenge.

Effective conservation programs historically have involved not only scientific study of the target species but complementary education of the public, who elect the officials that can champion a cause with legislation. Both approaches will be crucial to ensuring that productive populations of terrapins will persist, or in many cases recover, from Massachusetts to Texas. This book does an admirable job of covering both categories.

Among the conservation chapters is an excellent overview of the effect that overharvesting has had historically on these easy to find and catch animals (Chapter 13). Terrapins were the first reptile species in America whose steady and alarming decline in the wild can be attributed to commercial exploitation for food. An account of the astounding number of recipes for preparing terrapin and the list of restaurant menus on which they appeared provides sober testimony on how quickly a common, widespread species can decline in numbers when conservation programs and controls are not properly administered. Human disregard for the welfare of individual terrapins was commonplace, and the population status of these animals was not taken into consideration until their numbers and availability noticeably began to diminish. Even then, government programs that ultimately led to artificial propagation through terrapin farming were designed more to ensure a continuing supply of the popular dietary item than to lessen the rate of dwindling numbers and protect natural populations. Understanding the process of how public attitudes can drive a species from abundance to rarity in only a few decades is important. The terrapin story offers insight that can be applied to developing effective conservation practices for any overexploited species.

A conservation program that is scientifically based is key to the ultimate success of any species struggling under human-caused environmental threats, and the terrapin serves as a model for other species, especially turtles. Descriptions of the many dangers terrapins face because of commercial development and activities in coastal habitats are telling about the potential for continued declines. Collectively, these chapters set the stage for much-needed future research. Loss of terrapins as bycatch from both recreational and commercial crab trapping, as well as other fisheries activities, is clearly a key area of concern. Injuries and deaths caused by outboard motors in terrapin waters are documented and need attention. Road systems separating nesting habitats from aquatic areas are a bane to many turtle species, including terrapins. The value of examining geographic variation in the responses of terrapin populations to agricultural impacts, salt marsh development, and other human activities that vary regionally is also addressed and appropriately emphasized. Research leading to scientifically sound conservation approaches for beleaguered species such as terrapins must also include restorative efforts to replace and sustain what has been lost. Discussion of a variety of regulatory approaches, including harvest limitations, head-starting, and control of nesting area predation, provides the background necessary to pursue these conservation tactics.

Attention must be paid to all of these issues if effective conservation programs are to be implemented that will return terrapins to their natural state throughout their range and maintain them both ecologically and behaviorally. Finally, the importance of environmental education in concert with terrapin conservation programs is highlighted in Chapter 17.

The chapters on conservation threats that imperil the existence of terrapins throughout their range, coupled with recommendations for educational approaches, will serve as a vanguard for anyone wanting to aid in the protection of this prepossessing species. Continuation of the history of this venerable turtle will be marked by what is reported in this book, which will unequivocally be recognized as a milestone by those pursuing future studies on the diamond-backed terrapin's ecology and conservation.

ACKNOWLEDGMENTS

I appreciate the reviews of drafts of this chapter by Jeff Lovich, Willem Roosenburg, and Vic Kennedy.

PART I BIOLOGY AND ECOLOGY

2

Capture, Measurement, and Field Techniques

WILLEM M. ROOSENBURG

RUSSELL L. BURKE

Turtles are compelling animals to study because they are charismatic, many species are easy to catch, and their life histories are both diverse and unusual. Turtles also serve as model long-lived organisms and are the focus of several renowned long-term ecological studies. The turtle's shell allows the use of simple and inexpensive methods to mark and identify individuals (Cagle 1939) and is an excellent platform for attaching recording and transmitting devices. Simple scientific tools (balances, calipers, and a file) combined with hard work and persistence have resulted in many excellent population studies of a variety of turtle species (reviewed in Wilbur and Morin 1988), and more recent studies include the diamond-backed terrapin (*Malaclemys terrapin*). However, the population ecology of terrapins has unique and interesting challenges when compared with that of most freshwater turtles. To supplement two recent reviews of techniques for studying reptiles (McDiarmid et al. 2012; Graeter et al. 2013), we review here the specialized field techniques used with terrapins.

Comprehensive studies (those that catch more than just nesting females) require substantial investment in boats and traps and the tolerance of pesky or potentially dangerous bycatch that can include alligators, alligator gar, blue crabs, sharks, sting rays, and sea nettles. Thorough mark-recapture studies should employ a variety of capture techniques to minimize potential bias associated with different trapping methods (Roosenburg et al. 1997) and to sample a range of estuarine habitats, life cycle stages, and sexes (Roosenburg et al. 1999). Daily tidal fluctuation requires precautions to ensure that terrapins have access to air if they remain in traps for more than a few hours. Because of the variation in tidal amplitude (from <1 to >3 m) and the diversity of habitat/marsh structure throughout the terrapin's range, techniques that work well in some areas are not as effective in others. Finally, terrapins can be sympatric with other turtle species, including sea turtles, snapping turtles, kinosternids, and emydids, that may be caught in the same gear used for terrapins.

We describe a set of methods for catching terrapins and a standardized set of measurements to facilitate comparability of data among studies. We then describe successful techniques to catch, mark, measure, and track terrapins; highlight some methodological issues that require attention; and introduce some new techniques with great potential for terrapin research.

Trapping and Capture Methods

Checking traps and catching turtles can be the most exciting part of a population study. Traps that catch terrapins also reveal the diversity of their habitat by catching crabs, fish, other turtle species, and even sharks and rays. We describe here ways to catch terrapins (see Chapter 13 for an account of methods used in the early fisheries). Although our list is fairly comprehensive, we recommend asking local commercial fishermen both where and how to catch terrapins when initiating a study. Their experience can help identify the techniques that work best at your study site.

Fyke or Hoop Nets

Fyke nets are effective devices to catch riverine turtles in large open bodies of water or near nesting beaches (Vogt 1980). They consist of a series of mesh nets (a.k.a. hedgings or wings) attached to several hoops (fiberglass is preferred), nested within one another, and usually a double-throated mesh trap. They are most effectively configured with a central lead deployed perpendicular to the shoreline starting near shore, as well as two wings set at a 45° angle to the central lead, forming a funnel to the trap (cod end) (Fig. 2.1, A). Most mobile estuarine species, including terrapins, move parallel to the shoreline, and when they encounter an obstruction such as the central lead they move offshore to get around it, thus being directed to the cod end;

bait is not required. Fykes with two wings can be set parallel to the current in small tidal creeks, blocking the entire channel, but these will catch only with the tidal flow entering the funnel (Selman and Baccigalopi 2012; Selman et al. 2014).

Fyke nets require several modifications to catch terrapins. First, if the cod end is submerged, terrapins will drown when held underwater for more than 2 h in warm waters. Thus, a daily tidal amplitude of <1 m and a water depth of <2 m are ideal. Stakes or metal poles are preferred over the traditional anchors because they can hold the cod end up to help maintain an airspace. Tying 1 or 2 large

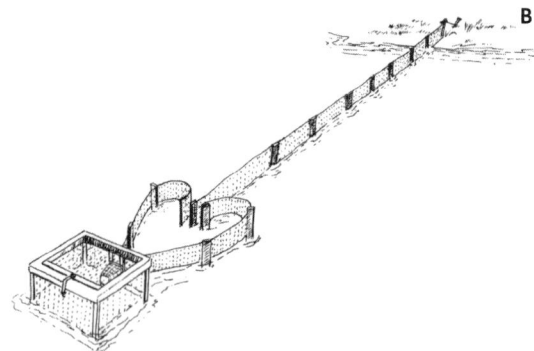

Fig. 2.1. (A) Fyke net. Stakes instead of anchors are used to hold the net in place. The float in the cod end of the net maintains a permanent air space so that terrapins have access to air. The net is set perpendicular to the shoreline, and the wings form a V shape that directs the catch to the cod end. (B) A bank trap (peeler pound) is similar in principle to a fyke net but with a net made of galvanized chicken wire. The wings are modified to a heart-shaped structure (the pound), and the cod end is a simple box that maintains an airspace above mean high tide. (B) Courtesy of Virginia Institute of Marine Science

floats in the cod end will also maintain the airspace. Second, a net mesh size of 2.5 cm should be used in areas with high blue crab densities because larger meshes entangle blue crabs, which damage the net, requiring mending. Dipping the webbing in net dip stiffens it and reduces crab damage. Floats and weights on the lead(s) keep the nets vertical in the water column, both preventing terrapins from swimming under or over the net and further reducing crab damage. Third, nets should be checked at least once a day to ensure that the airspace is available and to reduce stress on entrapped terrapins and other organisms; less frequent checking will increase bycatch mortality. Fyke nets can be easily moved between locations and should be cleaned and dried once every 5 to 10 days to prevent the buildup of estuarine fouling organisms, which reduces a net's effectiveness.

Bank Traps or Peeler Pounds

Bank traps or peeler pounds (Fig. 2.1, B) are essentially fyke nets made of galvanized chicken wire. They are not commercially available and require about 4 to 6 h to build. However, they are considerably cheaper than fyke nets and are immune to damage by blue crabs. They require a substantial effort to set and therefore are practical only in long-term trapping (>2 weeks) projects. Terrapin drowning is prevented by building the trap's box to be at least 30 cm above the mean high tide. Bank traps exploit the movement of animals parallel to the shoreline and thus do not need bait. Checking of bank traps requires a boat and a strong back or winch to lift the boxes.

Gill or Trammel Nets

Both gill and trammel nets entangle terrapins underwater and therefore must be attended while they are fished. They can catch terrapins in a variety of habitats (e.g., Seigel 1984; Simoes and Chambers 1999; Levesque 2000; Broyles 2010) but are most effective when terrapins are aggregated in open water near basking or nesting areas, or in tidal creeks during ebb tide (D. Owens, College of Charleston, pers. comm.). We have surrounded terrapin aggregations and then pulled the net toward shore like a seine or pounded the water with paddles or buckets to drive the terrapins toward the net. Nets should be removed within 2 h to prevent terrapins from drowning if they get entangled near the bottom. A 7.5 cm mesh is very effective for terrapins. These nets will generate bycatch of blue crabs and fish that can be difficult to untangle; in one instance, we caught 17 cow-nosed rays (*Rhinoptera bonasus*) along with 5 terrapins.

Turtle or Three-Ring Hoop Traps

Turtle traps work well with numerous species of freshwater turtles (McDiarmid et al. 2012) but are not as effective for terrapins. Traps typically are set in shallow water to maintain a permanent airspace and are baited with fish, crabs, or clams for terrapins (unbaited traps have caught terrapins in Barnegat Bay; H. Avery, Drexel University, pers. comm.). They can be successful in narrow, shallow marsh channels or while set submerged, but they must be checked every 3 h to avoid terrapin drowning. The traps catch better during the early spring when terrapins are emerging from their hibernacula and are attracted to bait.

Crab and Eel Pots

The ability of wire-mesh crab or eel pots to capture terrapins is of conservation concern (reviewed in Roosenburg 2004; see also Chapter 16). However, these pots can also catch terrapins for study (e.g., Davis 1942; Bishop 1983; Roosenburg at al. 1997; Hoyle and Gibbons 2000; Avissar 2006). The use of short pots that remain submerged requires checking every 3 h, but in areas of low tidal amplitude, tall crab pots (Roosenburg et al. 1997) can be used to catch terrapins and need to be checked only once per day. The tall traps can be built by cutting the bottom and top out of two pots and using hog rings to attach the two together. Tall crab pots (Fig. 2.2) should be tied to a stake to prevent strong wind or waves from knocking them over. Care should be taken to cut holes in the "church" that leads to the top section of the pot so that terrapins can retreat to the water during low tide. Crab

Fig. 2.2. Comparison of standard and tall crab pots. The tall crab pot contains a second "church" (or baffle) with upward and downward funnels that allow terrapins to move up and down with the rise and fall of the tide. Courtesy of Conservation Biology

pots checked regularly can be used in most habitats and can be baited with fish, crabs, or clams.

We have used round wire traps that are easier and cheaper to build than crab pots. They are approximately 1 m in diameter and 1.33 to 1.67 m tall and are made of 2.54 × 5.08 cm wire built on a frame of old fyke net hoops. They stand on end and have a single funnel on the side, near the bottom. Baited traps catch juvenile terrapins in small guts and tidal creeks late in the season.

Eel pots are another commercial gear that catches terrapins as bycatch. Cloth-funnel eel pots baited with razor clams capture terrapins at high rates in creek and nearshore river habitats, but they fish submerged and thus must be checked every 3 h (Radzio and Roosenburg 2005).

Otter Trawls

Otter trawls are tapered mesh nets with two wings attached to "doors" that spread the net while being towed. They require a motor boat to pull the net at a speed of 3 to 5 knots to prevent terrapins from avoiding or swimming out of the trawl. Larger trawls can be effective in open water with high tidal amplitude or where terrapins are concentrated

(Hurd et al. 1979); smaller trawls are most effective in small tidal creeks where their gape can span the width of the creek. The trawls should be emptied regularly (every 20 min) to prevent drowning terrapins or damaging them with accumulated debris and bycatch. Fishing with otter trawls requires attention to the position of the "doors," as their proper deployment is critical to the net fishing correctly.

Seines

In the absence of a boat with a powerful motor, hand seines can effectively catch terrapins in small tidal creeks (e.g., Grosse et al. 2011). This requires walking the banks of tidal creeks, where researchers can sink deeply in mud and have their shoes sucked off. Seines are most effective when pulled against the outgoing tide with a net that spans the width of the creek. A bag in the seine will reduce the escape of terrapins, and a mesh size between 1.25 cm and 2.5 cm will ensure the capture of most size classes. Dorcas et al. (2007), Gibbons et al. (2001), and Grosse et al. (2011) have seined in the narrow channels of salt marshes that lie behind the barrier islands of South Carolina and Georgia.

Terrestrial Drift Fences

The traditional herpetological technique using terrestrial drift fences (Graeter et al. 2013) can capture adult female terrapins when they come ashore to nest (Coleman 2011; Roberge 2012) and hatchlings as they emerge from nests (Muldoon and Burke 2012). Fences can be made of metal flashing, plastic UV-resistant screening, or even corrugated pipe (P. Baker, Wetlands Institute, pers. comm.). Typically, buckets (or smaller containers for hatchlings) are sunk in the sand along the fence to catch the animals as they search for passage across the barrier. The hot conditions on most nesting areas require that the fences be checked several times per day to prevent mortality due to overheating. This is particularly true for hatchlings, which are vulnerable to desiccation because of their smaller size. Wet sponges can be placed in the bottoms of buckets to reduce the risk of hatchling desiccation, and covers over buckets can protect terrapins from sun exposure (Roberge 2012).

Nest Rings and Predator Excluders

Nest rings are 50 to 100 cm diameter rings made of flashing (Fig. 2.3) that catch hatchlings emerging from the nest; they are useful to determine within-nest survivorship because they can account for all of the hatchlings within a nest (Roosenburg et al. 2014). Predator excluders (Riley and Litzgus 2013; Roosenburg et al. 2014) are devices designed to protect developing eggs from predators and can catch hatchlings if left in place long enough. Rings or excluders should be buried a minimum of 2.5 cm and centered on the nest. Hardware cloth (1.25 cm mesh) placed over the rings can prevent avian predation of emerging hatchlings. On hot days (>29°C), rings should be checked several times daily or shade should be provided to prevent hatchling mortality from overheating or desiccation. Rings and predator excluders may influence nest temperatures and/or nest moisture, but this has not been explored sufficiently (Riley and Litzgus 2013).

Rolling or Raking Wrack

Rolling or raking wrack uses a rake or cultivator to gently roll the accumulated wrack at the high-tide line in salt marshes. Wrack can also be lifted by hand (Pilter 1985). Hatchling and juvenile terrapins are active under the wrack, which is lifted above the substrate by emerging grasses, and this may be important terrestrial habitat in their life cycle. King (2007) found hatchling terrapins in both low and high *Spartina* marsh by searching under debris and vegetation. In Chesapeake Bay, this technique found more young of the previous year under

Fig. 2.3. Nest ring encircling emerged terrapin hatchlings.

wrack over *Spartina patens* than that over *S. alterniflora* (Roosenburg, pers. obs.). Raking wrack can catch juveniles and adults in some Florida marshes and can yield more than 60 terrapins per day (B. Turner, Florida Fish and Wildlife Conservation Commission, pers. comm.).

Winter Dredging or Scraping

Winter dredging was developed by commercial watermen in Chesapeake Bay and then adopted by scientists to catch terrapins (Haramis et al. 2011). A crab scrape modified to dig hibernating terrapins from the mud is pulled along the bottom by a powerful boat. The dredge is 1 to 1.3 m wide and has a twine net suspended from a metal frame. The bottom of the frame has 10 to 15 cm teeth that dig terrapins from the mud and into the net. Dredging requires knowledge of hibernacula where terrapins are concentrated (Haramis et al. 2011); catch rates can exceed 60 terrapins per hour. Haramis et al. (2011) also tested post-catch mortality and documented 100% survival of animals disturbed from their hibernaculum and returned to the water on the same day.

Miscellaneous Capture Techniques

Other techniques that catch terrapins include SCUBA, hand capture, and dip netting. In areas with clear water or in colder waters where terrapins do not move quickly, snorkeling and SCUBA diving can yield terrapins (Morreale 1992). This method can be supplemented with the use of dip nets to extend the reach of the catcher. However, SCUBA is ineffective in most estuaries throughout the terrapin's range because of low visibility in turbid waters. Collection of nesting terrapins by hand is employed by many researchers (e.g., Butler 2002; Feinberg and Burke 2003), and in many studies hand capture is the sole method of capture. This usually involves either patrolling nesting areas or watching from an inconspicuous location while females come ashore. This approach requires no equipment but a knowledge of nesting areas and times. Selman and Baccigalopi (2012) and R. Wood

(Wetlands Institute, pers. comm.) followed tracks left in the mud of tidal ponds and mud flats to catch terrapins by hand. Butler (2002) reported using cast nets for terrapins, but this required considerable skill and had limited success. Dip netting is used to capture many turtle species (Graeter et al. 2013) but is not commonly used for terrapins. However, Hart and McIvor (2008) dip-netted mangrove terrapins during a 2 h window around diurnal and nocturnal low tides. Hibernacula in shallows (<1 m) also can be searched for depressions on the bottom, and the terrapins can be excavated using a dip net from a boat (D. Marshall, Smith Island waterman, pers. comm.).

Census Techniques

Most of the capture techniques described above lend themselves well to standard mark-recapture analyses of populations (Hurd et al. 1979; Hoyle and Gibbons 2000; Tucker et al. 2001; Mitro 2003; Witczak et al. 2014). However, some projects do not require precise estimates of population size but instead focus on site occupancy (presence/absence) or relative abundance. Several techniques are available for these purposes.

Cyst Counts

Terrapins are the unique obligate final host for the small parasitic trematode *Pleurogonius malaclemys* (Hunter, 1961, 1967; Chodkowski et al. 2016). The parasite's intermediate host is a common terrapin prey species, the eastern mudsnail (*Ilyanassa obsoleta*). One of the juvenile stages is a <1 mm metacercarial cyst that is affixed to the outside of the snail (Fig. 2.4), where it remains until the snail is eaten by a terrapin. Mudsnails are easily collected along the beach at low tide, and the number of cysts correlates well with local terrapin population size in coastal Georgia (Byers et al. 2011), providing estimates of terrapin populations. Counting cysts is quick and has the potential for low-cost terrapin population estimates. Many questions remain throughout the terrapin's range, however, such as how far the trematodes disperse, the trematodes'

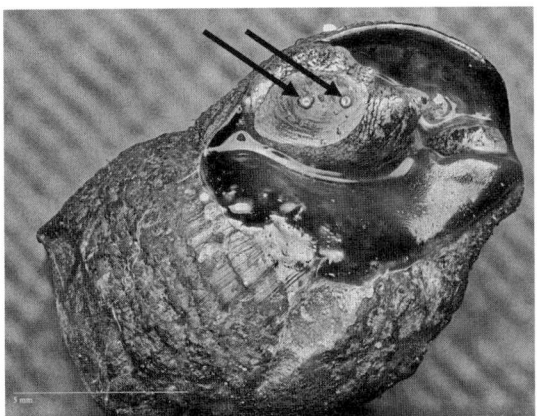

Fig. 2.4. Mud snail with *Pleurogonius malaclemys* metacercarial cysts, indicating that a terrapin population is nearby. The arrows indicate cysts on the snail's operculum and around the aperture.

range (to both north and south), and the strength of the relationship between cyst counts and population size. Burke's laboratory is exploring these questions in New York with a pilot program involving citizen scientists who are collecting snails and counting cysts throughout the Northeast.

Visual Counts

Visual counts of basking and swimming terrapins, taken from kayaks, canoes, and skiffs up to 8 m in length (Butler 2002; Harden et al. 2009) or from shore (Muehlbauer 1987; Estep 2005), can establish occupancy and identify terrapin concentrations. Harden et al. (2009) counted terrapins on standardized boat routes and compared the counts with terrapin population sizes estimated through mark-recapture. Head counts were highest at low tide and under partly cloudy conditions. Visual surveys estimated the number of terrapins and detected inter-annual trends over large areas at a fraction of the time and cost required for mark-recapture studies. Detailed studies are needed to document how head counts change seasonally and how weather and wave height affect detection probabilities. Preferably such studies would take place in an area with ongoing mark-recapture to explore the relationship between head-count data and actual population size.

Crawl and Nest Surveys

When only presence/absence data are needed, the simplest technique for identifying terrapin populations is to explore likely nesting areas and search for tracks and intact or depredated nests and their egg shells. Terrapin tracks are easily distinguished, but intact nests are more difficult to locate. Wind, rain, and dry sand will obscure tracks and nests within 24 h, making it virtually impossible to detect the presence of terrapins. Butler et al. (2004) documented crawls and nests on Florida beaches where they suspected terrapin presence. Where raccoons and foxes are common, nest predation rates are high. Depredated nests are easily identified by egg shells, which can remain on nesting beaches for more than a year. In areas where nest predators are absent (Roosenburg et al. 2014) or predators consume egg shells (Burke et al. 2009), detection probabilities decrease.

Marking Terrapins

Turtles are easy to mark (Graeter et al. 2013), and several identification technologies, each with its advantages and limitations, can be incorporated into demographic studies. We highly recommend using a double tagging system when possible, so that tag loss can be confirmed—for example, combining the use of marginal notching and PIT tags.

Noninvasive Methods

Imaging combined with pattern recognition software could identify individuals based on the distinctive color patterns on terrapins' shells and skin (e.g., Gamble et al. 2008). One researcher (M. Farina, Oceanside, NY, pers. comm.) identifies individual adult female terrapins by plastron scute seam patterns, photocopying plastrons and matching them visually to the catalog of previous captures. This system is limited to small populations, and scute seams may change through ontogeny; photocopying of adults works well, but it needs to be validated for efficacy in juveniles.

Scute Notching

Cagle (1939) developed the scute notching technique used in many turtle demographic studies. Using a triangular file, drill, or Dremel tool, one can notch or drill unique combinations of marginal scutes, based on a coding system. In older juveniles and adults the marks are permanent, and hatchling marks are recognizable as many as 8 to 10 years later; however, fast growth of hatchlings and juvenile turtles can obscure the marks. Scute marking is inexpensive and long lasting and can identify dead individuals. Drilling and marking is invasive, however, particularly if made deep enough to be long lasting, and large population studies (>4,000 individuals) require marking of at least 5 scutes. Furthermore, careful bookkeeping of used and unused ID codes is necessary to prevent duplicates. Finally, when marked animals incur carapace damage from boat injuries or other sources, the identification of notch codes can become confusing.

Passive Integrated Transponder Tags

Passive integrated transponder (PIT) tags are excellent for marking turtles (Buhlmann and Tuberville 1998; Szerlag-Egger and McRobert 2007). These tags come in several sizes, and the smallest can be implanted into hatchlings (Rowe and Kelly 2005). Such tags provide a 10- to 15-digit (depending on type) alphanumeric ID unique to the turtle. Locations of PIT tag injection vary among terrapin researchers. Some inject at the ventral base of the tail, in the muscle just above the plastron (K. Buhlmann, University of Georgia, pers. comm.); we inject in the loose flap of skin cranial to the rear leg along the plastron and have a <0.05% tag loss rate. We recommend Betadine to disinfect the injection site and New-Skin liquid bandage or cyanacrolite glue to seal the wound. Tags identify individuals from hatching to adulthood without the ambiguity caused by shell growth or damage, are minimally invasive, can often identify dead individuals, have no possibility of duplicate IDs, and enable recognition of individuals from other studies. An additional benefit is that these tags can identify wild terrapins that have ended up in the commercial trade. Disadvantages include the high cost ($2 per tag) and increased probability of tag loss if not properly injected.

Binary Coded Wire Tags

Binary coded wire tags (BCWT; Northwest Marine Technologies, Inc.) are miniature tags (1.1 × 0.25 mm) that can be injected into the terrapin. These tags are inscribed with a unique number that can be read only before injection or after removal. Therefore, detection is only presence/absence. We inject tags into the muscle tissue of the rear legs, using a 25 gauge syringe. These tags can mark large numbers of hatchlings in studies where cohort scute notches can assist in providing age data. The advantage of BCWT over PIT tags to mark hatchlings (Duncan 2013) is their lower cost ($0.33 each). However, the equipment necessary to locate inserted tags costs >$5,000. These tags were found to easily identify adults marked as hatchlings, and the cohort notch code confirmed age.

External Tags

Two types of external tags can mark terrapins. Sequentially numbered Monel tags (National Tag and Band Company) are miniature forms of sea turtle flipper tags. These small tags require the drilling of a 3 mm hole in one of the posterior marginals. However, the accompanying pliers do not close the tags well, and we recommend using a pair of needle-nose pliers to apply the tag. Floy or spaghetti tags (Floy Tag) are applied similarly but with a smaller hole that matches the diameter of the injector and holds the tag. These also bear a number that allows the identification of individuals.

Both types of tags are temporary because improper attachment and growing/breaking out of the shell result in tag loss. However, they provide a good secondary mark when coupled with PIT tags to facilitate clear identification of individuals. They also are visible, and thus tag return data can be obtained from individuals not associated with the study, e.g., commercial fishers or interested citizens who catch or find turtles.

Toe Clipping

Cagle (1939) suggested toe clipping for juvenile and hatchling turtles. Vogt (Instituto Nacional de Pesquisas da Amazonia, pers. comm.) toe-clipped hatchling mud turtles (*Kinosternon* sp.) and South American river turtles (*Podocnemis* sp.) that were later recognized when adults. No one has toe-clipped terrapins, but it may be an effective way to mark hatchlings. Studies testing the effects of toe clipping on turtle locomotion or behavior are needed, and the technique is not recommended for adults. However, the removed toes could be used to age adults (Thomas et al. 1997).

Paint

Fingernail polish or paint pens can mark terrapins temporarily to allow identification from a distance, such as during nesting (Roosenburg and Dunham 1997). Nail polish lasts about 7 to 10 days, which is shorter than the inter-nest interval (15 days; Roosenburg 1992). Paint markings or stick-on numbers also can identify individuals basking or on the beach.

Measuring Techniques

Turtle researchers have used a variety of measurements to quantify the morphology of turtles. Most researchers use calipers; however, cloth or plastic tape measures are low-cost alternatives. Digital scales and calipers greatly reduce the measurement error associated with analog scales and verniers, but they do not correct for differences among researchers' measuring of turtles. Although digital scales and calipers are considerably more expensive than analog tools, they greatly reduce measurement errors in data sets. Flexible rulers or tape can be used to measure terrapins, but the results are not comparable to straight-line measurements because of curvature of the shell. We recommend calipers, and below we describe measurements to standardize across population studies so as to improve comparisons throughout the terrapin's range.

Plastron and Carapace Length

Plastron length (PL) and carapace length (CL) are measured as the straight-line length of the plastron along the midsagittal plane. Measurements differ among researchers—primarily whether to measure into the anterior and posterior notches (= midline PL/CL) or flush with the ends of the shell (= maximum PL/CL). Use of both techniques ensures comparability among studies. Inserting the caliper tips into the notches is more consistent and provides less variability than maximum lengths. Calipers should be deployed parallel to the shell; angling the calipers dramatically affects the measurement. Use of flexible rulers and tape accounts for the curvature of the shell but is more prone to error created by variation of the doming of the shell (shorter individuals with a higher shell may have a longer carapace than a longer individual with a shorter shell). Doming can be accounted for by using calipers to measure shell height. Curved carapace measurements can also be useful for measurements of shell volume.

Carapace Width and Shell Height

Carapace width (CW) is measured at the widest point of the carapace. Pinch the terrapin between the beaks of the calipers to ensure they are perpendicular to the animal's sagittal plane. Shell height (SH) is measured at the deepest part of the terrapin. This is usually at the third vertebral scute on the carapace. The calipers should be perpendicular to the animal's longitudinal plane. Some problems may arise with individual terrapins that have knobs on vertebral scutes. Should one measure the maximum SH (including knobs) or SH without knobs? The individual study questions will determine which measurement is more appropriate (e.g., measure maximum SH if a study is related to bycatch reduction devices [BRDs]).

Head Width

Head width (HW) varies among terrapin populations, but it is unknown whether genetic differentiation or plasticity driven by local diet causes the variation. Head width is measured by gently pinching with calipers perpendicular to the sagittal plane at the surangular bone, which is the widest point of the head. Tricks to encourage terrapins to extend their heads include grabbing the head when extended and gently preventing retraction into the shell or holding the terrapin vertical, facing the ground. The latter method generally results in head extension within a few seconds. Pushing in on both forelimbs once the head is extended will prevent head retraction.

Mass

We recommend using portable digital top-loading balances or spring scales. A quick, vigorous shake can calm terrapins while the balance equilibrates.

Age

Growth ring (annulus) counting (Sexton 1959) can age terrapins, but uncertain annularity requires validation for every population (Litzgus and Brooks 1998; Wilson et al. 2003). Arrested growth in winter forms distinct ridges or growth rings on the shell. Although the ridges form on both the plastron and the carapace, the plastron is more reliable for aging. Rings on the carapace, though frequently deeper and larger, also are often less distinct and blend together more in older individuals, resulting in "false annuli." False annuli can sometimes be identified by comparing the preceding and following rings; if the middle ring is less distinct, then it may be false. We also suggest that age data include both the number of annuli counted (age in years) and the year of hatching.

The annularity of growth rings is confirmed when recaptures over several years document an increase in annuli consistent with the years passed. Recaptures over several years and recaptures of marked hatchlings have confirmed that counting of growth rings works well to estimate age in terrapins of Chesapeake Bay (Roosenburg, unpub. data) and South Carolina (Gibbons et al. 2001). However, we found very different results for Jamaica Bay, New York (Burke, unpub. data).

Permanent records of growth rings can be made using dental casts (Fig. 2.5) (Snider and Vano 1987). Examination of the casts under magnification facilitates precise observation of year-to-year changes in the number of growth rings and often reveals rings not visible to the naked eye. Using pencil and paper to make an imprint of the annuli also can reveal faint annuli and create a permanent record.

When growth slows at the onset of maturity, growth rings become compressed and difficult to count. Furthermore, the shedding of scutes eliminates younger growth rings, and after 10 to 15 years rings may no longer be used to precisely age a terrapin. At this point, experienced researchers can estimate an age range (e.g., <10, <15 years) from faint rings. We recommend carefully examining the rings on several plastron scutes to determine age, particularly when false annuli are suspected. Time of year also can affect the number of annuli. Animals caught in spring or early summer probably have not yet grown that year, in which case the ring at the midline seam should be included in the count. In contrast, the midline seam ring should not be included for terrapins caught in late summer; summer growth is lighter in color and sometimes softer than the rest of the plastron.

Reproductive Status

Demographic studies require identification of which females are reproducing and quantification of age of first reproduction, clutch frequency, interclutch interval, egg size, and egg and hatchling survivorship—data that are important for life history analyses and population modeling. Inguinal palpation (anterior to the hind leg) can detect the presence or absence of shelled eggs, which feel like small lumps or ping pong balls inside the terrapin. An experienced researcher may detect follicles, but eggs are distinct and often dimple inside the female. Palpation can underestimate the number of gravid females in earlier egg stages, with less calcification than is detected using radiographic methods (Keller 1998). X-rays provide data on reproductive status, number of eggs, and, if all individuals are X-rayed similarly, egg size and width (Gibbons and

Fig. 2.5. Cast from the carapace of an adult terrapin that can help evaluate accuracy of growth rings.

Greene 1979). To measure size, a radio-opaque object of known size (e.g., a US quarter) should be placed on the X-ray machine at the same height as the eggs. Settings for taking good pictures vary with machine and film, but for Chesapeake terrapins a setting of 75 mA with an exposure of 0.15 to 0.2 s has resulted in clear radiographs. The female should be X-rayed while lying on her back; Styrofoam wedges prevent the terrapin from rolling, and a large rubber band around the shell to partially immobilize the legs reduces movement. Ultrasound in the inguinal region can identify eggs and follicles and measure their sizes (e.g., Lee 2003). Ultrasound cannot accurately determine clutch size, but it can identify the size classes of follicles present in the ovary and therefore can help estimate the number of clutches per year. Ultrasound machines are expensive, but more affordable secondhand units can be found.

Nesting Ecology

Most researchers' initial encounter with terrapins occurs when walking a nesting beach and finding depredated nests. Study of the nesting ecology of terrapins can be simple and inexpensive and can reveal important life history information (e.g., Burger 1977). Because predators destroy many nests in most terrapin populations, surveys of nesting areas can provide a proxy of the local female population size and indicate preferred nesting habitat (e.g., Montevecchi and Burger 1975; Roosenburg and Dunham 1997; Feinberg and Burke 2003; Mitro 2003; Munscher et al. 2012; Mitchell and Walls 2013; Winters 2013). Excavation of intact nests can provide data on reproductive output and nesting behavior. However, before excavating, inspection for chalking or formation of the air pocket within the eggs can age the nest and determine the risk associated with excavation. Recently laid (<24 h) eggs are pink and translucent and can be safely handled and gently moved. Chalking begins after 24 h and forms a white opaque disc at the top of the egg, which envelopes the entire shell as the egg ages. If the chalky disc is larger than 10 mm, moving the egg increases the risk of rupturing membranes and killing the embryo. Eggs can still be excavated at this time but must be handled with care; they cannot be rotated or bounced, and thus their excavation should generally be avoided. Described below is a series of metrics that are useful in terrapin reproduction studies.

Finding Nests

Under most conditions, intact terrapin nests are inconspicuous, with only subtle field signs indicating their presence. Finding nests often requires either observing the female during oviposition or locating the nest after it has been depredated (e.g., Roosenburg 1992; Ner 2003). Sometimes terrapin nests can be discovered by tracks left by the female (e.g., Butler 2002; Butler et al. 2004) or tracks left by emerging hatchlings (e.g., Burger 1976a, 1977).

Clutch Size

Clutch size is determined by counting the eggs within the nest. Terrapins may abandon nesting after egg laying has started and sometimes continue to release eggs after leaving the nest. Because predators sometimes consume egg shells (Burke et al. 2009) or eat the eggs at some distance from the nest, use of egg shell fragments at a depredated nest to estimate clutch size (e.g., Feinberg and Burke 2003) is not recommended. As mentioned earlier, X-rays are an excellent mechanism to determine clutch size.

Egg Mass, Length, and Width

A small, portable electronic jeweler's balance (<$100) can weigh individual eggs in the field; a block of wood and a small bubble level can ensure that the balance is level. Sand can be removed carefully from eggs by wiping them with a soft cloth or brush. Egg length and width can be assessed using calipers to measure the longest and widest points. Hot conditions in the field may preclude making all of these measurements because eggs dehydrate rapidly.

Nest Depth to Top and Bottom

A ruler and a survey flag can be used to measure depths to the top of the first egg and to the bottom of the nest after the eggs have been removed. Excavate to the top egg, then lay the flag across the surface of undisturbed sand and measure the depth while holding the ruler perpendicular to the flag. Other nest cavity measurements are described by Montevecchi and Burger (1975).

Microhabitat Classification

Nest microhabitats can influence nest survival and incubation conditions (Burger and Montevecchi 1975; Burger 1976b; Roosenburg 1996; Scholz 2006; Chapter 7). A nest too close to mean high water may flood, and nests in areas with dense vegetation are frequently killed by roots that encase or penetrate the eggs to access water and nutrients (Lazell and Auger 1981; Feinberg and Burke 2003; Chapter 7). However, vegetation also provides shade that can reduce nest temperatures and be a mechanism for avoiding lethal incubation temperatures (Scholz 2006; Chapter 7). Microhabitat classification should therefore quantify vegetation around the nest and the solar radiation that contributes to heating the nest. Roosenburg (1996) classified nests with three discrete categories for vegetation (0, 1-50, >50 stems per 0.25 m^2) and light exposure (<4, 4-8, >8 h of direct sunlight per 0.25 m^2). Spherical densitometers quantify vegetation cover (Janzen 1994; Weisrock and Janzen 1999) but may not distinguish between high overhead canopy or shrubs and grasses in close proximity to the nest. Other metrics such as distance to nearest water and elevation above mean high tide can also be important metrics for hatchling survivorship (Fowler 1979; Whitmore and Dutton 1985).

Nest Monitoring, Marking, and Predation

Locating terrapin nests is critical for determining nest survivorship and monitoring nest predation (e.g., Feinberg and Burke 2003); however, making nests conspicuous to humans or nest predators is problematic. Commercially available survey flags can mark nests and are easily written on with a permanent marker. Surrounding the nest with 3 or 4 flags ensures successful ringing (see above) at a later date. However, in areas where predation rates are high, marking of nests in this way may help predators, particularly crows, identify them (Baker 1978; Yahner and Wright 1985; Rollinson and Brooks 2007; Roosenburg et al. 2014; but see Burke et al. 2005). Consequently, markers should be some distance from the nest (1-2 m), which can make locating them again 6 to 8 weeks later difficult. A penny or other metal object can be placed on the exact location of the nest and then located later with a metal detector. To deter avian (but not raccoon, snake, or fox) predation, a 25 × 25 cm, 12.5 mm mesh hardware cloth can be held in place by the flags marking a nest (Roosenburg et al. 2014). Counts of depredated nests combined with known predation rates can estimate the total number of nests (Feinberg and Burke 2003), but changes in predator behavior over the nesting season could skew results (Burke et al. 2009).

Biophysical Monitoring

The effects of temperature and soil moisture on sex, size, and fitness of the developing embryo (reviewed in Packard and Packard 1988; Janzen and Paukstis 1991) have stimulated the use of monitoring devices in terrapin nests. Initially, this monitoring required large data loggers with a tangle of thermocouple and other wires, limited by the proximity of multiple nests. Miniaturization and extended battery life have facilitated the placement of environmental loggers directly in nests or on turtles for long- and short-term monitoring. Thermochron iButtons sample temperatures at user-programmed time intervals, with the capacity to store a large number of readings. Before deployment, iButtons must be waterproofed. Finger cots, or finger-length latex sheaths, work well in nests and do not affect the performance of the iButtons (Roznik and Alford 2012). Harden et al. (2007), Williard and Harden (2011), and Akins et al. (2014) epoxied plastic-coated iButtons to terrapin carapaces to monitor body temperatures. Hobo Tidbits and Star-Oddi DST pressure loggers have been used to monitor depth and orientation of turtles (Chapter 15). Tidbits are larger and more expensive but more accurate than the iButtons (±0.1°C vs. ±0.5°C). Attaching devices to free-ranging terrapins requires recapture, making iButtons inexpensive (<$20 each) relative to Tidbits ($100) and depth sensors (>$400). Harden et al. (2007) radio-tracked terrapins and recovered 4 of 5 terrapins with iButtons. As external monitoring devices become smaller and cheaper, we anticipate more detailed studies of terrapin behavior, nesting environments, and habitat choice.

Tracking and Telemetry

A frustrating aspect of working with an estuarine species is that radio waves attenuate in saltwater, and acoustic telemetry requires line of sight between transmitter and receiver. Nevertheless, numerous studies (e.g., Spivey 1993; Butler 2002; Estep 2005; Williard and Harden 2011; Harden and Williard 2012; Roberge 2012; Winters 2013) have radio-tracked terrapins successfully, relying on signals transmitted when terrapins were at the surface and using sonic tags successfully with an array of stationary receivers (Estep 2005; Tulipani 2013; Winters 2013). Transmitters attach to the carapace with cable ties and fast-drying epoxy. The transmitter should be mounted so as to prevent snagging on external objects such as plants or nets and should not exceed 5% of the terrapin's weight. When mounting transmitters, tape should be placed over scute sutures so that shell growth is not adversely affected.

Both acoustic and radio tags can provide data on temperatures, depths, and swimming speeds (Chapter 15). Sonic tags are relatively inexpensive, but purchase, placement, and maintenance of receivers is expensive, particularly if a large area is to be monitored. Sonic tags can also monitor animals year-round and thus can locate hibernating terrapins.

String trailers unroll off a spool attached to the carapace of terrestrial turtles, leaving a trail of their movement (Graeter et al. 2013). Roosenburg et al. (1999; see also Tulipani 2013) trailed terrapins by sight using a fish bobber at the end of a 2 to 4 m string. Constant attendance can record location, but this also limits the use of bobbers to short-term (3-6 h) studies. The string and bobber should be removed at the end of the observation period because of the risk of entanglement. This technique can reveal short-term movements; for example, Burke and Reif (unpub. data) tracked post-oviposition females back to their marshes. Passive integrated transponders tags also can locate terrapins; Duncan (2014) located hatchlings in a salt marsh by using a Biomark BP Plus Portable Antenna. Similarly, stationary PIT tag antennas in key locations could track terrapin movement, as demonstrated in other animals (e.g., Gibbons and Andrews 2004; Charney et al. 2009).

Long-term and constant data on terrapin movements remain elusive. Fortunately, satellite transmitters and cellphone-based systems continue to get smaller and more accurate and are approaching usefulness for terrapins. Satellite transmitters send a signal to the Argos satellite system; locations are acquired by Argos or GPS and can be stored or sent directly to the researcher. Alternatively, in areas with good cellphone coverage, GPS data loggers with GSM (Global System for Mobile Communications) modules can obtain and transmit location fixes. Limitations include cumbersome and heavy power sources (batteries and/or solar), amount of time required at the water surface for transmission, low accuracy, attachment requirements, and price. Transmitters are now small and light enough to put on small migratory birds and juvenile sea turtles, and thus small enough for terrapins. However, terrapins often frequent turbid waters, and rapid growth of epiphytes or epifauna can foul solar cells. In the absence of solar cells, batteries for long-term tracking add prohibitive weight. Saltwater switches can save battery life, but also may foul. Antennas

must be above the water to transmit a signal, and a surface interval must be long enough to get a GPS fix. Required surface intervals under ideal conditions can be under 1 min, but with poor satellite coverage, obtaining a good location can be problematic. No data exist on terrapin surface intervals, so we cannot estimate success in good satellite coverage. Current technologies lack resolution and are thus of limited usefulness within a terrapin's reported home range (50-250 ha; Spivey 1993; Butler 2002), but they could be useful in studying dispersal, which has been observed as far as 55 km (Roosenburg, unpub. data). High resolution is possible with GPS-enabled Argos transmitters but will require developing a long-term, lightweight attachment system; technologies developed for sea turtle research (e.g., Mansfield et al. 2012) may offer a starting point. Finally, costs ($500-$5,000 per satellite tag, plus monthly costs for satellite time) stymie experimentation and pilot studies. Nonetheless, these technologies hold great promise for the future in understanding terrapin dispersal and other behaviors as transmitter technology evolves.

Molecular Methods and Population Biology

The increasing costs of labor, fuel, materials, and liability insurance challenge the traditional long-term methods for studying terrapin populations. However, recently developed molecular techniques can evaluate occupancy, parentage, mating systems, population structure, bottlenecks (reductions in population size), gene flow, and effective population size (Chapter 5). Newly developed eDNA technology could evaluate occupancy of terrapins in salt marshes throughout their range. eDNA and other molecular techniques require limited samples from populations yet can have robust management and conservation implications. High-resolution genetic techniques can measure gene flow among populations and distinguish between male- and female-biased gene flow (Sheridan et al. 2010). Genetic comparisons among terrapin populations have revealed low structuring (McCafferty et al. 2013; Hart et al. 2014; Petre et al. 2015; Chapter 5). However, developing analytical techniques, combined with sorely needed new genetic markers (possibly single nucleotide polymorphisms [SNPs]), could provide a more thorough understanding of gene flow and population structuring in terrapins. Such studies, combined with new tracking technologies, could improve our understanding of nest-site fidelity and lifetime movements. One compelling advantage of these tools is that they can provide a window into the past, identifying bottlenecks and population perturbations that traditional demography cannot elucidate. Nonetheless, molecular studies still require the collection of animals, and we recommend that modern terrapin demographers collect and store tissue samples from all individuals. We have successfully extracted DNA from the drill shavings from holes used to mount tags (see above). Although genetic studies are still costly, the labor necessary is greatly reduced and is shifted to the laboratory. We do not suggest that molecular studies are a panacea to replace long-term field studies, but they do provide a strong tool for the management and conservation of terrapins in understudied populations. Additionally, within- and among-population comparisons using traditional and genetic demographic techniques need to evaluate the consistency and accuracy of the results, but could reveal interesting patterns obscured by using the techniques independently (e.g., Anthonysamy et al. 2014). Such studies would be landmark for terrapins and the study of turtle populations.

REFERENCES

Akins, C., C.D. Ruder, S.J. Price, L.A. Harden, J.W. Gibbons and M.E. Dorcas. 2014. Factors affecting temperature variation and habitat use in free-ranging diamondback terrapins. Journal of Thermal Biology 44:63–69.

Anthonysamy, W.J.B., M.J. Dreslik, M.R. Douglas, N.K. Marioni and C.A. Phillips. 2014. Reproductive ecology of an endangered turtle in a fragmented landscape. Copeia 2014:437–446.

Avissar, N.G. 2006. Changes in population structure of diamondback terrapins (*Malaclemys terrapin terrapin*) in a previously studied creek in southern New Jersey. Chelonian Conservation and Biology 5:154–159.

Baker, B.W. 1978. Ecological factors affecting wild turkey nest predation in south Texas rangelands. Proceedings of the Annual Conference of the Southeast Association Fish and Wildlife Agencies 32:126–136.

Bishop, J.M. 1983. Incidental capture of diamondback terrapin by crab pots. Estuaries 6:426–430.

Broyles, E. 2010. Diamondback terrapins (*Malaclemys terrapin*) of Charleston, South Carolina: population estimate, sex ratios and distribution. MS thesis, College of Charleston, Charleston, SC. 63 pages.

Buhlmann, K.A. and T.D. Tuberville. 1998. Use of passive integrated transponder (PIT) tags for marking small freshwater turtles. Chelonian Conservation and Biology 3:102–104.

Burger, J. 1976a. Behavior of hatchling diamondback terrapins (*Malaclemys terrapin*) in the field. Copeia 1976:742–748.

Burger, J. 1976b. Temperature relationships in nests of the northern diamondback terrapin, *Malaclemys terrapin terrapin*. Herpetologica 32:412–418.

Burger, J. 1977. Determinants of hatching success in the diamondback terrapin, *Malaclemys terrapin*. American Midland Naturalist 97:444–464.

Burger, J. and W.A. Montevecchi. 1975. Nest site selection in the terrapin *Malaclemys terrapin*. Copeia 1975:113–119.

Burke, R.L., S.M. Felice and S.G. Sobel. 2009. Changes in raccoon (*Procyon lotor*) predation behavior affects turtle (*Malaclemys terrapin*) nest census. Chelonian Conservation and Biology 8:208–211.

Burke, R.L., C.M. Schneider and M.T. Dolinger. 2005. Cues used by raccoons to find turtle nests: effects of flags, human scent, and diamondback terrapin sign. Journal of Herpetology 29:312–315.

Butler, J.A. 2002. Population ecology, home range, and seasonal movements of the Carolina diamondback terrapin, *Malaclemys terrapin centrata*, in northeastern Florida. Report of the Florida Fish and Wildlife Conservation Commission. Tallahassee, FL. 72 pages.

Butler, J.A., C. Broadhurst, M. Green and Z. Mullin. 2004. Nesting, nest predation and hatchling emergence of the Carolina diamondback terrapin, *Malaclemys terrapin centrata*, in Northeastern Florida. American Midland Naturalist 152:145–155.

Byers, J.E., I. Altman, A.M. Grosse, T.C. Huspeni and J.C. Maerz. 2011. Using parasite trematode larvae to quantify an elusive vertebrate host. Conservation Biology 25:85–93.

Cagle, F.R. 1939. A system for marking turtles for future identification. Copeia 1939:170–173.

Charney, N.D., B.W. Letcher, A. Haro and P. Warren. 2009. Terrestrial passive integrated antennae for tracking small animal movements. Journal of Wildlife Management 73:1245–1250.

Chodkowski, N., J.D. Williams and R.L. Burke. 2016. Field surveys and experimental transmission of *Pleurogonius malaclemys* (Digenea: Pronocephalidae), an intestinal parasite of the diamondback terrapin *Malaclemys terrapin*. Journal of Parasitology 102:410–418.

Coleman, A.T. 2011. Biology and conservation of the diamondback terrapin *Malaclemys terrapin pileata* in Alabama. PhD dissertation, University of Alabama at Birmingham, Birmingham, AL. 207 pages.

Davis, C.C. 1942. A study of the crab pot as a fishing gear. Chesapeake Biological Laboratory Pub. 53. Solomons, MD. 20 pages.

Dorcas, M.E., J.D. Wilson and J.W. Gibbons. 2007. Crab trapping causes population decline and demographic changes in diamondback terrapin over two decades. Biological Conservation 137:334-340.

Duncan, N. 2013. Using a handheld PIT scanner and antenna system to successfully locate terrestrially overwintering hatchling turtles. Herpetological Review 44:233-235.

Duncan, N. 2014. Dispersal of newly-emerged diamond-backed terrapin (*Malaclemys terrapin*) hatchlings at Jamaica Bay, NY. MS thesis, Hofstra University, Hempstead, NY. 38 pages.

Estep, R.L. 2005. Seasonal movements and habitat pattern use of a diamondback terrapin (*Malaclemys terrapin*) population. MS thesis, College of Charleston, Charleston, SC. 152 pages.

Feinberg, J.A. and R.L. Burke. 2003. Nesting ecology and predation of diamondback terrapins, *Malaclemys terrapin*, at Gateway National Recreation Area, New York. Journal of Herpetology 37:517-526.

Fowler, L.E. 1979. Hatching success and nest predation in the green sea turtle, *Chelonia mydas*, at Tortuguero, Costa Rica. Ecology 60:946-955.

Gamble, L., S. Ravela and K. McGarigal. 2008. Multi-scale features for identifying individuals in large biological databases: an application of pattern recognition technology to the marbled salamander *Ambystoma opacum*. Journal of Applied Ecology 45:170-180.

Gibbons, J.W. and K.M. Andrews. 2004. PIT tagging: simple technology at its best. BioScience 54:447-454.

Gibbons, J.W. and J.L. Greene. 1979. X-ray photography: a technique to determine reproductive patterns of freshwater turtles. Herpetologica 35:86-88.

Gibbons, J.W., J.E. Lovich, A.D. Tucker, N.N. Fitzsimmons and J.L. Greene. 2001. Demographic and ecological factors affecting conservation and management of diamondback terrapins (*Malaclemys terrapin*) in South Carolina. Chelonian Conservation and Biology 4:66-74.

Graeter, G.J., K.A. Buhlmann, L.R. Wilkinson and J.W. Gibbons. 2013. Inventory and monitoring: recommended techniques for reptiles and amphibians, with application to the United States and Canada. Partners in Amphibian and Reptile Conservation Technical Pub. IM-1. Birmingham, AL. 321 pages.

Grosse, A.M., J.C. Maerz, J.A. Hepinstall-Cymerman and M.E. Dorcas. 2011. Effects of roads and crabbing pressures on diamondback terrapin populations in coastal Georgia. Journal of Wildlife Management 75:762-770.

Haramis, G.M., P.F. Henry and D.D. Day. 2011. Using scrape fishing to document terrapins in hibernacula in Chesapeake Bay. Herpetological Review 42:170-177.

Harden, L.A., N.A. Diluzio, J.W. Gibbons and M.E. Dorcas. 2007. Spatial and thermal ecology of the diamondback terrapin (*Malaclemys terrapin*) in a South Carolina salt marsh. Journal of the North Carolina Academy of Sciences 123:154-162.

Harden, L.A., S.E. Pittman, J.W. Gibbons and M.E. Dorcas. 2009. Development of a rapid-assessment technique for diamondback terrapin (*Malaclemys terrapin*) populations using head-count surveys. Applied Herpetology 6:237-245.

Harden, L.A. and A.S. Williard. 2012. Using spatial and behavioral data to evaluate the seasonal bycatch risk of diamondback terrapins (*Malaclemys terrapin*) in crab pots. Marine Ecology Progress Series 467:207-217.

Hart, K.M., M.E. Hunter and T.L. King. 2014. Regional differentiation among populations of diamondback terrapin (*Malaclemys terrapin*). Conservation Genetics 15:593-603.

Hart, K.M. and C.C. McIvor. 2008. Demography and ecology of mangrove diamondback terrapins in a wilderness area of Everglades National Park, Florida, USA. Copeia 2008:200-208.

Hoyle, M.E. and J.W. Gibbons. 2000. Use of a marked population of diamondback terrapins (*Malaclemys terrapin*) to determine impacts of recreational crab pots. Chelonian Conservation and Biology 3:735-737.

Hunter, W.S. 1961. A new monostome, *Pleurogonius malaclemys*, n. sp. (Trematoda: Pronocephalidae) from Beaufort, North Carolina. Proceedings of the Helminthological Society 28:111-114.

Hunter, W.S. 1967. Notes on the life history of *Pleurogonius malaclemys* Hunter 1961, (Trematoda: Pronocephalidae) from Beaufort, North Carolina, with a description of the cercaria. Proceedings of the Helminthological Society 34:33-40.

Hurd, L.E., G.W. Smedes and T.A. Dean. 1979. An ecological study of a natural population of diamond-back terrapins (*Malaclemys t. terrapin*) in a Delaware salt marsh. Estuaries 2:28-33.

Janzen, F.J. 1994. Vegetational cover predicts the sex ratio of hatchling turtles in natural nests. Ecology 75:1593-1599.

Janzen, F.J. and G.L. Paukstis. 1991. Environmental sex determination in reptiles: ecology, evolution, and experimental design. Quarterly Review of Biology 66:149-179.

Keller, C. 1998. Assessment of reproductive state in the turtle *Mauremys leprosa*: a comparison between inguinal palpation and radiography. Wildlife Research 25:520–531.

King, T.M. 2007. The diet of northern diamondback terrapins (Order Testudines; *Malaclemys terrapin terrapin*). MS thesis, C.W. Post Campus of Long Island University, Brookville, NY. 40 pages.

Lazell, J.D. Jr. and P.J. Auger. 1981. Predation on diamondback terrapin (*Malaclemys terrapin*) eggs by dunegrass (*Ammophila breviligulata*). Copeia 1981:723–724.

Lee, M.A. 2003. Reproductive biology and seasonal testosterone patterns of the diamondback terrapin, *Malaclemys terrapin*, in the estuaries of Charleston, South Carolina. MS thesis, College of Charleston, Charleston, SC. 146 pages.

Levesque, E.M. 2000. Distribution and ecology of the diamondback terrapin (*Malaclemys terrapin*) in South Carolina salt marshes. MS thesis, College of Charleston, Charleston, SC. 102 pages.

Litzgus, J.D. and R.J. Brooks. 1998. Testing the validity of counts of plastral scute rings in spotted turtle, *Clemmys guttata*. Copeia 1998:222–225.

Mansfield, K.L., J. Wyneken, D. Rittschof, M. Walsh, C.W. Lim and P.M. Richards. 2012. Satellite tag attachment methods for tracking neonate sea turtles. Marine Ecology Progress Series 457:181–192.

McCafferty, S.S., A. Shorette, J. Simundza and B. Brennessel. 2013. Paucity of genetic variation at an MHC class I gene in Massachusetts populations of the diamond-backed terrapin (*Malaclemys terrapin*): a cause for concern. Journal of Herpetology 47:222–226.

McDiarmid, R.W., M.S. Foster, C. Guyer, J.W. Gibbons and N. Chernoff (eds.). 2012. Reptile Biodiversity: Standard Methods for Inventory and Monitoring. University of California Press, Los Angeles. 424 pages.

Mitchell, J.C. and S.C. Walls. 2013. Nest site selection by diamond-backed terrapins (*Malaclemys terrapin*) on a Mid-Atlantic barrier island. Chelonian Conservation and Biology 12:303–308.

Mitro, M.G. 2003. Demography and viability analysis of a diamondback terrapin population. Canadian Journal of Zoology 81:716–726.

Montevecchi, W.A. and J. Burger. 1975. Aspects of the reproductive biology of the northern diamondback terrapin *Malaclemys terrapin terrapin*. American Midland Naturalist 94:166–178.

Morreale, S.J. 1992. The status and population ecology of the diamondback terrapin, *Malaclemys terrapin*, in New York. Final Report Submitted to the New York Department of Environment. Conservation Contract no. C002656. Hampton Bays, NY. 75 pages.

Muehlbauer, E. 1987. Field and laboratory studies of tidal activity in the turtle *Malaclemys terrapin terrapin*. MS thesis, New York University, New York, NY. 66 pages.

Muldoon, K.A. and R.L. Burke. 2012. Movements, overwintering, and mortality of hatchling diamond-backed terrapins (*Malaclemys terrapin*) at Jamaica Bay, New York. Canadian Journal of Zoology 90:651–662.

Munscher, E.C., E.H. Kuhns, C.C. Cox and J.A. Butler. 2012. Decreased nest mortality for the Carolina diamondback terrapin (*Malaclemys terrapin centrata*) following the removal of raccoons (*Procyon lotor*) from a nesting beach in northeastern Florida. Herpetological Conservation and Biology 7:167–184.

Ner, S.E. 2003. Distribution and predation of diamondback terrapin nests at six upland islands of Jamaica Bay and Sandy Hook Unit, Gateway National Recreation Area. MS thesis, Hofstra University, Hempstead, NY. 104 pages.

Packard, G.C. and M.J. Packard. 1988. The physiological ecology of reptilian eggs and embryos. Pages 523–607 in C. Gans and R.B. Huey (eds.), Biology of the Reptilia. Volume 16. Ecology B. Defense and Life History. Alan R. Liss, New York, NY.

Petre, C., W. Selman, B. Kreiser, S.H. Pearson and J.J. Wiebe. 2015. Population genetics of the diamondback terrapin, *Malaclemys terrapin*, in Louisiana. Conservation Genetics. DOI 10.1007/s10592-015-0735-z.

Pilter, R. 1985. *Malaclemys terrapin terrapin* (northern diamondback terrapin) behavior. Herpetological Review 16:82.

Radzio, T.A. and W.M. Roosenburg. 2005. Diamondback terrapin mortality in the American eel pot fishery and evaluation of a bycatch reduction device. Estuaries 28:620–626.

Riley, J.L. and J.D. Litzgus. 2013. Evaluation of predator-exclusion cages used in turtle conservation: cost analysis and effects on nest environment and proxies of hatchling fitness. Wildlife Research 40:499–511.

Roberge, T. 2012. Evaluating the reproductive ecology of the diamondback terrapin in Alabama saltmarshes: implications for the recovery of a depleted species. MS thesis, University of Alabama at Birmingham, Birmingham, AL. 91 pages.

Rollinson, N. and R.J. Brooks. 2007. Marking nests increases the frequency of nest depredation in a northern population of painted turtles (*Chrysemys picta*). Journal of Herpetology 41:174–176.

Roosenburg, W.M. 1992. Life history consequences of nest site choice by the diamondback terrapin, *Malaclemys terrapin*. PhD dissertation, University of Pennsylvania, Philadelphia, PA. 217 pages.

Roosenburg, W.M. 1996. Maternal condition and nest site choice: an alternate for the maintenance of environmental sex determination. American Zoologist 36:157-168.

Roosenburg, W.M. 2004. The impact of crab pot fisheries on terrapin (*Malaclemys terrapin*) populations: where are we and where do we need to go? Pages 23-30 in C. Swarth, W.M. Roosenburg and E. Kiviat (eds.), Conservation and Ecology of Turtles of the Mid-Atlantic Region: A Symposium. Bibliomania, Salt Lake City, UT.

Roosenburg, W.M., W. Cresko, M. Modesitte and M.B. Robbins. 1997. Diamondback terrapin (*Malaclemys terrapin*) mortality in crab pots. Conservation Biology 5:1166-1172.

Roosenburg, W.M. and A.E. Dunham. 1997. Allocation of reproductive output: egg and clutch-size variation in the diamondback terrapin. Copeia 1997: 290-297.

Roosenburg, W.M., K.L. Haley and S. McGuire. 1999. Habitat selection and movements of diamondback terrapins, *Malaclemys terrapin*, in a Maryland estuary. Chelonian Conservation and Biology 3:425-429.

Roosenburg, W.M., D.M. Spontak, S.P. Sullivan, E.L. Mathews, M.L. Heckman, R.J. Trimbath, R.P. Dunn, E.A. Dustman, L. Smith and L.J. Graham. 2014. Nesting habitat creation enhances recruitment in a predator free environment: *Malaclemys* nesting at the Paul S. Sarbanes Ecosystem Restoration Project. Restoration Ecology 22:815-823.

Rowe, C.L. and S.M. Kelly. 2005. Marking hatchling turtles via intraperitoneal placement of PIT tags: implications for long-term studies. Herpetological Review 36:408-410.

Roznik, E.A. and R.A. Alford. 2012. Does waterproofing Thermochron iButton dataloggers influence temperature readings? Journal of Thermal Biology 37:260-264.

Scholz, A.L. 2006. Impacts of nest site choice and nest characteristics on hatching success in the diamondback terrapins of Jamaica Bay, New York. MS thesis, Hofstra University, Hempstead, NY. 97 pages.

Seigel, R.A. 1984. Parameters of two populations of diamondback terrapins (*Malaclemys terrapin*) on the Atlantic coast of Florida. Pages 77-87 in R.A. Seigel, L.E. Hunt, J.L. Knight, L. Malaret and N.L. Zushlag (eds.), Vertebrate Ecology and Systematics—A Tribute to Henry S. Fitch. Museum of Natural History, University of Kansas, Lawrence, KS.

Selman, W. and B. Baccigalopi. 2012. Effectively sampling Louisiana diamondback terrapin (*Malaclemys terrapin*) populations, with description of a new capture technique. Herpetological Review 43:583-588.

Selman, W., B. Baccigalopi and C. Baccigalopi. 2014. Distribution and abundance of diamondback terrapins (*Malaclemys terrapin*) in Southwestern Louisiana. Chelonian Conservation and Biology 13:131-139.

Sexton, O.J. 1959. A method of estimating the age of painted turtles for use in demographic studies. Ecology 40:717-718.

Sheridan, C.M., J.R. Spotila, W.F. Bien and H.W. Avery. 2010. Sex-biased dispersal and natal philopatry in the diamondback terrapin, *Malaclemys terrapin*. Molecular Ecology 19:5497-5510.

Simoes, J.C. and R.M. Chambers. 1999. The diamondback terrapins of Piermont Marsh, Hudson River, New York. Northeastern Naturalist 6:241-248.

Snider, D.E. and A.R. Vano. 1987. Photographs and dental casts as permanent records for age estimates and growth studies of turtles. Herpetological Review 18:69-71.

Spivey, P.B. 1993. Home range, habitat selection, and diet of the diamondback terrapin (*Malaclemys terrapin*) in a North Carolina estuary. MS thesis, University of North Carolina at Wilmington, Wilmington, NC. 80 pages.

Szerlag-Egger, S. and S.P. McRobert. 2007. Northern diamondback terrapin occurrence, movement, and nesting activity along a salt marsh access road. Chelonian Conservation and Biology 6:295-301.

Thomas, R.B., D.W. Beckman, K. Thompson, K.A. Buhlmann, J.W. Gibbons and D.L. Moll. 1997. Estimation of age for *Trachemys scripta* and *Deirochelys reticularia* by counting annual growth layers in claws. Copeia 1997:842-845.

Tucker, A.D., J.W. Gibbons and J.L. Greene. 2001. Estimates of adult survival and migration for diamondback terrapins: conservation insight from local extirpation within a metapopulation. Canadian Journal of Zoology 79:2199-2209.

Tulipani, D. 2013. Foraging ecology and habitat use of the northern diamondback terrapin (*Malaclemys terrapin terrapin*) in southern Chesapeake Bay. PhD dissertation, College of William and Mary, Williamsburg, VA. 238 pages.

Vogt, R.C. 1980. New methods for trapping aquatic turtles. Copeia 1980:368-371.

Weisrock, D.W. and F.J. Janzen, 1999. Thermal and fitness-related consequences of nest location in painted turtles (*Chrysemys picta*). Functional Ecology 13:94-101.

Whitmore, C.P. and P.H. Dutton. 1985. Infertility, embryonic mortality and nest-site selection in leatherback and green sea turtles in Suriname. Biological Conservation 34:251-272.

Wilbur, H.M. and P.J. Morin. 1988. Life history evolution in turtles. Pages 387-439 in C. Gans and R.B. Huey (eds.), The Biology of the Reptilia. Volume 16. Ecology B. Defense and Life History. Alan R. Liss, New York, NY.

Williard, A.S. and L.A. Harden. 2011. Seasonal changes in thermal environment and metabolic enzyme activity in the diamondback terrapin (*Malaclemys terrapin*). Comparative Biochemistry and Physiology A 158:477–484.

Wilson, D.S., C.R. Tracy and C.R. Tracy. 2003. Estimating age of turtles from growth rings: a critical evaluation of the technique. Herpetologica 59:178-194.

Winters, J.M. 2013. The effects of bulkheading on diamondback terrapin nesting in Barnegat Bay, New Jersey. PhD dissertation, Drexel University, Philadelphia, PA. 115 pages.

Witczak, L.R., J.C. Guzy, S.J. Price, J.W. Gibbons and M.E. Dorcas. 2014. Temporal and spatial variation in survivorship of diamondback terrapins (*Malaclemys terrapin*). Chelonian Conservation and Biology 13:146-151.

Yahner, R.H. and A.L. Wright. 1985. Depredation on artificial ground nests: effects of edge and plot age. Journal of Wildlife Management 49:508-513.

Evolutionary History and Paleontological Record

DANA J. EHRET

BENJAMIN K. ATKINSON

The fossil record of the diamond-backed terrapin (*Malaclemys terrapin*) is sparse. Records include late Pleistocene carapace fragments from Edisto Beach in South Carolina (Dobie and Jackson 1979; Ehret and Atkinson 2012), the Aucilla River in Florida (Webb and Simons 2006; Ehret and Atkinson 2012), and the Brunswick River in Georgia (Ehret and Atkinson 2012), as well as a Holocene shell and postcranial elements from Bermuda (Parham et al. 2008) (Fig. 3.1). Extant diamond-backed terrapins occur along the Atlantic and Gulf coasts of the United States from Cape Cod, Massachusetts, to Texas (Ernst and Lovich 2009), with a naturally occurring population in Bermuda (Davenport et al. 2005; Parham et al. 2008). Unlike other emydids, *Malaclemys* is specifically adapted to salt marshes, estuaries, tidal creeks, and mangroves (Wood 1977; Parham et al. 2008; Ernst and Lovich 2009).

We review here the fossil record of *Malaclemys* and its phylogenetic position, introduce a new occurrence from Louisiana, refute a fossil claim from the Pleistocene of Maryland (Parris and Daeschler 1995), and discuss why more material has not been recovered or identified. The recent appearance of known terrapin fossils (Late Pleistocene [Rancholabrean] and Holocene) limits our knowledge of the paleontological history of the genus. Molecular analyses (mitochondrial and nuclear) suggest that *Malaclemys* is closely related to both *Graptemys* and *Trachemys*; however, conflicting phylogenetic trees complicate the known evolutionary history of the deirochelyines (Seidel and Jackson 1990; Bickham et al. 1996; Lamb and Osentoski 1997; Stephens and Wiens 2003; Spinks and Shaffer 2009; Wiens et al. 2010).

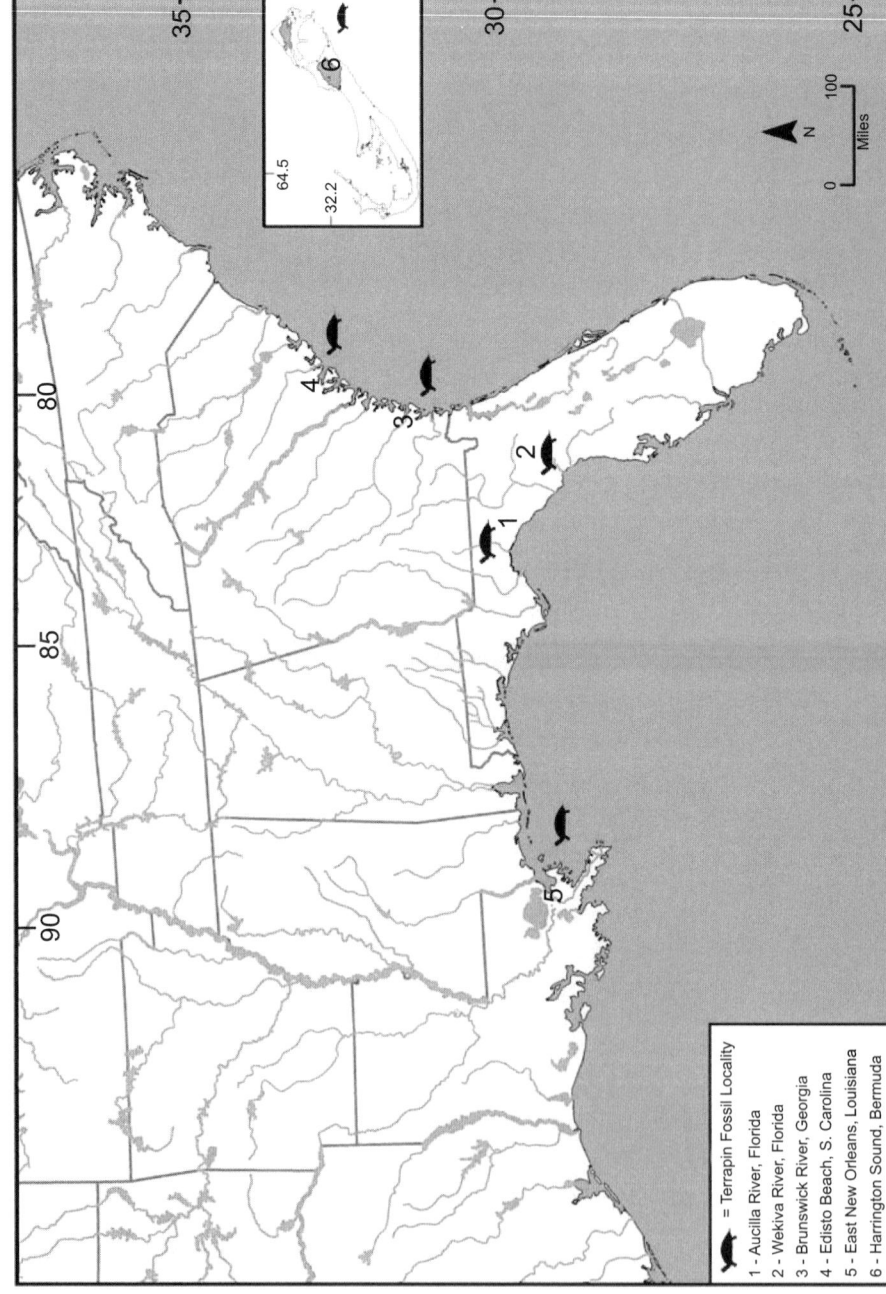

Fig. 3.1. Map of known *Malaclemys terrapin* fossil localities. Bermuda, located outside the map area, is shown in the inset

Methods and Materials

Most of the fossils that have been referred to the diamond-backed terrapin are disarticulated carapacial bones. (The nearly complete, more recent shell and postcranial material recovered from Bermuda is an exception.) We used the following combined criteria to distinguish *Malaclemys* fossils from other emydids: relative thinness of the shell elements with respect to overall size; the presence of well-defined and somewhat raised concentric annuli, rib scars, and scute sulci patterns; and the deep and wide nuchal notch that identifies the nuchal bone. Dobie and Jackson (1979), Parham et al. (2008), and Ehret and Atkinson (2012) provide a more complete treatment of previously described specimens.

We and our colleagues collected several of the fossils discussed in this chapter through a combination of surface prospecting and snorkeling/diving. Other specimens described before Ehret and Atkinson's 2012 publication were collected by a combination of surface prospecting and dredging (Table 3.1). Specimens from Edisto Beach, Colleton County, South Carolina, were surface-collected along the shoreline and were probably washed in from the continental shelf along the Atlantic Ocean (Fig. 3.2, C, F, G, I) (Dobie and Jackson 1979; Roth and Laerm 1980; Ehret and Atkinson 2012). Materials recovered from Andrews Island, Glynn County, Georgia, were found in dredge spoil piles from the South Brunswick River, Georgia (Fig. 3.2, D, M-O) (Ehret and Atkinson 2012). One specimen, UF 256391, was collected while snorkeling along the bottom of the Wekiva River, Levy County, Florida (Fig. 3.2, E) (Ehret and Atkinson 2012). Materials from the Aucilla River, Taylor County, Florida, were recovered by scuba diving as part of the 1987-1994 Aucilla River Project of the Florida Museum of Natural History (Fig. 3.2, A-B, H, J-L) (Webb and Simons 2006). One specimen, MMNS 5394, was recovered from dredge spoil in Orleans County, Louisiana, in the late 1970s; this previously undescribed fragment was recently brought to our attention (Fig. 3.2, P).

Materials discussed here are deposited in the Division of Vertebrate Paleontology of the Florida Museum of Natural History (UF), Gainesville, Florida; the Chelonian Research Institute (PCHP), Oviedo, Florida; the Mississippi Museum of Natural History (MMNS), Jackson, Mississippi; the Bermuda Aquarium, Museum, and Zoo (BAMZ), Flatts FL 04, Bermuda; and the American Museum of Natural History (AMNH), New York. Anatomical abbreviations include: C, costal bone (e.g., C1 = 1st costal); N, neural bone; P, peripheral bone; V, vertebral scute; Pl, pleural scute; and M, marginal scute.

Description of Fossil Material

Family Emydidae (Bell 1825)
Subfamily Deirocheylinae (Agassiz 1857)
Genus Malaclemys (Schoepff 1793)
Malaclemys terrapin
Referred fossil materials: See Table 3.1;
    Figs. 3.2, 3.3

Malaclemys fossils have been found in deposits from the coastal plain of the southeastern United States (Fig. 3.1; Table 3.1) (Dobie and Jackson 1979; Parham et al. 2008; Ehret and Atkinson 2012). The only exception is the specimen collected from Bermuda, which was recovered in a cave located at Paynter's Hill (Parham et al. 2008).

Specimens collected in Florida, Georgia, and South Carolina are considered Rancholabrean (Late Pleistocene) in age, meaning they are 240,000 years old or younger (Ehret and Atkinson 2012). Previous work on the Edisto Beach, Aucilla River, and Wekiva River localities indicated Rancholabrean ages based on associated mammalian fossils (Dobie and Jackson 1979; Roth and Laerm 1980; Webb and Simons 2006; Ehret and Atkinson 2012). We infer that *Malaclemys* specimens from the South Brunswick River are also Rancholabrean, based on other associated turtle material, which includes *Trachemys* cf. *T. scripta*, *Chelydra* cf. *C. serpentina*, Cheloniidae indet., and *Geochelone crassicutata* (Ehret and Atkinson 2012). While most of these species remain extant, except for *G. crassiscutata*, the fossils are mineralized and dredged from an estuary where a majority of the taxa are no longer present.

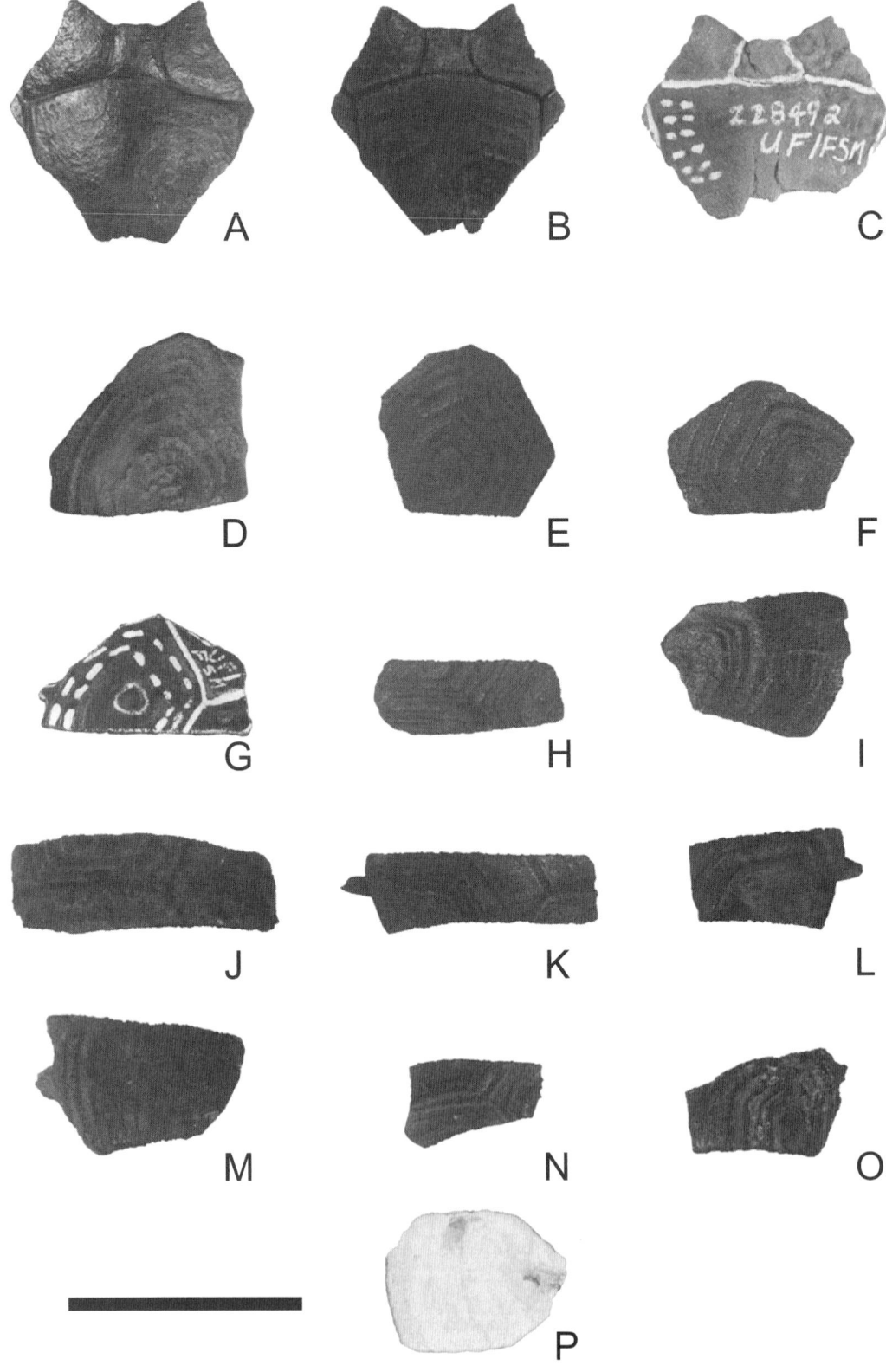

Table 3.1. Catalog numbers, localities, and skeletal identifications of *Malaclemys terrapin* fossils identified and discussed in the chapter

Catalog number	Locality	Elements present
BAMZ 2006-237-0001	Paynter's Hill, Bermuda	Carapace, plastron, and postcranial elements
UF 92451	Taylor County, FL	Nuchal
UF 22849	Colleton County, SC	Nuchal
UF 247192	Taylor County, FL	Nuchal
UF 256391	Levy County, FL	1st left costal
UF 256392	Glynn County, GA	1st left costal
PCHP 12754	Colleton County, SC	1st right costal
UF 131346	Taylor County, FL	2nd left costal
UF 132617	Taylor County, FL	2nd right costal
MMNS 5394	Orleans County, LA	3rd right costal fragment
UF 227628	Taylor County, FL	4th left costal
UF 227629	Taylor County, FL	5th left costal
UF 256393	Glynn County, GA	5th left costal
UF 227627	Taylor County, FL	5th right costal
PCHP 12753	Edisto Beach, SC	5th right costal fragment
UF 256394	Glynn County, GA	6th left costal
UF 256395	Glynn County, GA	7th right costal
UF 256396	Glynn County, GA	Two costal fragments

Note: The BAMZ material is illustrated in Fig. 3.3; the remaining material is illustrated in Fig. 3.2.

The Bermuda materials were recovered from the surface of a cave floor and therefore could not be incorporated into a stratigraphic framework (Parham et al. 2008). Instead, conservative radiometric dates for samples were cited as between AD 1427 and 1620, assuming a marine diet (Parham et al. 2008), placing the specimen firmly in the late Holocene (11,700 years BP [before the present]).

The Louisiana specimen was recovered from a chronologically mixed sample that was pipe-dredged from ~25 m depth within the East New Orleans back levee (Hollander and Dockery 1977; G. Phillips, MMNS, pers. comm.) The sample contained three temporally mixed assemblages consisting of distinctive sedimentary layers (from youngest to oldest): Holocene$_2$ Marine Mollusk Sand (see Hollander and Dockery 1977) and Holocene$_1$ Calcareous Sand (containing shells, crab concretions, archaeological bone, and charcoal), both of which are roughly 4,000 to 5,000 years old,

and a Pleistocene (Rancholabrean) vertebrate facies (containing marine fish, freshwater turtles [*Trachemys*, *Pseudemys*, *Deirochelys*, and *Apalone*], a manatee [*Trichechus*], and some terrestrial mammals; G. Phillips, pers. comm.). Based on its preservation and appearance, the *Malaclemys* specimen was probably part of the Holocene$_1$ assemblage; however, without conducting radiometric dating we conservatively refer the specimen to the Holocene.

The newly recognized specimen collected from the East New Orleans back levee dredge material (Orleans County, LA) represents a right C3 (MMNS 5394; Fig. 3.2, P). Only the distal third of the costal is present. The rib end is worn flush with the distal edge of the bone. The anterior edge is chipped distally. The fossil is 23 mm wide at the distal margin and 31 mm long. The thickness of the bone is 5.7 mm at the distal edge where the rib end would have projected. This specimen is thicker than other specimens; we believe it belonged to a large, old

Fig. 3.2. *Facing page:* Fossil *Malaclemys terrapin* specimens. (A) UF 247192, nuchal; (B) UF 92451, nuchal; (C) UF 22849, nuchal; (D) 256392, left first costal; (E) UF 256391, left first costal; (F) PCHP 12754, right first costal; (G) UF 22849, left first costal; (H) UF 131346, left second costal; (I) PCHP 12753, right fifth costal; (J) UF 227628, left fourth costal; (K) UF 227629, left fifth costal; (L) UF 227627, right fifth costal; (M) UF 256393, left fifth costal; (N) UF 256394, left sixth costal; (O) UF 256395, right seventh costal; (P) MMNS 5394, costal fragment. Scale bar = 5 cm. See Table 3.1 for localities.

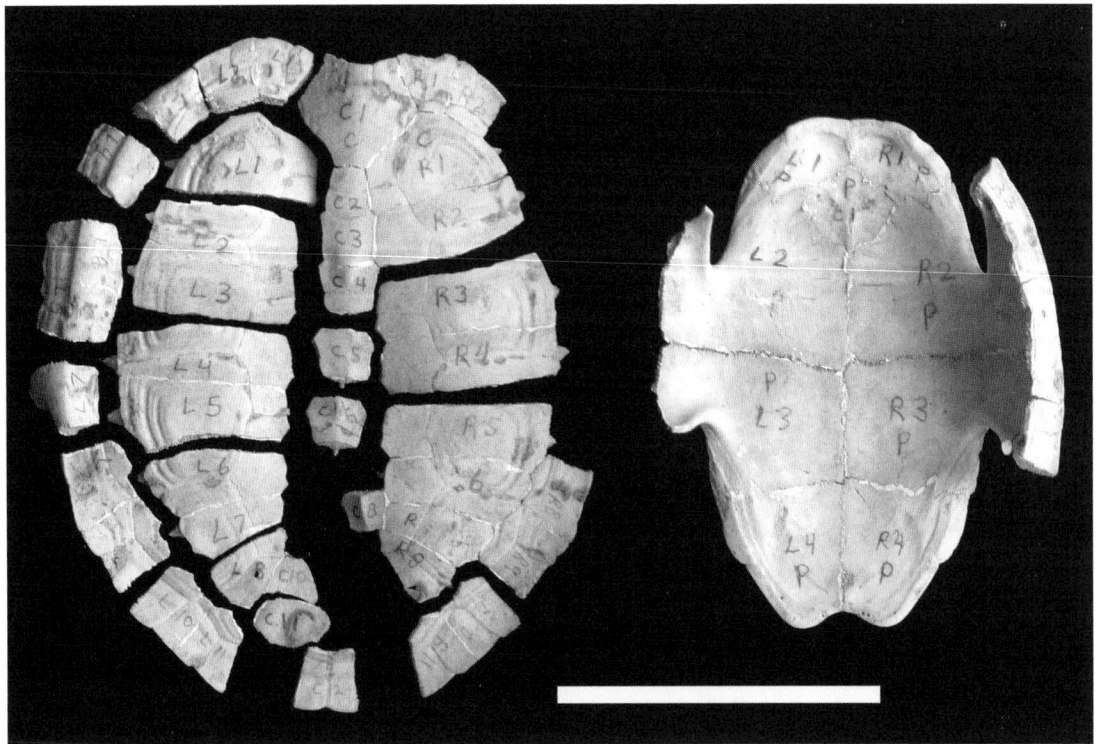

Fig. 3.3. Fossil *Malaclemys terrapin* specimen from Bermuda: BAMZ 2006-237-001. Scale bar = 5 cm. Image modified from Parham et al. 2008.

female, based on extant comparative material. Four scute annuli are still visible on the dorsal surface, though most of the bone is worn smooth— also indicative of a bigger, older terrapin. Although not as old geologically as other specimens discussed here and by Ehret and Atkinson (2012), the fossil is fully mineralized and is clearly not from a recent individual.

Finally, an entoplastron currently associated with (i.e., within the same box as) the holotype of *Terrapene eurypygia* (Cope 1869; AMNH 1484) from Talbot County, Maryland, was referred, on the basis of sulcus patterns, to *Malaclemys* by Parris and Daeschler (1995), who suspected that the bone did not belong with the holotype. This specimen was figured by Cope (1869) in his original description of *Cistudo eurypygia* and in subsequent reviews by Hay (1902, 1908). We examined the fossil entoplastron and agree with Parris and Daeschler (1995) that it is incorrectly associated with AMNH 1484. The gular scute sulci are posteriorly situated on the bone, which is not consistent with *Malaclemys*.

Identification of the entoplastron as Emydidae *incertae sedis* is more appropriate.

Discussion

The fossil record for *Malaclemys* is not as robust as that for many other emydid genera (e.g., *Deirochelys*, *Trachemys*, *Pseudemys*, and *Terrapene*; Hay 1908; Auffenberg 1958; Jackson 1977, 1978, 1988; Holman 1995; Bentley and Knight 1998). However, there is now definitive evidence that *Malaclemys* established populations in Atlantic coastal waters from South Carolina to Florida and along the Gulf Coast of Florida westward to Louisiana by the late Pleistocene (Fig. 3.1; Table 3.1). Additional evidence shows that *Malaclemys* inhabited the Mississippi River delta of Louisiana, as well as Bermuda, in the Holocene. Parham et al. (2008) demonstrated that a naturally occurring population of *Malaclemys* inhabited Bermuda before the presence of humans on the island (however, see Chapter 4). The presence of fossils from Edisto

Beach, South Carolina, supports Parham et al.'s (2008) hypothesis that terrapins could have been swept offshore into the Gulf Stream current and carried toward Bermuda in the late Pleistocene or Holocene.

A lack of pre-Rancholabrean *Malaclemys* fossil specimens (or any emydid materials that exhibit *Malaclemys*-like characteristics) precludes inference about geographic origins with regard to a specific drainage within the Atlantic or Gulf coast regions. Wood (1977) hypothesized that subspecies of *Malaclemys* independently gave rise to different species of *Graptemys* during the late Pleistocene. This hypothesis would imply that *Malaclemys* evolved before the late Pleistocene and that *Graptemys* is polyphyletic. However, the fossil record does not support this, and genetic evidence has rejected a polyphyletic origin of *Graptemys*, negating this hypothesis (Lamb and Osentoski 1997; Stephens and Wiens 2003; Spinks and Shaffer 2009; Wiens et al. 2010). Lamb and Osentoski (1997) calculated an estimated divergence time for *Malaclemys* and *Graptemys* from a common ancestor in the late Miocene (~7 to 11 million years ago) based on cytochrome *b* sequence data. This calculation suggests that there is much more work to be done on Mio-Pleistocene emydids. Furthermore, abundant disarticulated Neogene and Quaternary emydid material in museum collections from the southeastern Gulf Coastal Plain suggests that more terrapin fossils will be identified (Jackson 1975, 1977, 1988; Bentley and Knight 1998; Meylan et al. 2001; Phillips 2006; Ehret and Bourque 2011).

Populations of *Malaclemys* inhabited the Atlantic Coast from South Carolina to Florida and along the Gulf Coast before 12,000 years ago, with the possibility of their presence dating back to the Miocene (up to 10 million years ago). The lack of *Malaclemys* (or *Malaclemys*-like) fossils from freshwater deposits where other emydid fossils are prevalent suggests that the common ancestor invaded brackish water and adapted into the salt-tolerant *Malaclemys* lineage that could inhabit estuaries and salt marshes. This hypothesis entertains an evolutionarily parsimonious scenario of rapid speciation, instigated by rapid shifts in sea level. For example, Donoghue (2006) estimated a period of rapid sea-level rise 18,000 to 7,000 radiocarbon years ago, when the shoreline at the Aucilla River mouth in Florida was retreating northward across the continental shelf at a rate of approximately 18 m/year. Ephemeral environments might have promoted species able to adapt quickly to the rapidly changing conditions. The ancestral *Malaclemys* population is likely to have lived close to the coastlines of east-central North America and adapted to ecosystem shifts during transgressive/regressive cycles at some point during the Mio-Pleistocene. Sliders, both yellow-bellied (*Trachemys scripta scripta*) and red-eared (*T. s. elegans*) subspecies, have been documented inhabiting brackish and sometimes even saline environments (Gibbons and Coker 1978; Moll and Moll 2004; Ehret and Parker 2005; Ernst and Lovich 2009). Therefore, it is conceivable that a population of deirochelyine turtles could have transitioned into a brackish environment and evolved into the genus *Malaclemys*, which then spread along the Atlantic and Gulf coasts.

Morphological and genetic studies of the Emydidae indicate a close taxonomic relationship and evolutionary history for *Malaclemys*, *Graptemys*, and *Trachemys* (McDowell 1964; Wood 1977; Dobie 1981; Bickham et al. 1996; Lamb and Osentoski 1997; Stephens and Wiens 2003; Spinks and Shaffer 2009; Wiens et al. 2010). A sister-taxon relationship between *Malaclemys* and *Graptemys* is supported by both mitochondrial and nuclear DNA analyses (Bickham et al. 1996; Lamb and Osentoski 1997; Stephens and Wiens 2003; Wiens et al. 2010; Chapter 4). A *Malaclemys*-*Graptemys* relationship also has been suggested based on congruent morphological features (McDowell 1964). Obvious parallel and possibly homologous morphological adaptations include: the presence of neural protuberances in juvenile Florida Gulf coast *Malaclemys* and the genus *Graptemys* (Wood 1977; Lamb and Osentoski 1997); the macrocephaly of adult females of both genera; and the substantial sexual dimorphism exhibited by these taxa, with adult females ultimately growing much larger than adult males (Ernst and Lovich 2009; Lindeman 2013).

The ecological restriction of diamond-backed terrapins to coastal salt marshes, mangroves, and estuaries limits the potential for finding their fossils. Fluctuations in sea level further limit the time that the fossil record is exposed and accessible (Reed 1990; Webb 1990; Emslie 1998; Donoghue 2006). It is quite likely that deposits containing *Malaclemys* fossils are currently offshore along both the Atlantic and Gulf coasts and are inaccessible. Furthermore, the fragile nature of terrapin fossils makes preservation less likely than for other, thicker-shelled, emydids. Future prospecting for fossil *Malaclemys* should target deposits where salt marshes and estuaries occurred during prior glacial/interglacial cycles, including estuaries that reflect earlier shorelines during the late Miocene through Pleistocene, particularly in the Gulf Coastal Plain. Another potential source for diamond-backed terrapin fossils is museum collections. Neogene emydids are understudied; the plethora of fossil materials already housed in research collections will provide surprises for many years to come. We hope paleontological investigations will continue to provide important supplemental data to extant morphological and molecular studies, shedding further light on this extraordinary species.

ACKNOWLEDGMENTS

We thank J. M. Butler, A. Kerner, and C. Kirby for the donation of fossil terrapin materials; J. Bourque, M. A. Nickerson, R. Hulbert, S. D. Webb, P. C. H. Pritchard, R. C. Wood, and G. Phillips for providing access to fossil and recent terrapin materials; M. Outerbridge for use of fossil images; and J. Parham for useful comments and discussion.

REFERENCES

Agassiz, L. 1857. Contributions to the Natural History of the United States of America. Volume 1. Little, Brown and Company, Boston, MA. 452 pages.

Auffenberg, W. 1958. Fossil turtles of the genus *Terrapene* in Florida. Bulletin of the Florida State Museum 3(2):1-92.

Bell, T. 1825. A monograph of the tortoises having a moveable sternum, with remarks on their arrangement and affinities. Zoological Journal of London 2:299-310.

Bentley, C.C. and J.L. Knight. 1998. Turtles (Reptilia: Testudines) of the Ardis local fauna Late Pleistocene (Rancholabrean) of South Carolina. Brimleyana 25:3-33.

Bickham, J.W., T. Lamb, P. Minx and J.C. Patton. 1996. Molecular systematics of the genus *Clemmys* and the intergeneric relationships of emydid turtles. Herpetologica 52:89-97.

Cope, E.D. 1869. Synopsis of the extinct Batrachia, Reptilia and Aves of North America. Transactions of the American Philosophical Society, New Series 14(1):1-252.

Davenport, J., A. Glasspool and L. Kitson. 2005. Occurrence of diamondback terrapins, *Malaclemys terrapin*, on Bermuda: native or introduced? Chelonian Conservation and Biology 4:956-959.

Dobie, J.L. 1981. The taxonomic relationship between *Malaclemys* Gray, 1844 and *Graptemys* Agassiz, 1857 (Testudines: Emydidae). Tulane Studies in Zoology and Botany 23:85-102.

Dobie, J.L. and D.R. Jackson. 1979. First fossil record for the diamondback terrapin, *Malaclemys terrapin* (Emydidae), and comments on the fossil record of *Chrysemys* (Emydidae). Herpetologica 35:139-145.

Donoghue, J.F. 2006. Geography and geomorphology of the Aucilla River region. Pages 31-48 in S.D. Webb (ed.), First Floridians and Last Mastodons: The Page-Ladson Site in the Aucilla River. Springer Press, Dordrecht, The Netherlands.

Ehret, D.J. and B.K. Atkinson. 2012. The fossil record of the diamondback terrapin *Malaclemys terrapin* (Testudines: Emydidae). Journal of Herpetology 46:351-355.

Ehret, D.J. and J. Bourque. 2011. An extinct map turtle *Graptemys* (Testudines: Emydidae) from the Pleistocene of Florida. Journal of Vertebrate Paleontology 31:575-587.

Ehret, D.J. and D. Parker. 2005. *Trachemys scripta elegans*: geographic distribution. Herpetological Review 36:78.

Emslie, S.D. 1998. Avian community, climate, and sea-level changes in the Plio-Pleistocene of the Florida peninsula. Ornithological Monographs 50:1-113.

Ernst, C.H. and J.E. Lovich. 2009. Turtles of the United States and Canada. 2nd edition. Johns Hopkins University Press, Baltimore, MD. 827 pages.

Gibbons, J.W. and J.W. Coker. 1978. Herpetofaunal colonization patterns of Atlantic coast barrier islands. American Midland Naturalist 99:212-233.

Hay, O.P. 1902. Bibliography and catalogue of fossil vertebrata of North America. Bulletin of the United States Geological Survey 179:1-868.

Hay, O.P. 1908. The Fossil Turtles of North America. Carnegie Institution of Washington, Pub. no. 75. Carnegie Institution, Washington, DC. 568 pages + 113 plates.

Hollander, E.E. and D.T. Dockery III. 1977. Molluscan assemblages of the Mid-Holocene New Orleans Barrier Trend. Compass 55:1-29.

Holman, J.A. 1995. Pleistocene Amphibians and Reptiles in North America. Oxford University Press, New York, NY. 256 pages.

Jackson, D.R. 1975. A Pleistocene *Graptemys* from the Santa Fe River of Florida. Herpetologica 30:213-219.

Jackson, D.R. 1977. Fossil freshwater emydid turtles of Florida. PhD dissertation, University of Florida, Gainesville, FL. 128 pages.

Jackson, D.R. 1978. Evolution and fossil record of the chicken turtle *Deirochelys* with a re-evaluation of the genus. Tulane Studies in Zoology and Botany 20:35-55.

Jackson, D.R. 1988. A re-examination of fossil turtles of the genus *Trachemys* (Testudines: Emydidae). Herpetologica 44:317-325.

Lamb, T. and M.F. Osentoski. 1997. On the paraphyly of *Malaclemys*: a molecular genetic assessment. Journal of Herpetology 31:258-265.

Lindeman, P.V. 2013. The Map Turtle and Sawback Atlas: Ecology, Evolution, Distribution, and Conservation. University of Oklahoma Press, Norman, OK. 488 pages.

McDowell, S.B. 1964. Partition of the genus *Clemmys* and related problems in the taxonomy of the aquatic Testudinidae. Proceedings of the Zoological Society of London 143:239-279.

Meylan, P.A., W.A. Auffenberg and R.C. Hulbert. 2001. Reptilia 1: turtles and tortoises. Pages 118-136 in R.C. Hulbert (ed.), The Fossil Vertebrates of Florida. University Press of Florida, Gainesville, FL.

Moll, D. and E.O. Moll. 2004. The Ecology, Exploitation, and Conservation of River Turtles. Oxford University Press, New York, NY. 420 pages.

Parham, J.F., M.E. Outerbridge, B.L. Stuart, D.B. Wingate, H. Erlenkeuser and T.J. Papenfuss. 2008. Introduced delicacy or native species? A natural origin of Bermudian terrapins supported by fossil and genetic data. Biology Letters 4:216-219.

Parris, D.C. and E. Daeschler. 1995. Pleistocene turtles of Port Kennedy Cave (Late Irvingtonian), Montgomery County, Pennsylvania. Journal of Paleontology 69:563-568.

Phillips, G.E. 2006. Paleofaunistics of nonmammalian vertebrates from the late Pleistocene of the Mississippi-Alabama black prairie. MS thesis, North Carolina State University, Raleigh, NC. 277 pages.

Reed, D.J. 1990. The impact of sea-level rise on coastal salt marshes. Progress in Physical Geography 14:465-481.

Roth, J.A. and J. Laerm. 1980. A Late Pleistocene vertebrate assemblage from Edisto Island, South Carolina. Brimleyana 3:1-29.

Schoepff, J.D. 1793. *Testudo terrapin*. Page 64 in Historia Testudinum Iconibus Illustrata. Fascicles III and IV, pages 33-80, plates 11-16, 17B-20. Ioannis Iacobi Palm, Erlangae.

Seidel, M.E. and D.R. Jackson. 1990. Evolution and fossil relationships of slider turtles. Pages 68-73 in J.W. Gibbons (ed.), Life History and Ecology of the Slider Turtle. Smithsonian Institution Press, Washington, DC.

Spinks, P.Q. and H.B. Shaffer. 2009. Conflicting mitochondrial and nuclear phylogenies for the widely disjunct *Emys* (Testudines: Emydidae) species complex, and what they tell us about biogeography and hybridization. Systematic Biology 58:1-20.

Stephens, P.R. and J.J. Wiens. 2003. Ecological diversification and phylogeny of emydid turtles. Biological Journal of the Linnean Society 79:577-610.

Webb, S.D. 1990. Historical biogeography. Pages 70-100 in R. L. Myers and J. J. Ewel (eds.), Ecosystems of Florida. University of Central Florida Press, Orlando, FL.

Webb, S.D. and E. Simons. 2006. Vertebrate paleontology. Pages 215-246 in S.D. Webb (ed.), First Floridians and Last Mastodons: The Page-Ladson Site in the Aucilla River. Springer Press, Dordrecht, The Netherlands.

Wiens, J.J., C.A. Kuczynski and P.R. Stephens. 2010. Discordant mitochondrial and nuclear gene phylogenies in emydid turtles: implications for speciation and conservation. Biological Journal of the Linnean Society 99:445-461.

Wood, R.C. 1977. Evolution of the emydine turtles *Graptemys* and *Malaclemys* (Reptilia, Testudines, Emydidae). Journal of Herpetology 11:415-421.

4

Taxonomy: A History of Controversy and Uncertainty

JEFFREY E. LOVICH

KRISTEN M. HART

What's in a name? That which we call a rose/By any other word would smell as sweet." When these words were spoken by Juliet in Shakespeare's *Romeo and Juliet*, she was proclaiming the inadequacy of words to convey deeper meanings. Such is often the case in zoological nomenclature. The proclivity of humans to name, catalog, and describe things in our best efforts to bestow order on the universe—extending as far back in human history as the Biblical story of Adam's naming of all the animals—is often complicated by levels of diversity and complexity that can defy simple classification schemes that are limited to words. The common and scientific names applied to the diamond-backed terrapin, *Malaclemys terrapin*, have been subjected to such complications. The species displays a remarkable degree of pattern and morphological variation across its range (Ernst and Lovich 2009), contributing to a taxonomic and nomenclatural history punctuated by confusion and controversy, especially over how many races or subspecies should be recognized. The purpose of this chapter is to review this history, describe the traits used to discriminate among the putative taxa, discuss current research findings, and suggest ways to stabilize the intraspecific taxonomy of this iconic species.

The common name "terrapin" has an interesting and perhaps little appreciated history (Ernst and Lovich 2009). Although associated with turtles, it is rarely used in "formal" lists of common names. In a recent checklist of the 335 turtles of the world (van Dijk et al. 2014), only 5 of the 94 genera (*Batagur, Malaclemys, Mauremys, Pelomedusa,* and *Pelusios*)—collectively occupying four continents: Africa, Asia, Europe, and North America—incorporate the word "terrapin" in the common names of at least one species ($N=10$). That the name "terrapin" would achieve near global usage is even more remarkable given its humble origin as a Native American word. According to the Merriam-Webster online dictionary (www.merriam-webster.com/dictionary /terrapin), "terrapin" is an alteration of the word "torope" used by Algonquian Native Americans of Virginia to denote turtles. The dictionary further states that the first known use of the word "terrapin" was in 1613. Early European colonists of eastern North America adopted the word, allowing it to enter the English lexicon, as well as other languages. Despite the certainty of attribution associated with terrapins in antiquity, later recognition of terrapin names applied to subspecies or races was less than definitive.

We reluctantly concede to the common name "diamond-backed" terrapin as has been suggested (Crother et al. 2012), although some contributors to this volume, including us, prefer "diamondback." Much ado has been made of "standardizing" common names of animals, including amphibians and reptiles (Crother et al. 2012). We recognize that what some consider a common name is not necessarily used commonly and can vary from region to region, culture to culture, and even publication to publication. For example, one of us has used both hyphenated (Ernst and Lovich 2009) and non-hyphenated (Lovich and Gibbons 1990) versions of the name for terrapins. Part of the reason for the now-global dominance of the English language is that it has been subjected to minimal standardization and has the flexibility to accommodate user-driven change, including indigenization (Bruthiaux 2002). Standardizing common names of animals restricts that flexibility in ways best reserved for Latin binomial nomenclature.

Detailed reviews of the taxonomic and nomenclatural history of *M. terrapin* and its various subspecies are available from several sources, including Ernst and Bury (1982), Fritz and Havaš (2007), and van Dijk et al. (2014). Instead of repeating that information, in the systematic accounts that follow we borrow from those and other reviews, especially a recent presentation on the topic (Ernst and Lovich 2009). We add new information only as necessary, including more details on external characteristics and the distribution of Florida subspecies (Butler et al. 2006) and terra-

pins from Bermuda. The reader is referred to these earlier publications for details beyond the scope of this chapter. Our intent here is to summarize briefly the characteristics and taxonomic history of each of the recognized subspecies of terrapins and then discuss the body of evidence in support of their continued recognition. In addition, we review the controversial history and status of terrapins in Bermuda, even though their taxonomic affinities remain unclear.

Position of the Genus *Malaclemys* in Turtle Phylogenies

The relationship of the monotypic genus *Malaclemys* to other turtle genera has been subject to various interpretations. On the basis of three skull characteristics, McDowell (1964) lumped the genus *Graptemys* with *Malaclemys*, an arrangement first suggested by Boulenger (1889) under the junior synonym *Malacoclemmys terrapen* (not *"terrapin"* as shown in Ernst and Bury 1982). In his treatment, McDowell (1964) did not even recognize *Graptemys* as a subgenus, stating that differences between *Malaclemys* and some *Graptemys* were no more significant than differences between a pair of *Graptemys* species. Paradoxically, he also stated that dividing his consolidated genus *Malaclemys* into two groups would not adequately express the structural diversity in the group and that he did not have "sufficient material to divide the genus into its many natural groupings." His arrangement was ultimately rejected by other researchers.

Wood (1977) and Dobie (1981) disagreed with McDowell (1964), and Wood (1977) hypothesized that various populations of *Malaclemys* gave rise to many or most species of *Graptemys*. Dobie (1981) maintained that both genera arose from *Pseudemys*-like stock or that *Malaclemys* was derived from *Graptemys* stock, the opposite suggestion to Wood's (1977). Modern molecular analyses rejected both proposals, redefining our understanding of the relationships between these and other taxa (Lamb and Osentoski 1997; Stephens and Wiens 2003). For example, Lamb et al. (1994) hypothesized that a *Graptemys/Malaclemys* ancestor was derived from a *Trachemys*-like species (not *Pseudemys*) in the early

Middle Miocene. The sister taxon relationship between *Trachemys* and *Graptemys* is supported by other studies (Bickham et al. 1996; Shaffer et al. 1997; Spinks et al. 2009; Thomson and Shaffer 2010; Wiens et al. 2010). Available evidence suggests that *Malaclemys* and *Graptemys* diverged 7 to 11 million years ago during the Late Miocene (Lamb and Osentoski 1997).

Traditionally Recognized Subspecies of *Malaclemys*

Since the diamond-backed terrapin was formally described by Schoepff in 1793, seven subspecies have been named over 162 years, or about one every 23 years, including the nominate form, *M. t. terrapin*. Three were recognized before 1900 (Fig. 4.1), and four more were described in the years that followed. The last subspecies described was *M. t. tequesta* from the Atlantic Coast of Florida in 1955 (Schwartz 1955). With the exception of *M. t. pileata*, more northern subspecies were described before 1900, while the remaining southern subspecies were described thereafter. In particular, the majority of Florida subspecies were recognized later than the others.

The ranges of the traditionally recognized subspecies correspond closely to six major physiographic regions along the US Atlantic and Gulf coasts (Thornbury 1965), as documented by Hartsell and Ernst (2004). In the following descrip-

tions, we discuss the major biological, geological, and oceanographic features that characterize the ranges of each putative subspecies. Type specimen localities are taken from Ernst and Bury (1982), and the reader is referred there for additional information on synonymies, etc. Various measures of adult or maximum body size (carapace or plastron length) are given, as available in the literature, to further characterize each putative taxon.

Malaclemys terrapin terrapin (Schoepff 1793)

As its common name suggests, the northern diamond-backed terrapin is the northernmost subspecies, ranging along the Atlantic Coast from Cape Cod (Wellfleet Bay) (Lazell 1969; Holden 1979) to Cape Hatteras or a little farther south (Hildebrand and Prytherch 1947). The type locality for this subspecies is unknown, but the type for the species was restricted to "coastal waters of Long Island [New York]." Its median keel does not possess knobs, and the sides of its carapace diverge posteriorly (although this latter character is not consistent). The carapace color varies from black to light brown, with distinct concentric rings, and the plastron is orangish to greenish gray. The modal carapace length of males from near Lewes, Delaware, was 11.0 cm, but some were reported up to 18.0 cm (Hurd et al. 1979). Females in the same population attained a maximum carapace length of about 21.0 cm. Males and females in Chesapeake Bay reach carapace lengths of 17.1 cm and 26.5 cm, respectively (W. Roosenburg, Ohio University, pers. comm.).

The habitat occupied by this subspecies closely corresponds to the Embayed Section of the Atlantic Coast, a physiographic region that stretches from Cape Cod, Massachusetts, to just south of the Neuse River, North Carolina. The region is characterized by productive large bays and estuaries. The northern portion is influenced by the cold Labrador Current and is thus the coldest portion of the range occupied by *M. terrapin*. South of about Long Island, New York, terrapin populations are more affected by the warm Gulf Stream Current.

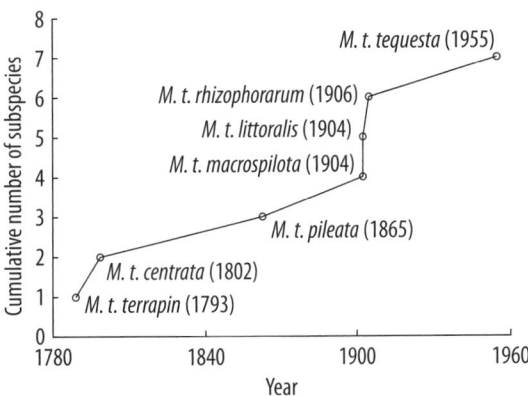

Fig. 4.1. Accumulation of subspecies designations for *Malaclemys terrapin* over time. Year of formal description is given after each taxon.

Malaclemys t. centrata (Latreille, in Sonnini de Manoncourt and Latreille 1802)

The Carolina diamond-backed terrapin ranges from Cape Hatteras to northern Florida. The type locality was restricted to the "vicinity of Charleston, South Carolina." There are few differences between *M. t. centrata* and *M. t. terrapin*, according to Hildebrand (1929). As in the latter subspecies, the median keel of *M. t. centrata* does not possess knobs (Figs. 4.2, 4.3). In addition, in *M. t. centrata*, the sides of the carapace are parallel and the posterior marginals are curled upward. In comparison with *M. t. terrapin*, *M. t. centrata* is reputed to have a larger head and blunter snout (Coker 1906; Hildebrand 1929). Males and females of this taxon attain plastron lengths of 12.0 cm and 18.5 cm, respectively (Hildebrand 1932). Coker (1906) stated that terrapins with plastrons 20.3 cm in length were rare in "Carolina," sug-

gesting that the subspecies may nevertheless attain that size. Mean plastron lengths of males and females at Kiawah Island, South Carolina, were 10.2 cm and 14.7 cm, respectively (Gibbons and Lovich 1990).

The Sea Island Section of the coast occurs from the Neuse River, North Carolina, south to northern Florida, corresponding well with the range of *M. t. centrata*. The region is characterized by fewer estuaries than the Embayed Section to the north but has a border of barrier islands with highly diverse and productive salt marshes on the continental side. The southern part of this section has a subtropical climate (Gibbons and Harrison 1981) nurtured by proximity to the Gulf Stream.

Malaclemys t. tequesta (Schwartz 1955)

The eastern Florida diamond-backed terrapin occurs along the Atlantic Coast of Florida from Volusia

Fig. 4.2. Male *Malaclemys terrapin centrata* from the Edisto River near Seabrook Island, South Carolina. This is an exceptionally patterned specimen, as most are darker and show less well developed patterns on their carapace. Photograph by Jeff Lovich

Fig. 4.3. Male *Malaclemys terrapin centrata* from near Seabrook Island, South Carolina. This specimen is typical of the carapace shading, which can range from gray to brown to black, as seen in this area. Photograph by Jackie Guzy; used with permission

County south to Miami and perhaps the upper Keys (Butler et al. 2006). The type locality is "Miami Beach, Dade County, Florida." As in other terrapin subspecies to the south and west (Lamb and Avise 1992), the median keel possesses posterior-facing knobs, most pronounced in males of this subspecies. The carapace is dark or tan, with little trace of concentric light circles, but the centers of the large carapace scutes may be lighter than surrounding areas. The yellowish plastron may have dark blotches or smudges. The head is silver to gray with patterns consisting of dots and short stripes. Exemplifying the variation seen in terrapin subspecies, specimens with traits of *M. t. tequesta* have been observed in areas where one would normally expect to find *M. t. rhizophorarum* (Butler et al. 2006). Those authors also noted the difficulty in distinguishing between *M. t. tequesta* and *M. t. centrata* from northern Florida. Mean adult male and female carapace lengths are 12.4 cm and 17.3 cm, respectively (Butler et al. 2006).

Terrapins of this subspecies live in the Floridian Section, extending from the Southern Sea Island Section down to the tip of Florida. However, they also share the section with *M. t. rhizophorarum*. This section is characterized by relatively recently submerged formations of carbonate rocks. Offshore there are sandbars with coastward lagoons and small limestone islands or keys. Salt marshes are present in some areas, and extensive groves of mangroves occur at the extreme southern end of the distribution of *M. t. tequesta*. The climate is subtropical throughout the species' range in the Gulf of Mexico.

Malaclemys t. rhizophorarum (Fowler 1906)

The mangrove diamond-backed terrapin (Fig. 4.4) is generally thought to be restricted to the Florida Keys, although some (Butler et al. 2006) have reported it from a wider area extending from the

Fig. 4.4. Diamond-backed terrapin (*Malaclemys terrapin rhizophorarum*) from Big Sable Creek, Everglades National Park, Florida. Note the bulbous protrusions on the vertebral scutes. Photograph by Kristen Hart

Keys to Naples or even Fort Meyers, Florida, on the Gulf Coast. The type locality is "Boca Grande Key, [Lee County] Florida." The subspecies' median keel on the carapace usually possesses bulbous knobs (although it can also be keeled, flat, or smooth), and the shell is strongly oblong, but the sides of the carapace are sometimes parallel, possibly related to age and/or sex (Butler et al. 2006). The dark gray, brown, or black carapace lacks light centers, and the posterior marginal may be flared somewhat. The yellow to orange plastron is characterized by scute seams outlined with black pigmentation. Skin color is typically light gray, the head has black spots and a light dorsal smudge, and the hind legs may be striped. The latter trait holds for Key West and Big Sable Creek, Florida, but not necessarily elsewhere (Butler et al. 2006).

Although the name *M. tuberculifera* is a senior synonym, predating *Malaclemys t. rhizophorarum* by 62 years, the long-standing common usage of the latter was recommended by Ernst and Hartsell (2000a, b). Wood (1994) thought that two morphologically distinct populations of *Malaclemys* occurred within the Florida Keys: a southern one

(*M. t. rhizophorarum*) and a northern one, for which he proposed the name *M. t. fordorum*. However, he provided no description of his new taxon. Thus, it is a *nomen nudum* and invalid. Ernst and Lovich (2009) reported no distinction between *M. t. rhizophorarum* from the northern and southern Keys. Male and female maximum carapace lengths at Big Sable Creek in Everglades National Park, Florida, were 14.2 cm and 20.1 cm, respectively (Hart and McIvor 2008). Mean carapace lengths of males and females were 11.9 cm and 16.8 cm, respectively (Butler et al. 2006). See the description of the Floridian Section of the US coastline given above for a general portrait of the habitat this subspecies shares, in part, with *M. t. tequesta*.

Malaclemys t. macrospilota (Hay 1904)

The ornate diamond-backed terrapin ranges from Florida Bay (Monroe County) to the panhandle of Florida (Walton County). The type locality is "Charlotte Harbor, [Charlotte County] Florida." The carapace of adults is medium to dark gray, and their scutes sometimes have an orange or yellow

Fig. 4.5. Female diamond-backed terrapin at Rockefeller Wildlife Refuge, near Vermillion, Louisiana. The flared marginal and lighter scute centers suggest characters diagnostic for the Mississippi diamond-backed terrapin, *Malaclemys terrapin pileata*. Photograph by Will Selman; used with permission

center. The carapace sides may be parallel or not. The median carapace keel may or may not have terminal, often bulbous, knobs (Butler et al. 2006). The plastron may be orange or mottled with black pigment. Skin color is light to dark gray with dark bars or spots. Mean carapace lengths of males and females were reported as 12.5 cm and 18.1 cm, respectively (Butler et al. 2006).

The East Gulf Coastal Plain Section of the US coast extends from southwestern Florida to Mobile Bay, Alabama, roughly matching the distribution of *M. t. macrospilota*. This region is characterized by a Pleistocene coastal plain consisting of weak rock types including limestone and clay shales. Numerous rivers drain into the Gulf of Mexico, but none are as large as those typical of the Embayed or Sea Island sections. Vegetation types range from mangroves in the south to tidal marshes in the north.

Malaclemys t. pileata (Wied-Neuwied 1865)

The Mississippi diamond-backed terrapin ranges from the Florida panhandle near the western Choctawhatchee Bay in Okaloosa County to western Louisiana near the mouth of the Mississippi River (Hildebrand and Prytherch 1947). The type locality is "New Orleans, Louisiana." The sides of the carapace in this subspecies are usually parallel, with strongly flared marginal edges around the carapace rim (Fig. 4.5). The median carapace keel possesses terminal tuberculate knobs, especially on the fourth vertebral scute, but not necessarily in older females. Males usually have a black carapace with orange marginals, while females have black, yellowish brown, or yellow carapaces with yellow, orange, red, or black marginals (Butler et al. 2006). The scutes of the oval carapace lack light centers. The dorsal surfaces of the head, neck, and limbs are dark brown or black, and the upturned edges of the marginals are orange or yellow. Males and females mature at a plastron length of about 9.9 cm and 17.0 cm, respectively (Cagle 1952). In Mississippi, mean carapace lengths of males and females are 12.1 cm and 16.8 cm, respectively (Butler et al. 2006).

The Mississippi Alluvial Plain Section covers most of the range of *M. t. pileata*. Numerous rivers draining the mainland from north to south, some penetrating deep into the North American continent, created alluvial and deltaic plains with scattered barrier islands. Large salt marshes are associated with river deltas.

Malaclemys t. littoralis (Hay 1904)

The Texas diamond-backed terrapin ranges from western Louisiana to Corpus Christi, Texas, including offshore islands (Hildebrand 1929). The type locality is "Rockport, [Aransas County] Texas." The deep carapace possesses terminal knobs on the median keel and scutes without light centers. The plastron is pale or white. The dorsal surface of the head is usually whitish, and the neck and legs are greenish gray with heavy black spotting. Females may exceed 20.3 cm in plastron length, but male size is similar to that of *M. t. centrata* (Hildebrand 1932). The mean carapace length of 11 egg-bearing females from Comfort Island, Bernard Parish, Louisiana, was 19.5 cm (Burns and Williams 1972). Maximum male and female carapace lengths for terrapins collected at South Deer Island, Texas, were 14.8 cm and 20.1 cm, respectively (Hogan 2003).

The westernmost coastal section, the West Gulf Coastal Plain, is partially congruent with the range of *M. t. littoralis*. This section grades from a geology similar to that of the adjacent Mississippi Alluvial Plain Section in the east to a series of fault systems, igneous intrusions, and salt domes to the west. Vegetation along the coast is similar to that of other sections on the Gulf Coast.

Terrapins in Bermuda

The island of Bermuda is located more than 1,000 km east-southeast of the nearest known habitat for terrapin populations that are unquestionably native: Cape Hatteras, North Carolina. Because of Bermuda's isolation, it has relatively few species of native amphibians or nonmarine reptiles (Bacon et al. 2006). Terrapins were first reported in Bermuda between 1950 and 1953 (Parham et al. 2008), but by the mid-1990s, significant numbers of terrapins (as well as red-eared sliders [Outerbridge 2008], *Trachemys scripta elegans*, a species native to the lower Mississippi River valley [Ernst and Lovich 2009]) were observed in artificial ponds on a golf course and a nearby natural, brackish lake (Davenport et al. 2005). Additionally, a live juvenile terrapin was found in a tide pool at Crawl Point, Hamilton Parish, on May 10, 1974, on the opposite side of the island from the golf course ponds (see photograph in supplementary material of Parham et al. 2008).

Are Terrapins Native to Bermuda?

Davenport et al. (2005) concluded that there were three possible explanations for how terrapins arrived in Bermuda: (1) they could have been imported as pets and subsequently released, as was undoubtedly the case for the red-eared slider; (2) they could have been imported during the heyday of terrapin exploitation as a food source (see Ernst and Lovich 2009 for a recent review of that literature); or (3) they could have arrived on their own, without human intervention. A turtle species adapted to brackish water environments would appear to be an excellent candidate for dispersal in the ocean, but terrapins cannot withstand permanent exposure to seawater (Hartsell and Ernst 2004). (For additional reviews on osmoregulation in terrapins, see Chapter 9 and Ernst and Lovich 2009.)

Davenport et al. (2005) were unable to find any written records or evidence in support of the first two explanations and concluded that it was plausible that terrapins arrived on their own, perhaps from Florida. However, there is little doubt that the red-eared sliders that were syntopic with terrapins were released by humans in Bermuda, so human agency is an established means of turtle transportation to the island. In support of Davenport et al.'s (2005) conclusion that terrapins are native to

Bermuda, there is evidence that other extinct and extant reptiles (e.g., *Hesperotestudo bermudae* and *Eumeces longirostris*) probably arrived in Bermuda from Florida on eddies spun off by the Gulf Stream (Meylan and Sterrer 2000), a distance of more than 2,000 km. Unlike *H. bermudae*, fossils of *M. terrapin* are unknown from Bermuda (but are known from South Carolina, as discussed in Dobie and Jackson 1979; see also Chapter 3), other than a skeleton with some scutes still attached that was found in a dry cave in Hamilton Parish near the extant terrapin populations (Davenport et al. 2005; Parham et al. 2008). Because the shell was not buried, it could not be incorporated into a stratigraphic paleontological framework (Parham et al. 2008).

In the supplementary material associated with their publication online, Parham et al. (2008) described the cave and the position of the specimen as follows:

The cave is one of several deep fissure openings along the cliff face on the south side of Paynter's Hill, approximately 50 m above sea level. A steeply sloped dry soil talus was present inside the cave immediately below a vertical chimney hole near the entrance, which was littered with bones of the cahow (or Bermuda Petrel) *Pterodroma cahow* and Audubon's shearwater *Puffinus lherminieri*. The terrapin bones were located towards the bottom of the talus in an upright position, as if the specimen died there. The skeletal remains were only partly disarticulated with the central segments of the carapace collapsed inwards, the skull was missing, and the lower dentary was at the posterior end. The appendicular skeleton was incomplete, but most of the major bones in both the pectoral and pelvic girdles were present. However, virtually all of the carpals, metacarpals, tarsals, metatarsals and phalanges were missing. Additionally, a number of scales were present, although all were separated from the underlying bones. The straight mid-line carapace and plastron lengths measured 205 mm and 144 mm respectively.

Using the shell, Parham et al. (2008) performed radiometric and genetic analyses to ascertain its age and provenance. The radiometric analysis suggested an estimated age for the shell of AD 1427 to 1620. The latter date would place it 11 years after

human settlement of the island (1609), opening the possibility of human transport. The genetic analysis suggested that the Bermudian specimen is from the vicinity of North or South Carolina, which could argue for a natural or human-mediated mechanism of dispersal. Nevertheless, Parham et al. (2008) concluded that terrapins should be considered a native species in Bermuda.

In contrast, Lever (2003) reported a personal communication from D. B. Wingate (one of the co-authors of Parham et al. 2008), dated 2000, that stated the following:

[The Bermudian terrapin] remains localized in Tucker's Town lakes and ponds with occasional records elsewhere and we are still not sure whether it is native or introduced. I found a skeleton in a cave once with scutes still intact. A carbon date gave an ambiguous reading, i.e., the margin of error spanned the period of Bermuda's settlement after which time this popular gourmet food could have been introduced deliberately to Bermuda from the Chesapeake area.

Subsequent research appears to have altered this preliminary conclusion (Parham et al. 2008).

It is clear to us that additional genetic analysis or discovery of older fossils (Chapter 3) would be desirable to shed more light on the provenance of Bermuda terrapins. Until then, they remain an interesting isolated population worthy of conservation attention.

Description of Bermuda Terrapins

Three adult females were described by Davenport et al. (2005) as having extremely smooth carapaces without noticeable scute annuli or grooves. The dorsal keel was also insignificant. As with other terrapin populations, coloration was variable, ranging from light brown with ivory showing on new scute growth to olive green shells with recent scute growth being light green. Carapace shape ranged from elliptical to oval, and the marginal scutes showed minimal upcurling at the posterior. Skin color ranged from gray or pale gray to light green with dark gray spots. "Moustache" patterns seen in other terrapin populations were not present. The largest specimen was 21.5 cm in carapace length. The juvenile terrapin from Hamilton Parish described and photographed in Parham et al.'s (2008) supplementary material had a decidedly mahogany-colored carapace, unusual for the species. The head was off-white with dark speckles behind a light patch between the eyes. The centers of the vertebral and pleural scutes were lighter than the rest of the carapace.

Photographs presented by Davenport et al. (2005) reflect the colors and patterns typically seen within the range of variation of terrapins from Kiawah Island, South Carolina (based on the experience of author Lovich) and appear to be assignable to *M. t. centrata*. This conclusion is also suggested by Davenport et al. (2005).

Summary Comments on Traditionally Recognized Terrapin Subspecies

Collectively, the range of all *M. terrapin* subspecies extends over a remarkable diversity of habitats and climates from 28° to 42° north latitude, a distance of almost 1,600 km. Total occupied range is estimated to be 262,846 km², a figure consistent with the individual range sizes (100,000–1,000,000 km²) for the majority (44%) of the world's turtle species (Buhlmann et al. 2009). The only common denominator for terrapins throughout this vast area is occupancy of brackish water habitats. This essentially continuous diversity of habitats provides abundant opportunity for local adaptation, although such adaptation may be hindered by gene flow (Chapter 5), confounded by genetic drift, opposed by natural selection due to temporal environmental variability, and constrained by lack of genetic variation or by the genetic architecture of underlying traits (Kawecki and Ebert 2004).

Adding further complexity to identification of the subspecies were the US government's efforts in the past century to translocate and deliberately hybridize terrapins in an attempt to propagate terrapins in captivity (Hildebrand 1933). Stocks of several subspecies were mixed, as reflected in the results of more recent microsatellite analysis (Hauswaldt and Glenn 2005). Given their wide and continuous range of variation in geographic space

(e.g., Diniz-Filho and De Campos Telles 2002) and the mixing of stocks by humans, is it any wonder that diamond-backed terrapins exhibit such a bewildering array of patterns and morphologies?

Molecular Ecological Contributions to Understanding Terrapins

The changing use of various genetic markers through time (see FitzSimmons and Hart 2007) is reflected in our molecular understanding of terrapin populations. Previous attempts at genetic stock identification in terrapins included assessment of restriction enzymes to assay variation in mitochondrial DNA (mtDNA; Lamb and Avise 1992). Results of this seminal study indicated a population division at Cape Canaveral, Florida, and a suggestion of an Atlantic/Gulf population divide. More recent studies using microsatellites, or simple sequence tandem repeats (SSTRs), have shown population-level patterns of gene flow. Hauswaldt and Glenn (2005) found no genetic structuring in terrapins sampled within several proximal creeks in the Charleston estuary, South Carolina. Hart (2005) also reported high levels of gene flow in terrapins sampled at four sites near Beaufort, North Carolina, and Sheridan et al. (2010) reported similarly high levels of gene flow for terrapins sampled in Barnegat Bay, New Jersey. Drabeck et al. (2014) found no evidence of population structure in Louisiana, even at sample locations 120 km apart. These studies confirm high levels of gene flow at local and regional geographic scales (i.e., distances of 10s to 100s km), a pattern similar to that seen in some other turtle species (Averill-Murray and Hagerty 2014). (See Chapter 5 for further details.)

Most recently, using microsatellite DNA analysis to investigate levels of gene flow among and genetic variability within many geographically separate terrapin collections, Hart et al. (2014) quantified levels of genetic variability (allelic diversity, genotypic frequencies, and heterozygosity) across the range of the species to reveal three zones of genetic discontinuity. They described four discrete genetic clusters or possible metapopulations (Northeast Atlantic, Coastal Mid-Atlantic, Florida, and Texas/Louisiana) and reported misalignment between geographic boundaries and the accepted terrapin subspecies limits. Specifically, Hart et al. (2014) found that the Northeast Atlantic cluster was geographically smaller than the extent of the *M. t. terrapin* range, the Coastal Mid-Atlantic cluster was larger than the extent of the *M. t. centrata* range, and the Florida cluster extended around the southwest coast of the state and so was larger than the previous *M. t. rhizophorarum* range (note that no *M. t. tequesta* samples were analyzed). Although their Texas/Louisiana grouping included portions of both *M. t. pileata* and *M. t. littoralis* ranges, no samples were available for analysis from northwest Florida (Gulf Coast) or southern Texas. An important finding in this new study was that the levels of differentiation reported are similar to those described as management units in other species.

Characterizations of gene flow for terrapin populations have also revealed sex biases in dispersal. Hart (2005) detected male-biased gene flow in North Carolina and Florida Everglades (mangrove) terrapins, and Sheridan et al. (2010) reported the same male bias in gene flow for terrapins sampled in New Jersey. The consistency of this finding in sites across the range of the species indicates that males may be more valuable than previously thought for maintaining genetic structure in populations. Given the vulnerability of males to crab-trap mortality (Dorcas et al. 2007), this may have significant effects on gene flow.

How Many Subspecies of Diamond-backed Terrapins Should Be Recognized?

Terrapin taxonomy is firmly rooted in the traditional recognition of seven subspecies, even though this is at odds with the molecular analyses to date. We believe there is still value in retaining the general concept of terrapin subspecies—if only for understanding how morphology, patterns, and genes differ across their extensive range—but not necessarily based on a seven-taxon model. The mismatch between phenotype and genotype across much of the range renders recognition of all the traditional subspecies of heuristic value only.

From the discussion in this chapter, it can be seen that the number of natural "groups" recognized by various authors ranges from two to seven. The traditional classification scheme was to recognize seven subspecies (*M. t. terrapin, M. t. centrata, M. t. tequesta, M. t. rhizophorarum, M. t. macrospilota, M. t. pileata,* and *M. t. littoralis*). In contrast, Lamb and Avise (1992) recognized two groups that were separated at approximately Cape Canaveral, Florida. Based on the accumulating weight of evidence that the species is oversplit, we conclude that four discrete populations (Northeast Atlantic, Coastal mid-Atlantic, Florida, and Texas/Louisiana) are now supported (Hart et al. 2014). The existence of a single Gulf Coast population (west of Florida) was partially reinforced by the findings of Drabeck et al. (2014) for Louisiana. It is important to emphasize again that these four groups are misaligned with the traditional seven subspecies and the major physiographic regions along the US Atlantic and Gulf coasts (Hartsell and Ernst 2004). This complicates efforts to make taxonomic changes, a topic beyond the scope of this chapter.

Pennock and Dimmick (1997) and Berry et al. (2002) reviewed the various designations used to recognize and conserve groups of organisms at units below that of species, including distinct population segments (DPS), evolutionarily significant units (ESU), and management units (MU). Precise definitions of these units are complicated by disagreement on what actually constitutes a species, in both concept and operation (Frost and Hillis 1990; de Queiroz 1998, 2007). The DPS has a specific legal definition under the US Endangered Species Act. Units can be designated based on discreteness of the group in relation to the rest of the species, the significance of the group to the species, and the conservation status of the group (see also Fallon 2007). Discreteness can be measured by physical separation or by physiological, ecological, or behavioral factors. Significance can be assessed relative to the distribution of the group or survival in a unique habitat relative to the remainder of the species. Evolutionarily significant units are based on recognition of significant adaptive variation within species that may include unique adaptations or reproductive isolation (Pennock and Dimmick 1997), later redefined in terms of the evolutionary species concept. The MU definition

was developed to identify genetically unique sub-
divisions within an ESU. The four terrapin groups
identified by Hart et al. (2014) conform to the definition of an MU, and we recommend use of that designation until formal taxonomic studies confirm
and codify such usage. The divisions/zones of
genetic discontinuity identified by Hart et al. (2014)
may be useful for defining region-specific conservation actions.

We recognize that this is a new way of dealing
with an old problem in terrapin taxonomy and that
it may be neither quickly nor universally adopted.
New information will doubtless accumulate over
time, further refining our understanding of terrapin genetic structure. Additional sampling of terrapin DNA in locales in the southeast United States
(Georgia, northern Florida, southern Florida, and
all along the Gulf Coast) could help to resolve current boundaries of genetic regional groupings. In
addition, use of existing microsatellite markers to
discern location of origin for terrapins sold in commercial seafood markets would be valuable for
determining pathways of terrapins to markets and
identifying populations vulnerable to harvest.
Future research employing new molecular tools
such as next-generation sequencing technologies
may hold promise for further resolving terrapin
taxonomic units, as well as uncovering patterns of
gene flow not yet currently recognized. Until then,
the four-group arrangement we recommend is a
parsimonious working hypothesis, given the information available.

ACKNOWLEDGMENTS

An earlier version of this chapter benefited greatly
from comments provided by Carl Ernst and Whit
Gibbons. Any use of trade, product, or firm
names is for descriptive purposes only and does
not imply endorsement by the US Government.

REFERENCES

Averill-Murray, R.C. and B.E. Hagerty. 2014. Transloca-
tion relative to spatial genetic structure of the Mojave
Desert Tortoise, *Gopherus agassizii*. Chelonian
Conservation and Biology 13:35-41.

Bacon, J.P., J.A. Gray and L. Kitson. 2006. Status and
conservation of the reptiles and amphibians of the
Bermuda Islands. Applied Herpetology 3:323-344.

Berry, K.H., D.J. Morafka and R.W. Murphy. 2002.
Defining the desert tortoise: our first priority for a
coherent conservation strategy. Chelonian Conservation and Biology 4:249-262.

Bickham, J.W., T. Lamb, P. Minx and J.C. Patton. 1996.
Molecular systematics of the genus *Clemmys* and
the intergeneric relationships of emydid turtles.
Herpetologica 52:89-97.

Boulenger, G.A. 1889. Catalogue of the Chelonians,
Rhynchocephalians, and Crocodiles in the British
Museum (Natural History). Taylor and Francis,
London. 311 pages.

Bruthiaux, P. 2002. Predicting challenges to English as a
global language in the 21st century. Language
Problems and Language Planning 26:129-157.

Buhlmann, K.A., T.S.B. Akre, J.B. Iverson, D. Karapatakis, R.A. Mittermeier, A. Georges, A.G. J. Rhodin, P.P.
van Dijk and J.W. Gibbons. 2009. A global analysis of
tortoise and freshwater turtle distributions with
identification of priority conservation areas. Chelonian Conservation and Biology 8:116-149.

Burns, T.A. and K.L. Williams. 1972. Notes on the
reproductive habits of *Malaclemys terrapin pileata*.
Journal of Herpetology 6:237-238.

Butler, J.A., R.A. Seigel and B.K. Mealey. 2006. *Malaclemys
terrapin*—diamondback terrapin. Chelonian Research
Monographs 3:279-295.

Cagle, F.R. 1952. A Louisiana terrapin population. Copeia
1952:74-76.

Coker, R.E. 1906. The natural history and cultivation of
the diamond-back terrapin with notes on other forms
of turtles. North Carolina Geological Survey Bulletin
14:1-67.

Crother, B.I., J.F. Boundy, F.T. Burbrink, J.A. Campbell,
K. de Queiroz, D.R. Frost, D.M. Green, R. Highton,
J.B. Iverson, F. Kraus, R.W. McDiarmid, J.R. Mendelson
III, P.A. Meylan, R.A. Pyron, T.W. Reeder, M.E. Seidel,
S.G. Tilley and D.B. Wake. 2012. Scientific and
standard English names of amphibians and reptiles
of North America North of Mexico, with comments
regarding confidence in our understanding. 7th ed.
Society for the Study of Amphibians and Reptiles
Herpetological Circular no. 39. Topeka, KS.
92 pages

Davenport, J., A.F. Glasspool and L. Kitson. 2005.
Occurrence of diamondback terrapins, *Malaclemys
terrapin*, on Bermuda: native or introduced? Chelonian
Conservation and Biology 4:956-959.

de Queiroz, K. 1998. The general lineage concept of
species, species criteria and the process of speciation: a conceptual unification and terminological
recommendations. Pages 57-75 in D.J. Howard and

S.H. Berlocher (eds.), Endless Forms: Species and Speciation. Oxford University Press, New York, NY.

de Queiroz, K. 2007. Species concepts and species delimitation. Systematic Biology 56:879-886.

Diniz-Filho, J.A.F. and M.P. De Campos Telles. 2002. Spatial autocorrelation analysis and the identification of operational units for conservation in continuous populations. Conservation Biology 16:924-935.

Dobie, J.L. 1981. The taxonomic relationship between *Malaclemys* Gray, 1844 and *Graptemys* Agassiz, 1857 (Testudines: Emydidae). Tulane Studies in Zoology and Botany 23:85-102.

Dobie, J.L. and D.R. Jackson. 1979. First fossil record for the diamondback terrapin, *Malaclemys terrapin* (Emydidae), and comments on the fossil record of *Chrysemys nelsoni* (Emydidae). Herpetologica 35:139-145.

Dorcas, M.E., J.D. Willson and J.W. Gibbons. 2007. Crab trapping causes population decline and demographic changes in diamondback terrapins over two decades. Biological Conservation 137:334-340.

Drabeck, D.H., M.W.H. Chatfield and C.L. Richards-Zawacki. 2014. The status of Louisiana's diamondback terrapin (*Malaclemys terrapin*) populations in the wake of the Deepwater Horizon oil spill: insights from population genetic and contaminant analyses. Journal of Herpetology 48:125-136.

Ernst, C.H. and R.B. Bury. 1982. *Malaclemys terrapin* (Schoepff) diamondback terrapin. Catalogue of American Amphibians and Reptiles, 299.1-299.4.

Ernst, C.H. and T.D. Hartsell. 2000a. An earlier name for the mangrove diamondback terrapin, *Malaclemys terrapin rhizophorarum* (Reptilia: Testudines: Emydidae). Proceedings of the Biological Society of Washington 113:887-889.

Ernst, C.H. and T.D. Hartsell. 2000b. *Malaclemys littoralis rhizophorarum* Fowler, 1906: precedence of names in wide use over disused synonyms or homonyms in accordance with Article 23.9 of the Code (Reptilia: Testudines, Case 3108). Bulletin of Zoological Nomenclature 57:6-10.

Ernst, C.H. and J.E. Lovich. 2009. Turtles of the United States and Canada. 2nd edition. Johns Hopkins University Press, Baltimore, MD. 827 pages.

Fallon, S.M. 2007. Genetic data and the listing of species under the U.S. Endangered Species Act. Conservation Biology 21:1186-1195.

FitzSimmons, N.N. and K.M. Hart. 2007. Genetic studies of freshwater turtles and tortoises: a review of the past 70 years. Chelonian Research Monographs 4:15-46.

Fowler, H.W. 1906. Some cold blooded vertebrates from the Florida Keys. Proceedings of the Academy of Natural Sciences of Philadelphia 58:77-113.

Fritz, U. and P. Havaš. 2007. Checklist of chelonians of the world. Vertebrate Zoology (Museum für Tierkunde Dresden) 57:149-368.

Frost, D.R. and D.M. Hillis. 1990. Species in concept and practice: herpetological applications. Herpetologica 46:86-104.

Gibbons, J.W. and J.R.I. Harrison. 1981. Reptiles and amphibians of Kiawah and Capers Islands, South Carolina. Brimleyana 5:145-162.

Gibbons, J.W. and J.E. Lovich. 1990. Sexual dimorphism in turtles with emphasis on the slider turtle (*Trachemys scripta*). Herpetological Monographs 4:1-29.

Hart, K.M. 2005. Population biology of diamondback terrapins (*Malaclemys terrapin*): defining and reducing threats across their range. PhD dissertation, Duke University, Durham, NC. 259 pages.

Hart, K.M., M.E. Hunter and T.L. King. 2014. Regional differentiation among populations of the diamondback terrapin (*Malaclemys terrapin*). Conservation Genetics 15:593-603.

Hart, K.M. and C.C. McIvor. 2008. Demography and ecology of mangrove diamondback terrapins in a wilderness area of Everglades National Park, Florida, USA. Copeia 2008:200-208.

Hartsell, T.D. and C.H. Ernst. 2004. A review of environmental conditions along the coastal range of the diamondback terrapin, *Malaclemys terrapin*. Herpetological Bulletin 89:12-20.

Hauswaldt, J.S. and T.C. Glenn. 2005. Population genetics of the diamondback terrapin (*Malaclemys terrapin*). Molecular Ecology 14:723-732.

Hay, W.P. 1904. A revision of *Malaclemmys*, a genus of turtles. Bulletin of the US Bureau of Fisheries 24:1-20.

Hildebrand, S.F. 1929. Review of experiments on artificial culture of diamond-back terrapin. Bulletin of the US Bureau of Fisheries 45:25-70.

Hildebrand, S.F. 1932. Growth of diamond-back terrapins: size attained, sex ratio and longevity. Zoologica 9:551-563.

Hildebrand, S.F. 1933. Hybridizing diamond-back terrapins. Journal of Heredity 24:231-238.

Hildebrand, S.F. and H.F. Prytherch. 1947. Diamond-back terrapin culture. Fish and Wildlife Service Leaflet 216:1-5.

Hogan, J.L. 2003. Occurrence of the diamondback terrapin (*Malaclemys terrapin littoralis*) at South Deer Island in Galveston Bay, Texas, April 2001-May 2002. US Geological Survey Open File Report, no. 03-022, pages 1-24. https://pubs.usgs.gov/of/2003/ofr03-022.

Holden, T. 1979. Terrapin decline at Wellfleet. Massachusetts Audubon 18:10-11.

Hurd, L.E., G.W. Smedes and T.A. Dean. 1979. An ecological study of a natural population of diamondback terrapins (*Malaclemys t. terrapin*) in a Delaware salt marsh. Estuaries 2:28-33.

Kawecki, T.J. and D. Ebert. 2004. Conceptual issues in local adaptation. Ecology Letters 7:1225-1241.

Lamb, T. and J.C. Avise. 1992. Molecular and population genetic aspects of mitochondrial DNA variability in the diamondback terrapin, *Malaclemys terrapin*. Journal of Heredity 83:262-269.

Lamb, T., C. Lydeard, R.B. Walker and J.W. Gibbons. 1994. Molecular systematics of map turtles (*Graptemys*): a comparison of mitochondrial DNA restriction site versus sequence data. Systematic Biology 43:543-559.

Lamb, T. and M.F. Osentoski. 1997. On the paraphyly of *Malaclemys*: a molecular genetic assessment. Journal of Herpetology 31:258-265.

Lazell, J.D., Jr. 1969. Terrapin terminals. Massachusetts Audubon 53(4). 3 pages. No pagination.

Lever, C. 2003. Naturalized Reptiles and Amphibians of the World. Oxford University Press, New York, NY. 344 pages.

Lovich, J.E. and J.W. Gibbons. 1990. Age at maturity influences adult sex ratio in the turtle *Malaclemys terrapin*. Oikos 59:126-134.

McDowell, S.B. 1964. Partition of the genus *Clemmys* and related problems in the taxonomy of the aquatic Testudinidae. Proceedings of the Zoological Society of London 143:239-279.

Meylan, P.A. and W. Sterrer. 2000. *Hesperotestudo* (Testudines: Testudinidae) from the Pleistocene of Bermuda with comments on the phylogenetic position of the genus. Zoological Journal of the Linnean Society 128:51-76.

Outerbridge, M.E. 2008. Ecological notes on feral populations of *Trachemys scripta elegans* in Bermuda. Chelonian Conservation and Biology 7:265-269.

Parham, J.F., M.E. Outerbridge, B.L. Stuart, D.B. Wingate, H. Erlenkeuser and T.J. Papenfuss. 2008. Introduced delicacy or native species? A natural origin of Bermudian terrapins supported by fossil and genetic data. Biology Letters 4:216-219.

Pennock, D.S. and W.W. Dimmick. 1997. Critique of the evolutionarily significant unit as a definition for "distinct population segments" under the U.S. Endangered Species Act. Conservation Biology 11:611-619. DOI 10.1098/rsbl.2007.0599.

Schoepff, J.D. 1793. *Testudo terrapin*. Page 64 in Historia Testudinum Iconibus Illustrata. Fascicles III and IV, pages 33-80, plates 11-16, 17B-20. Ioannis Iacobi Palm, Erlangae.

Schwartz, A. 1955. The diamondback terrapins (*Malaclemys terrapin*) of peninsular Florida. Proceedings of the Biological Society of Washington 68:157-164.

Shaffer, H.B., P. Meylan and M.L. McKnight. 1997. Tests of turtle phylogeny: molecular, morphological, and paleontological approaches. Systematic Biology 46:235-268.

Sheridan, C.M., J.R. Spotila, W.F. Bien and H.W. Avery. 2010. Sex-biased dispersal and natal philopatry in the diamondback terrapin, *Malaclemys terrapin*. Molecular Ecology 19:5497-5510.

Sonnini de Manoncourt, C.S. and P.A. Latreille. 1802. Histoire naturelle des reptiles, avec figures dessinées d'après nature, I. Imprimerie de Crapelet, Paris. 280 pages.

Spinks, P.Q., R.C. Thompson, G.A. Lovely and H.B. Shaffer. 2009. Assessing what is needed to resolve a molecular phylogeny: simulations and empirical data from emydid turtles. BMC Evolutionary Biology 9:1-17.

Stephens, P.R. and J.J. Wiens. 2003. Ecological diversification and phylogeny of emydid turtles. Biological Journal of the Linnean Society 79:577-610.

Thomson, R.C. and H.B. Shaffer. 2010. Sparse supermatrices for phylogenetic inference: taxonomy, alignment, rogue taxa, and the phylogeny of living turtles. Systematic Biology 59:42-58.

Thornbury, W.D. 1965. Regional Geomorphology of the United States. John Wiley and Sons, New York, NY. 609 pages.

van Dijk, P.P., J.B. Iverson, A.G.J. Rhodin, H.B. Shaffer and R. Bour. 2014. Turtles of the world, 7th edition: annotated checklist of taxonomy, synonymy, distribution with maps, and conservation status. Pages 329-479 in A.G.J. Rhodin, P.C.H. Pritchard, P.P. van Dijk, R.A. Saumure, K.A. Buhlmann, J.B. Iverson and R.A. Mittermeier (eds.), Conservation Biology of Freshwater Turtles and Tortoises: A Compilation Project of the IUCN/SSC Tortoise and Freshwater Turtle Specialist Group. Chelonian Research Monographs. Chelonian Research Foundation, Lunenburg, MA.

Wied-Neuwied, M.A.P. 1865. Verzeichniss der Reptilien, welche auf einer Reise in nördlichen America beobachtet wurden. Nova Acta Academiae Caesareae Leopoldino-Carolinae Germanicae Naturae Curiosorum 32. 146 pages + 2 figures.

Wiens, J.J., C.A. Kuczynski and P.R. Stephens. 2010. Discordant mitochondrial and nuclear gene phylogenies in emydid turtles: implications for speciation and conservation. Biological Journal of the Linnean Society 99:445-461.

Wood, R.C. 1977. Evolution of the emydine turtles *Graptemys* and *Malaclemys* (Reptilia, Testudines, Emydidae). Journal of Herpetology 11:415-421.

Wood, R.C. 1994. The distribution, status, ecology, and taxonomy of diamondback terrapins, *Malaclemys terrapin*, in the Florida Keys. Abstract in G.E. Heinrich (ed.), A Symposium on the Status and Conservation of Florida Turtles. Eckerd College, St. Petersburg, FL. No pagination.

5

Molecular Ecology and Phylogeography

PAUL E. CONVERSE

SHAWN R. KUCHTA

The diamond-backed terrapin (*Malaclemys terrapin*) is one of the most charismatic and recognizable turtle species in North America, populating brackish water habitats along the Gulf and Atlantic coasts from Corpus Christi, Texas, to Cape Cod, Massachusetts, with a small disjunct population in Bermuda (Chapters 3 and 4). Historically, the terrapin has been divided into seven subspecies: the Texas diamond-backed terrapin (*Malaclemys terrapin littoralis*), the Mississippi diamond-backed terrapin (*M. t. pileata*), the ornate diamond-backed terrapin (*M. t. macrospilota*), the mangrove diamond-backed terrapin (*M. t. rhizophorarum*), the eastern Florida diamond-backed terrapin (*M. t. tequesta*), the Carolina diamond-backed terrapin (*M. t. centrata*), and the northern diamond-backed terrapin (*M. t. terrapin*). All subspecies were delimited based on regional variation in size, color, and shape (described in Chapter 4). The recognition of seven morphologically distinct subspecies suggests that substantial amounts of phylogeographic and population genetic structure should be present within the terrapin.

Elements of the natural history and ecology of the terrapin add weight to this thesis. Ecological studies show that terrapins generally exhibit low vagility, reside in small home ranges, and demonstrate high nest site philopatry (Auger 1989; Lovich and Gibbons 1990; Roosenburg 1996; Spivey 1998; Roosenburg et al. 1999). Indeed, diamond-backed terrapins have been recaptured in the same creek or river system for decades (Gibbons et al. 2001). On the other hand, not all females demonstrate perfect nest site fidelity, and some individuals are known to disperse over long distances (up to 8 km; Sheridan et al. 2010); the dynamics of juvenile dispersal are largely unknown. Thus, the ecological data are mixed in their predictions about levels of expected population genetic structure.

Unfortunately, during the nineteenth and twentieth centuries, before patterns of terrapin diversity were well studied, populations along the East Coast underwent severe demographic declines as a consequence of demand for terrapin meat (Coker 1906). The result was regional and local extirpations of terrapin populations. As natural populations declined, government and private terrapin farms were created to meet the demand for meat (Barney 1922; Hildebrand and Hatsel 1926; Hildebrand 1929; Chapter 13). As described by Coker (1920), farm-raised terrapins were collected from poorly documented locations, often locally but also from other parts of the terrapin's range. For example, as terrapin populations in Chesapeake Bay diminished, terrapins from North Carolina were translocated to replenish the dwindling stock; terrapins from South Carolina were moved to North Carolina and into Chesapeake Bay. Additionally, Gulf terrapins were translocated into south Atlantic populations (Coker 1920; Hildebrand 1933). Indeed, terrapins were moved around their range with such frequency that Coker (1920) quipped, "Had these terrapin carried handbags, they might have displayed an array of hotel stickers to shame the traveler returned from Europe." When demand for terrapin meat subsided, farms purportedly simply released their terrapins. The ability of released terrapins to interbreed with local populations and the numbers and success of the translocated terrapins remain unknown.

Thus, the terrapin has a complicated history with humans, which creates significant challenges to understanding the patterns of genetic variation. Most genetic work has focused primarily on conservation needs and management strategies. In this chapter we survey studies of genetic variation in the diamond-backed terrapin, addressing the terrapin's phylogeographic and population genetic structure and the abundance and distribution of its genetic diversity and examining the relationship between terrapin genetic studies and ecological findings. We end with recommendations for future genetic work.

Phylogeographic Studies

The diamond-backed terrapin has extensive phenotypic variation among its seven putative subspecies (Chapter 4), and accordingly, there was an expectation of underlying high levels of genetic structure across its range. Lamb and Avise (1992) conducted the first analysis of genetic structure throughout the terrapin's range, using mitochondrial DNA (mtDNA) and 18 restriction enzymes. They recovered 74 restriction fragments from 53 terrapins, with six haplotypes. Compared with other vertebrates, this level of divergence is unusually low (uncorrected $P = 0.001\%$), and geographic structure is weak, with only one phylogeographic break, near Cape Canaveral, Florida (referred to in the literature as the "Atlantic/Gulf divide"; Fig. 5.1).

In contrast to taxonomic conclusions, the genetic study by Lamb and Avise (1992) did not recover any of the subspecies as diagnosable entities. However, the genetic divide near Cape Canaveral did correspond with a noticeable change in terrapin morphology: terrapins to the north lack knobs on their medial keels, while terrapins in the south and in the Gulf possess knobby keels.

One possible cause of the discordance between subspecific designations and terrapin phylogeography may be that terrapins have low levels of mtDNA variation. Early work suggested that chelonians exhibit low rates of nuclear and mitochondrial evolution relative to other vertebrates (Avise et al. 1992; Bromham 2002; Shaffer et al. 2013). The slowdown in terrapin mtDNA evolutionary rate was inferred from concordant phylogeographic breaks among taxa (Avise et al. 1992; Lamb and Avise 1992). For instance, the eastern oyster (*Crassostrea virginica*) and seaside sparrow (*Ammodramus maritimus*) exhibit a phylogeographic break in the Gulf, while the horseshoe crab (*Limulus polyphemus*) exhibits a phylogeographic break near Cape Canaveral, suggesting that these taxa share a common vicariant history with the terrapin (Avise et al. 1992). Relative to these other taxa, however, the terrapin demonstrates a paucity of mtDNA variation and thus would require a much slower rate of mtDNA evolution than other vertebrates (~14-fold slower) if they shared this vicariant event (Avise et al. 1992). However, Avise et al. (1992) also noted that terrapins may simply have been isolated near Cape Canaveral more recently.

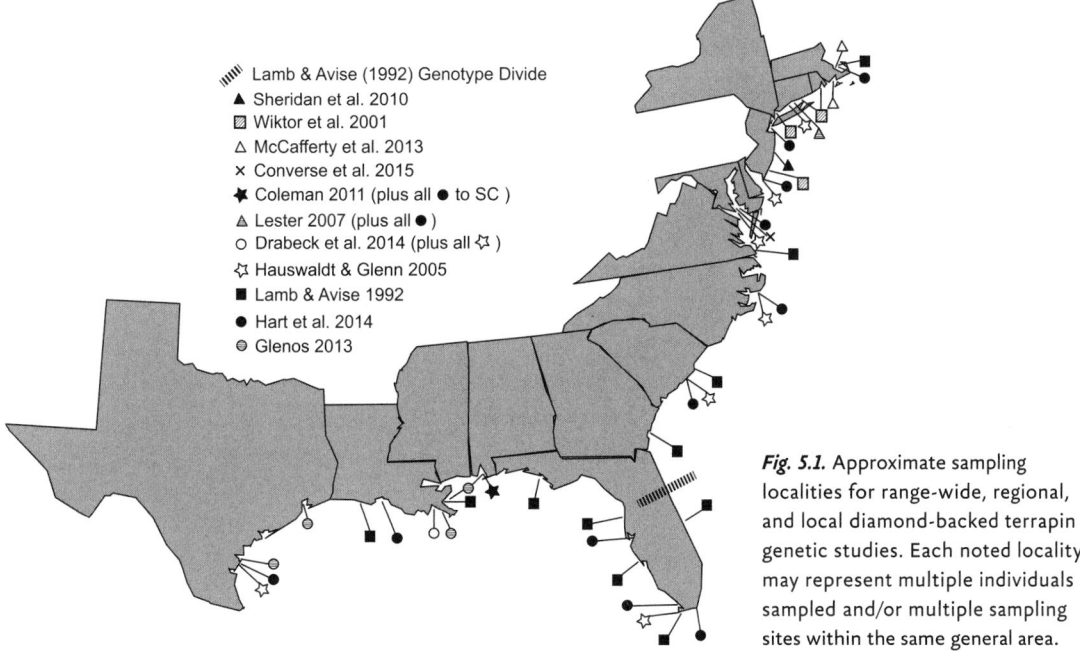

Fig. 5.1. Approximate sampling localities for range-wide, regional, and local diamond-backed terrapin genetic studies. Each noted locality may represent multiple individuals sampled and/or multiple sampling sites within the same general area.

Legend:
- ⟋⟍ Lamb & Avise (1992) Genotype Divide
- ▲ Sheridan et al. 2010
- ▨ Wiktor et al. 2001
- △ McCafferty et al. 2013
- × Converse et al. 2015
- ✦ Coleman 2011 (plus all ● to SC)
- ▲ Lester 2007 (plus all ●)
- ○ Drabeck et al. 2014 (plus all ✿)
- ✿ Hauswaldt & Glenn 2005
- ■ Lamb & Avise 1992
- ● Hart et al. 2014
- ◉ Glenos 2013

Perhaps as a consequence of the low genetic variation found by Lamb and Avise (1992), no further phylogeographic work was conducted on the terrapin for more than a decade, until revisited by Hauswaldt (2004) and Hauswaldt and Glenn (2005). Hauswaldt (2004) surveyed the mtDNA control region, a rapidly evolving marker in most vertebrates, across the terrapin's range. She recovered 28 mtDNA haplotypes, with unique haplotypes in Maryland, Texas, and Florida. The phylogeographic break found by Lamb and Avise (1992) was recovered, and haplotypes for the Atlantic and Gulf coasts formed separate clades. However, limited phylogenetic structure was recovered in both clades, and even the Atlantic/Gulf divide lacked strong support (bootstrap < 70%; Hillis and Bull 1993).

In a parallel study, Hauswaldt and Glenn (2005) quantified genetic diversity across the terrapin's range, using six nuclear microsatellites (*TerpSH*; Hauswaldt and Glenn 2003). Population structure, population assignment, and bottleneck detection were the primary foci of the work. Most sampling took place along the Atlantic coast (New York, New Jersey, Maryland, North Carolina, and three locations in South Carolina), with additional sampling in the Gulf (Florida and Texas). Across the terrapin's range, the researchers found high levels of heterozygosity (H_o; 0.70–0.87) and large numbers of alleles at most loci (A; 7.5–13), except in Florida ($H_o = 0.56$; $A = 6.5$). Using an analysis of molecular variance (AMOVA; Excoffier et al. 1992), Hauswaldt and Glenn (2005) found that little variance could be explained by grouping all sampling localities (16.8%), while partitioning populations into Gulf and Atlantic groups explained the most, though still little, variance (25.1%). In addition, Hauswaldt and Glenn (2005) demonstrated that individuals in some populations could be correctly assigned to their population of origin by using Bayesian assignment tests. Terrapins in the Gulf had higher instances of correct assignment (71%–100%) than populations in the Atlantic (10%–75%). Terrapins along the Atlantic were more similar to Texas populations than Florida populations, consistent with the observation that terrapin translocations occurred during the early twentieth century. Surprisingly, the program BOTTLENECK (Piry et al. 1999) found no evidence of demographic bottlenecks in any population, including populations known to have been historically overharvested.

More recently, Hart et al. (2014) used 12 nuclear microsatellites (*Gmu*; King and Julian 2004) to revisit phylogeographic structure along the terrapin's range. Populations in the North Atlantic, Gulf, and Florida showed low levels of heterozygosity (0.32–0.47), while mid-Atlantic populations showed moderate levels (0.66–0.71). Genetic clusters (populations) were inferred by using the Bayesian clustering program STRUCTURE (Pritchard et al. 2000). STRUCTURE operates by dividing individuals into clusters such that conformity to Hardy-Weinberg equilibrium is maximized and linkage disequilibrium is minimized. Hart et al. (2014) initially detected the presence of two genetic clusters. The first consisted of terrapins from Atlantic populations, and the second consisted of terrapins from Florida. STRUCTURE plots indicated that populations sampled from the Gulf (Louisiana and Texas) were admixed with samples from the Atlantic and Florida. Within the Atlantic cluster, two subclusters were found: Massachusetts vs. the remaining Atlantic populations. Further analysis showed that Louisiana and Texas formed a cluster, while Florida formed the final cluster, for a total of four clusters. Hart et al. (2014) also found support for these clusters with the use of neighbor-joining phenograms, AMOVAs, and F_{ST} calculations. In contrast to previous studies, evidence of population contraction was detected with BOTTLE-NECK, with each locality exhibiting some signature of a bottleneck. Hart et al. (2014) concluded that precipitous drops in population sizes had occurred.

Regional and Population Level Studies

Atlantic Coast

While phylogeographic studies did not recover large genetic discontinuities or incipient species, investigation of patterns of genetic variation within and among populations has been an ongoing research interest. Such studies aim to reveal smaller-scale patterns of differentiation, quantify directions and levels of gene flow, probe demographic history, and direct conservation and management.

Terrapins were historically harvested at unsustainable levels (Garber 1990), and the genetic effects of this exploitation are a common thread running through terrapin research. To date, no range-wide, coordinated conservation plan for terrapins is in place; rather, conservation efforts are left to individual states, with inconsistent enforcement (Glenos 2013; Hart et al. 2014). To investigate the problem of illegal harvests, Lester (2007) conducted a covert study to identify the source populations of terrapins sold at the Fulton Fish Market in New York City. She obtained blood samples from 63 terrapins in 2004 and, using assignment tests and 12 *Gmu* microsatellites, determined that terrapins had been collected along much of the East Coast, from South Carolina to New York. The majority originated from Chesapeake Bay (34/63), especially from Sandy Island Cove, Maryland. Maryland closed its legal terrapin fishery in 2007 due to its adverse effects on local populations (Roosenburg et al. 2008). How illegal markets currently affect terrapin populations remains unclear.

Many studies have examined genetic variation within and among sets of populations, with an eye toward documenting population structure and habitat use (e.g., Callens et al. 2011). Elucidating how habitat features affect patterns of gene flow has important implications for metapopulation dynamics and conservation planning. In a creative study, Sheridan (2010) used six *Gmu* microsatellite markers and computational modeling to compare and contrast barriers to dispersal in Barnegat Bay, New Jersey. First, she measured genetic differentiation among sampling localities by using F_{ST} and G'_{ST} (F_{ST} for multiple alleles) statistics. She then estimated gene flow among populations using a diversity of models. Simple models quantified genetic differentiation as a function of straight-line and shoreline distances. Complex models included landscape features and assigned costs to traverse them. Open, deep and shallow waters, emergent wetlands, development, roads, and riprap (rocks used to armor shorelines) were investigated for their propensity to facilitate or impede gene flow. Sheridan (2010) found that open water and developed landscapes were the biggest impediments to gene flow, while shallow water and emergent wetlands facilitated genetic exchange. These results are consistent with ecological studies of terrapin movement (Roosenburg et al. 1999).

Populations located at the edge of a species range commonly have smaller population sizes and lower levels of genetic diversity than more centrally located populations (Kirkpatrick and Barton 1997). McCafferty et al. (2013) investigated this phenomenon in terrapins by examining populations near the northern edge of the range in Massachusetts. Populations in this region not only are at the terminus of the terrapin's range but also may have undergone bottlenecks as a consequence of harvesting. To investigate this, McCafferty et al. (2013) combined nuclear sequence data (class 1 major histocompatibility complex, *MHC1*) with six *Gmu* microsatellite loci. Despite their study populations' being at the northern edge of the terrapin's range, McCafferty et al. (2013) found moderate levels of heterozygosity (0.62–0.71), contrary to the lower levels typically associated with range margins and population bottlenecks. Using STRUCTURE, the researchers found that microsatellites demarcated two populations, while F_{ST} statistics documented low to moderate levels of population differentiation (0.03–0.11). The first STRUCTURE cluster included samples from Wellfleet Harbor and Sandy Neck, Massachusetts, while samples in Sippican Harbor, Massachusetts, constituted a second cluster. In contrast to microsatellites, no variation was recovered at the nuclear locus, *MCH1*. McCafferty et al. (2013) propose that this is the consequence of the locus operating under purifying selection, though recent range expansion or low effective population size could lead to the same lack of variation. A 1972 translocation of approximately 100 terrapins from New Jersey to Buzzards Bay (Lazell 1976) may also explain the observed population structure.

Not all population level studies have recovered variation among populations. In the Hudson River of New York, Piermont Marsh harbors a terrapin population that is small and broadly distributed, with few high-quality nesting sites. This population may be isolated from neighboring populations, and Wiktor et al. (2001) hypothesized that terrapins in Piermont Marsh might be genetically differentiated from other populations

in the region. To test this hypothesis, they compared terrapins from Piermont Marsh with terrapins from Rhode Island, Connecticut, and New Jersey, using two nuclear inter-simple sequence repeat (ISSR) loci. In contrast to their predictions, they found little genetic differentiation between Piermont Marsh and other populations in the North Atlantic. Wiktor et al. (2001) concluded that the Piermont Marsh population must be connected to others via high levels of gene flow, most likely from populations in the Hudson River. An alternative explanation is that the Piermont Marsh population was founded relatively recently. That is, it could currently be isolated demographically, but this isolation might not yet have left its genetic signature.

Most recently, Converse et al. (2015) quantified historical and contemporary gene flow among terrapin populations in Chesapeake Bay. Using 12 *Gmu* microsatellite loci and STRUCTURE, they found four terrapin populations that were weakly structured but harbored moderate to high levels of heterozygosity (0.69–0.79). Using MIGRATE-n (Beerli 2008) and BAYESASS (Wilson and Rannala 2003) to estimate historical and contemporary gene flow, respectively, Converse et al. (2015) showed large increases in gene flow into Chesapeake Bay, as well as large increases between the Patuxent River and Kent Island, potentially due to terrapin translocation events. By contrast, levels of gene flow decreased over time among most populations within the Bay. They attributed this latter pattern to habitat loss and the impact of a large crabbing industry, which adversely affects terrapin dispersal, especially of juveniles and males (Roosenburg et al. 1997; Roosenburg and Green 2000).

Gulf Coast

Until recently, populations in the Gulf were understudied by molecular ecologists. The work of Hart (2005), Hart et al. (2014), and Hauswaldt and Glenn (2005) suggested that Gulf Coast populations contained lower levels of genetic diversity than populations in the mid-Atlantic, but these studies included limited sampling in the Gulf region. Coleman (2011) was the first to include a wide sampling of Gulf populations. Using 12 *Gmu* microsatellites and STRUCTURE, he identified three clusters of populations: Florida, South Carolina, and the Gulf (Texas, Louisiana, Alabama). STRUCTURE suggested little admixture among the clusters. While earlier studies found mixed support for population bottlenecks, Coleman (2011) found strong support for bottlenecks in populations in the Gulf, Florida, and Atlantic by using *M*-ratios (all $M < 0.40$). Moderate levels of heterozygosity (0.50–0.65) and significant deviations from Hardy-Weinberg equilibrium in the Alabama and South Carolina populations provided support for genetic bottlenecks. His findings differed from those of Hauswaldt and Glenn (2005), who found much higher levels of heterozygosity (0.56–0.80) and no evidence for bottlenecks in Gulf populations.

Work on Gulf Coast terrapins continued with Glenos (2013), whose sampling was most dense in Texas but included Louisiana and Alabama. She did not find significant differentiation between the two sampling localities within Texas, suggesting that terrapins along the Texas coast constitute a single population. On the other hand, she found terrapins from Alabama and Louisiana to be genetically separated from the Texas population. Consistent with Coleman's (2011) finding, her 12 *Gmu* microsatellites exhibited low levels of heterozygosity (0.39–0.52), few private alleles, and a low number of alleles (3.0–5.5) in all populations.

The value of diversity studies is always highlighted by environmental disasters, such as the Deepwater Horizon oil spill in 2010 that deeply affected coastal ecology in Louisiana. Drabeck et al. (2014) studied the effects of the oil spill on terrapin populations. Using 4 *TerpSH* and 12 *Gmu* microsatellites, they found no meaningful population structure among Louisiana terrapins, nor did they detect population contractions with the program BOTTLENECK, in contrast to the findings of Coleman (2011) and Hart et al. (2014). However, consistent with Lamb and Avise (1992) and Hart et al. (2014), Drabeck et al. (2014) found Florida, Gulf, and Atlantic terrapin populations to be diagnosable entities. The AMOVAs showed that most genetic variance (7.47%) was explained by combining Gulf and Atlantic terrapin populations while excluding Florida populations. Hauswaldt and Glenn (2005) similarly found evidence for terrapin translocations during the twentieth century, as Gulf Coast terrapins were more similar to coastal Atlantic populations than adjacent populations in the Florida Keys. This may be due to the undesirable commercial traits of Florida terrapins—considered insipid in taste and too small to sell or hybridize, while terrapins from Texas were preferred for shipment to the Atlantic seaboard (Hay 1917; Hildebrand 1933).

Further work on Louisiana terrapins was conducted by Petre (2014). Twenty-six sampling localities (analyzed as eight sites) along the Louisiana coastline were examined for population structure, genetic diversity, population connectivity, and bottlenecks, using 13 *Gmu* microsatellites. While STRUCTURE did not detect any meaningful population structure, isolation-by-distance was detected along Louisiana's coastline, and MIGRATE-n found that a stepping stone model best explained patterns of gene flow. Petre (2014) found fairly high heterozygosity (0.74–0.77) and allele number (5.8–10.4) among populations. Despite the high levels of genetic diversity, BOTTLENECK found that two of the eight sites exhibited the genetic signature of population contraction.

In summary, while many ecological studies document high levels of population structure, phylogeographic and regional genetic studies have found more limited genetic variation among terrapin populations. This could be because genetic and ecological studies quantify structure on different time scales, with ecological studies documenting real-time population structure that has yet to manifest itself in the genome. It remains possible that the terrapin recently underwent a rapid range expansion across much of its range, resulting in limited phylogeographic structure. Terrapin populations along the coasts may also have experienced sporadic episodes of increased gene flow, including human-mediated gene flow.

Insights into Reproductive Ecology Provided by Genetic Studies

Large documented declines in terrapin populations in the early twentieth century were a major impetus for studying the patterns of terrapin genetic diversity. By contrast, studies of terrapin reproductive biology received less attention. Two themes within reproductive biology have been studied by molecular ecologists: natal philopatry and multiple paternity.

Natal philopatry is the propensity for offspring to return to their birth location to nest. Whether emerging terrapin hatchlings exhibited natal philopatry was unknown until recently, though this behavior had been documented in the Mississippi map turtle (*Graptemys pseudogeographica kohni*), a closely related species (Freedberg et al. 2005). High levels of natal philopatry are important for studies of genetic differentiation because they represent a mechanism by which genetic structure among populations can evolve. If mothers and their daughters return to the same nesting locality each season, genetic evidence should complement ecological findings of mothers' nest site fidelity. Sheridan et al. (2010) documented natal philopatry in terrapins in Barnegat Bay, New Jersey, using six *Gmu* microsatellites and spatial autocorrelation analysis. They calculated genetic correlations among terrapins and then surveyed for relationships with geographic distance. The only demographic group with a positive genetic autocorrelation was nesting females (at 0–50 m), indicating that they exhibit spatial structure that could result from nest site philopatry. Sheridan et al. (2010) also found that many juvenile males were transient at their study site. High levels of juvenile movement may be partly responsible for the lack of population genetic structure found in terrapins.

A clutch of diamond-backed terrapin hatchlings share a mother, but not necessarily a father. Multiple paternity in terrapins was not well understood until advances in genetic techniques allowed researchers to estimate paternity accurately, but it was highly suspected because terrapins can store sperm for up to 4 years (Hildebrand and Hatsel 1926). The first study of multiple paternity in terrapins was conducted by Hauswaldt (2004), who surveyed six *TerpSH* microsatellites in 439 hatchlings from 26 mothers in Oyster Bay, New York. She found that a majority of nests (82%) were sired by one male. This result was somewhat surprising and suggested that multiple paternity is uncommon in the terrapin relative to other chelonians (e.g., Galbraith et al. 1993; Valenzuela 2000; Ireland et al. 2003).

Like any trait, multiple paternity can vary by region or population, or even over time in the same location. Following Hauswaldt (2004), Sheridan (2010) surveyed six *Gmu* microsatellites in 1,608 hatchlings and 1,558 adults from four sites in Barnegat Bay, New Jersey, and Poplar Island in Chesapeake Bay. She recorded higher levels of multiple paternity than did Hauswaldt (2004), with a mean of 29%. However, the occurrence of multiple paternity varied widely among nesting locations, from 12.5% to 45.7%. This variation was correlated with population sex ratio: populations with moderate female-biased sex ratios (4.7:1) demonstrated the highest levels of multiple paternity, whereas populations with extreme female-biased sex ratios (9:1) had lower levels.

Discussion

Studies of genetic variation have taught us much about terrapin biology. On the broadest scale, range-wide phylogeographic analyses by Lamb and Avise (1992), Hauswaldt (2004), Hauswaldt and Glenn (2005), and Hart et al. (2014) show that genetic structure across the range of the terrapin is limited and do not support the existence of seven subspecies—at least if one views subspecies as incipient evolutionary lineages. Thus, the subspecific designations in terrapins are in need of reevaluation.

Genetic studies of the terrapin are made more difficult by limited sequence variation and the specter of frequent human translocations. In general, mtDNA exhibits higher levels of structure than nuclear markers (microsatellites excluded), making mtDNA useful for phylogeography. In chelonians, however, mtDNA is thought to mutate more slowly than in other vertebrates (Avise et al. 1992; Bromham 2002; Shaffer et al. 2013), resulting in less structure and weaker inference. Ultimately, many nuclear loci from next-generation sequencing resources will be needed to elucidate the evolutionary history of the terrapin.

At the range-wide and regional scales, a lack of genetic structure may be a consequence of the terrapin's rapidly expanding its range from a refugium. Because the terrapin's range closely follows the Gulf and Atlantic coasts, it is essentially structured as a linear stepping stone, which should produce isolation-by-distance. STRUCTURE assumes no isolation-by-distance (Pritchard 2000), but isolation-by-distance is known to be present at regional and range-wide scales in terrapins (Hauswaldt and Glenn 2005; Hart et al. 2014). Violation of the assumptions of F_{ST} statistics (equal population sizes, island migration model, etc.) may also obfuscate population structure estimates. For this reason, it is important that future work take full advantage of the latest advances in population genetics, especially coalescent theory, a powerful stochastic model for evaluating demographic history (Wakeley 2008). For example, Bayesian skyline plots could be used to examine variation in effective population sizes through time. Alternatively, historical events could be investigated using approximate Bayesian computations, which allow complex models to be constructed and tested against genetic data (Csilléry et al. 2010). Models including admixture, population divergence dates, population sizes, and translocation could be assembled and tested against one another.

Among population genetic studies, a shared finding is that terrapin genetic diversity is lowest in southern populations, especially in Florida. This is in contrast to the findings for many taxa in the eastern United States, especially terrestrial taxa, which display the highest levels of genetic diversity in the south and lowest levels in the north (Avise 2000; Soltis et al. 2006). In terrapins, the legacy of terrapin harvests may have contributed to lowered genetic diversity. However, historical factors most likely contributed as well. For example, if the terrapin had a Pleistocene refugium along the Atlantic Coast and only recently (within the last several thousand years) expanded into the Gulf, Gulf populations would be expected to show low levels of genetic diversity.

By contrast, genetic diversity is highest in the mid-Atlantic region, possibly due to translocations into the region. Translocation is supported by Bayesian population assignment tests, which often assign Atlantic terrapins to Gulf populations (Hauswaldt and Glenn 2005). In addition, Drabeck et al. (2014) found that Atlantic populations are genetically more similar to Gulf populations than neighboring populations. Alternatively, the mid-Atlantic could be the origin of the terrapin's historical range. If the terrapin originated in this portion of its range, genetic diversity would be expected to be higher there, while peripheral populations (North Atlantic, Florida, and the Gulf) would be expected to exhibit lower genetic diversity.

A common finding among genetic studies of terrapins is that Florida populations stand out as distinctive. Hart et al. (2014), Drabeck et al. (2014), and Hauswaldt and Glenn (2005) all found Florida terrapins to be genetically differentiated from neighboring populations in the Atlantic and Gulf. The cause of this differentiation is not entirely clear, but terrapin populations in Florida exhibit much lower levels of heterozygosity (0.363; Hart et al. 2014) and fewer private alleles than populations in other regions, consistent with a severe genetic bottleneck, inbreeding, or small population sizes. One possible cause is the Suwanne Seaway, which separated peninsular Florida from the North American continent by sea level rise during the Miocene and Pliocene (Felder and Staton 1994). This seaway would have isolated terrapin populations in Florida while facilitating gene flow between Gulf and Atlantic populations. The Suwanne Seaway could also account for the Atlantic/Gulf genetic divide first documented by Lamb and Avise (1992).

It is noteworthy that the specific microsatellite markers used in terrapin studies influence the recovered patterns of diversity. Two sets of microsatellite primers are commonly used: *TerpSH* (Hauswaldt and Glenn 2003) and *Gmu* (King and Julian 2004). These two sets of markers differ in mutation rate. Hauswaldt (2004) estimated the *TerpSH* loci to have an average mutation rate of 2.72×10^{-3} mutations per site per generation, while the *Gmu* loci developed by King and Julian (2004) have a mean mutation rate of 4.34×10^{-4} mutations per site per generation (Converse et al. 2015)—approximately 10-fold slower. Accordingly, studies that use the *TerpSH* microsatellites consistently recover higher levels of diversity (Hauswaldt and Glenn 2005; Drabeck et al. 2014) than do studies that use the *Gmu* microsatellites (Hart et al 2014; Drabeck et al. 2014). Inference of terrapin phylogeographic structure is also affected by marker choice, with mtDNA recovering two phylogeographic groups (the Atlantic/Gulf divide; Lamb and Avise 1992; Hauswaldt 2004) and microsatellites recovering four population clusters (Gulf, Atlantic, Massachusetts, Florida; Hart et al. 2014). While incongruence between mitochondrial and nuclear loci is not uncommon, terrapin studies have routinely recycled the same markers. Development of new high-resolution genetic markers is sorely needed.

Somewhat disturbingly, evidence for population bottlenecks is highly variable among studies, even though there is good historical evidence for demographic bottlenecks across the terrapin's range (Garber 1990). Louisiana terrapins provide an excellent example of the discordance among studies. Hart et al. (2014) and Coleman (2011) reported bottlenecks in this region, Petre (2014) found evidence that varied by site, and Drabeck et al. (2014) found no evidence of a bottleneck. In some cases, outbreeding due to terrapin translocations among populations may have obfuscated genetic evidence of a bottleneck. More work on population bottlenecks and their detection is warranted.

Finally, although many ecological studies show terrapins to exhibit high site fidelity and low dispersal, which should promote the evolution of population genetic structure, population genetic studies consistently recover low levels of genetic structure. Several hypotheses have been put forth to explain this discrepancy between ecology and genetics. Hauswaldt and Glenn (2005) hypothesize that terrapins form mating aggregations in the early spring and disperse far from their home range to participate. This would homogenize local population structure such as that generated by nest site fidelity. Ecological work by Seigel (1980) and Butler (2002) supports this interpretation. However, work by Sheridan (2010) suggests that terrapins do not travel long distances to form mating aggregations. Converse et al. (2015) hypothesize that mating aggregations homogenize populations at the local level, while male-biased dispersal homogenizes populations at larger spatial scales. Because most populations are strongly female-biased, low levels of male migration can result in high levels of admixture and thereby reduce population structure. We recommend that future ecological work focus on population sex ratios, juvenile dispersal, and adult dispersal in relation to mating aggregations, as these elements of terrapin biology have the potential to harmonize ecological and molecular findings.

Future genetic work should also take full advantage of advances in genomics to more accurately test models of population structure and population history. Thankfully, a large volume of high-quality loci is on the horizon, as scientists at the University of Maryland have recently sequenced the terrapin's entire mitochondrial and nuclear genomes (M. Pop, Center for Bioinformatics and Computational Biology, University of Maryland, pers. comm.). With a reference genome, a study could now easily include tens of thousands of single nucleotide polymorphisms or microsatellite loci. Hypervariable sequence regions unique to terrapins also could be identified and used. Indeed, it is now possible to sequence whole genomes from throughout the terrapin's range.

The diamond-backed terrapin has a complicated history, and it is important that future studies incorporate the latest advances in genomics, computational resources, and population genetic theory. Our understanding of the history of the terrapin is incomplete, and a range-wide phylogeographic study that takes advantage of next-generation sequencing resources is needed to define management units, identify possible Pleistocene refugia, and better document population history. Populations in Florida are of particular interest because they are genetically distinct from both Gulf and Atlantic populations, perhaps as a consequence of the Suwanne Seaway. On a finer scale, landscape genetic studies that document contemporary movement patterns would inform conservation efforts. For instance, models of gene flow that include electrical circuit theory (McRae and Beier 2007) could be used to investigate the effects of crab pots on population connectivity. Studies of contemporary gene flow could include historical estimates, as these provide important context for interpreting patterns of connectivity (Epps et al. 2013; Converse et al. 2015). Finally, more studies on natal philopatry and juvenile dispersal are badly needed because these elements of terrapin biology have an overarching effect on population genetic structure. By taking advantage of the latest developments in DNA sequencing and population genetic theory, and by conducting important work on natural history and ecology in parallel, we can ensure that our understanding of the evolutionary history of the diamond-backed terrapin becomes more detailed than ever.

REFERENCES

Auger, P.J. 1989. Sex ratio and nesting behavior in a population of *Malaclemys terrapin* displaying temperature-dependent sex-determination. PhD dissertation, Tufts University, Medford, MA. 174 pages.

Avise, J.C. 2000. Phylogeography: The History and Formation of Species. Harvard University Press, Cambridge, MA. 464 pages.

Avise, J.C., B.W. Bowen, T. Lamb, A.B. Meylan and E. Bermingham. 1992. Mitochondrial DNA evolution at a turtle's pace: evidence for low genetic variability and reduced microevolutionary rate in the Testudines. Molecular Biology and Evolution 9:457-473.

Barney, R.L. 1922. Further notes on the natural history and artificial propagation of the diamond-back terrapin. Bulletin of the US Bureau of Fisheries 38:91-111.

Beerli, P. 2008. MIGRATE version 3.6.5: a maximum likelihood and Bayesian estimator of gene flow using the coalescent. http://popgen.scs.edu/migrate.html.

Bromham, L. 2002. Molecular clocks in reptiles: life history influences rate of molecular evolution. Molecular Biology and Evolution 19:302-309.

Butler, J.A. 2002. Population ecology, home range, and seasonal movements of the Carolina diamondback terrapin, *Malaclemys terrapin centrata*, in northeastern Florida. Florida Fish and Wildlife Conservation Commission Project Report NG96-007. Final Report. Tallahassee, FL. 69 pages.

Callens, T., P. Galbuser, E. Matthysens, E.Y. Durand, M. Githiru, J.R. Huyghe and L. Lens. 2011. Genetic signature of population fragmentation varies with mobility in seven bird species of a fragmented Kenyan cloud forest. Molecular Ecology 20:1829-1844.

Coker, R.E. 1906. The natural history and cultivation of the diamond-back terrapin with notes on other forms of turtles. North Carolina Geological Survey Bulletin 14:1-67.

Coker, R.E. 1920. The diamondback terrapin: past, present and future. Scientific Monthly 11:171-186.

Coleman, A.T. 2011. Biology and conservation of the diamondback terrapin, *Malaclemys terrapin pileata*, in Alabama. PhD dissertation, University of Alabama at Birmingham, Birmingham, AL. 221 pages.

Converse, P.E., S.R. Kuchta, W.M. Roosenburg, P.F.P. Henry, G.M. Haramis and T.L. King. 2015. Spatiotemporal analysis of gene flow in Chesapeake Bay diamondback terrapins (*Malaclemys terrapin*). Molecular Ecology 24: 5864-5876.

Csilléry, K., M.G.B. Blum, O.E. Gaggiotti and O. François. 2010. Approximate Bayesian computation (ABC) in practice. Trends in Ecology and Evolution 25:410-418.

Drabeck, D.H., M.W.H. Chatfield and C.L. Richards-Zawacki. 2014. The status of Louisiana's diamondback terrapin (*Malaclemys terrapin*) populations in the wake of the Deepwater Horizon oil spill: insights from population genetic and contaminant analyses. Journal of Herpetology 48:125-136.

Epps, C.W., S.K. Wasser, J.L. Keim, B.M. Mutayoba and J.S. Brashares. 2013. Quantifying past and present connectivity illuminates a rapidly changing landscape for the African elephant. Molecular Ecology 22:1574-1588.

Excoffier, L., P.E. Smouse and J. M. Quattro. 1992. Analysis of molecular variance inferred from metric distance among DNA haplotypes: application to human mitochondrial DNA restriction data. Genetics 131:479-491.

Felder, D.L. and J. L. Staton. 1994. Genetic differentiation in trans-Floridian species complexes of *Sesarma* and *Uca* (Crustacea: Decapoda: Brachyura). Journal of Crustacean Biology 14:191-209.

Freedberg, S., M.A. Ewert, B.J. Ridenhour, M. Neiman and C.E. Nelson. 2005. Nesting fidelity and molecular evidence for natal homing in the freshwater turtle, *Graptemys kohni*. Proceedings of the Royal Society B: Biological Sciences 272:1345-1350.

Galbraith, D.A., B.N. White, R.J. Brooks and P.T. Boag. 1993. Multiple paternity in clutches of snapping turtles (*Chelydra serpentina*) detected using DNA fingerprints. Canadian Journal of Zoology 71:318-324.

Garber, S.W. 1990. The ups and downs of the diamond-back terrapin. Conservationist 44:44-47.

Gibbons, J.W., J.E. Lovich, A.D. Tucker, N.N. Fitz-Simmons and J.L. Greene. 2001. Demographic and ecological factors affecting conservation and management of the diamondback terrapin (*Malaclemys terrapin*) in South Carolina. Chelonian Conservation and Biology 4:66-74.

Glenos, S.M. 2013. A comparative assessment of genetic variation of diamondback terrapin (*Malaclemys terrapin*) in Galveston Bay, Texas, in relation to other northern Gulf Coast populations. MS thesis, University of Houston, Clear Lake, TX. 71 pages.

Hart, K.M. 2005. Population biology of diamondback terrapins: defining and reducing threats across their geographic range. PhD dissertation, Duke University, Durham, NC. 259 pages.

Hart K.M., M.E. Hunter and T.L. King. 2014. Regional differentiation among populations of the Diamond-back terrapin (*Malaclemys terrapin*). Conservation Genetics 15:593-603.

Hauswaldt, J.S. 2004. Population genetics and mating pattern of the diamondback terrapin (*Malaclemys terrapin*). PhD dissertation, University of South Carolina, Columbia, SC. 216 pages.

Hauswaldt, J.S. and T.C. Glenn. 2003. Microsatellite DNA loci from the diamondback terrapin (*Malaclemys terrapin*). Molecular Ecology Notes 3:174-176.

Hauswaldt, J.S., and T.C. Glenn. 2005. Population genetics of the diamondback terrapin (*Malaclemys terrapin*). Molecular Ecology 14:723-732.

Hay, W.P. 1917. Artificial propagation of the diamond-back terrapin. Bulletin of the US Bureau of Fisheries 24:1-20.

Hildebrand, S.F. 1929. Review of experiments on artificial culture of diamondback terrapin. Bulletin of the US Bureau of Fisheries 45:25-70.

Hildebrand, S.F. 1933. Hybridizing diamond-back terrapins. Journal of Heredity 113:231-238.

Hildebrand, S.F. and C. Hatsel. 1926. Diamond-back terrapin culture at Beaufort, N.C. US Bureau of Fisheries Economic Circular 60:1-20.

Hillis, D.M. and J.J. Bull. 1993. An empirical test of bootstrapping as a method for assessing confidence in phylogenetic analysis. Systematic Biology 42:182-192.

Ireland, J.S., A.C. Broderick, F. Glen, B.J. Godley, G.C. Hays, P.L.M. Lee and D.O.F. Skibinski. 2003. Multiple paternity assessed using microsatellite markers, in green turtles Chelonia mydas (Linnaeus, 1758) of Ascension Island, South Atlantic. Journal of Experimental Marine Biology and Ecology 291:149-160.

King, T.L. and S.E. Julian. 2004. Conservation of microsatellite DNA flanking sequence across 13 emydid genera assayed with novel bog turtle (*Glyptemys muhlenbergii*) loci. Conservation Genetics 5:719-725.

Kirkpatrick, M. and N.H. Barton. 1997. Evolution of a species' range. American Naturalist 150:1-23.

Lamb, T. and J.C. Avise. 1992. Molecular and population genetic aspects of mitochondrial DNA variability in the diamondback terrapin, *Malaclemys terrapin*. Journal of Heredity 83:262-269.

Lazell, J.D. 1976. This Broken Archipelago: Cape Cod and the Islands, Amphibians and Reptiles. Quadran-gle/New York Times Book Company, New York, NY. 260 pages.

Lester, L.A. 2007. Tracking terrapins through genetic analysis: multilocus assignment tests shed light on origin of turtles sold in markets. MEM thesis, Duke University, Durham, NC. 38 pages.

Lovich, J.E. and J.W. Gibbons. 1990. Age at maturity influences adult sex ratio in the turtle *Malaclemys terrapin*. Oikos 59:126-134.

McCafferty, S.S., A. Shorette, J. Simundza and B. Brennessel. 2013. Paucity of genetic variation at an MHC class I gene in Massachusetts populations of the diamond-backed terrapin (*Malaclemys terrapin*): a cause for concern? Journal of Herpetology 47:222-226.

McRae, B.H. and P. Beier. 2007. Circuit theory predicts gene flow in plant and animal populations. Proceedings of the National Academy of Sciences 104:19,885-19,890.

Petre, C.L. 2014. The conservation genetics of two emydid turtles: *Emydoidea blandingii* and *Malaclemys terrapin*. MS thesis, University of Southern Missis-sippi, Hattiesburg, MS. 62 pages.

Piry, S., G. Luikart, and J.-M. Cornuet. 1999. BOTTLE-NECK: a computer program for detecting recent reductions in the effective population size using allele frequency data. Journal of Heredity 90:502-503.

Pritchard, J.K., M. Stephens and P. Donnelly. 2000. Inference of population structure using multilocus genotype data. Genetics 155:945-959.

Roosenburg, W.M. 1996. Maternal condition and nest site choice: an alternative for the maintenance of environmental sex determination? American Zoologist 36:157-168.

Roosenburg, W.M., J. Cover and P.P. van Dijk. 2008. Legal issues: legislative closure of the Maryland terrapin fishery: perspectives on a historical accomplishment. Turtle and Tortoise Newsletter 12:27-30.

Roosenburg, W.M., W. Cresko, M. Modesitte and M.B. Robbins. 1997. Diamondback terrapin (*Malaclemys terrapin*) mortality in crab pots. Conservation Biology 11:1166-1172.

Roosenburg, W.M. and J.P. Green. 2000. Impact of a bycatch reduction device on diamondback terrapins and blue crab capture in crab pots. Ecological Applications 10:882-889.

Roosenburg, W.M., K.L. Haley and S. McGuire. 1999. Habitat selection and movements of diamondback terrapins, *Malaclemys terrapin*, in a Maryland estuary. Chelonian Conservation and Biology 3:425-429.

Seigel, R.A. 1980. Nesting habitats of diamondback terrapins (*Malaclemys terrapin*) on the Atlantic Coast of Florida. Transactions of the Kansas Academy of Science 83:239-246.

Shaffer, H.B. and 58 coauthors. 2013. The western painted turtle genome, a model for the evolution of extreme physiological adaptations in a slowly evolving lineage. Genome Biology 14:R28. DOI 10.1186/gb-2013-14-3-r28. 22 pages.

Sheridan, C.M. 2010. Mating system and dispersal patterns in the diamondback terrapin (*Malaclemys terrapin*). PhD dissertation, Drexel University, Philadelphia, PA. 204 pages.

Sheridan, C.M., J.R. Spotila, W.F. Bien and H.W. Avery. 2010. Sex-biased dispersal and natal philopatry in the diamondback terrapin, *Malaclemys terrapin*. Molecular Ecology 19:5497-5510.

Soltis, D.E., A.B. Morris, J.S. McLachlan, P.S. Manos and P.S. Soltis. 2006. Comparative phylogeography of unglaciated eastern North America. Molecular Ecology 15:4261-4293.

Spivey, P.B. 1998. Home range, habitat selection, and diet of the diamondback terrapin (*Malaclemys terrapin*) in a North Carolina estuary. MS thesis. University of Georgia, Athens, GA. 80 pages.

Valenzuela, N. 2000. Multiple paternity in side-neck turtles, *Podocnemis expansa*: evidence from microsatellite DNA data. Molecular Ecology 9:99-105.

Wakeley, J. (ed.). 2008. Coalescent Theory: An Introduction. Roberts and Company Publishers, Greenwood Village, CO. 326 pages.

Wiktor, D.M., M. Hill and R.M. Chambers. 2001. Genetic diversity of diamondback terrapins (*Malaclemys terrapin*) from Piermont Marsh, Hudson River, NY. Pages 1-20 in J.R. Waldman and W.C. Nieder (eds.), Final Reports of the Tibor T. Polgar Fellowship Program, 2000. Volume VIII. Hudson River Foundation, New York, NY.

Wilson, G.A. and B. Rannala. 2003. Bayesian inference of recent migration rates using multilocus genotypes. Genetics 163:1177-1191.

6

Life History with Emphasis on Geographic Variation

JEFFREY E. LOVICH
J. WHITFIELD GIBBONS
KATHRYN M. GREENE

Every organism is defined by a set of vital rates that evolve to enhance lifetime reproductive success and survival of individuals and their progeny. These rates vary due to the complex but sometimes predictable interactions among individuals, populations, and their environments. Collectively, these rates are controlled by life history traits including age and size at maturity, longevity, clutch size, offspring size, clutch frequency, and survivorship throughout the organism's life cycle. A significant body of literature is devoted to life history theory, which is beyond the scope of this chapter; we refer the reader to the seminal works of Bernardo, Cole, Congdon, Roff, Stearns, and others on this topic. Here, we review the life history traits of diamond-backed terrapins across all subspecies (Chapter 4) to understand both their unique adaptations and their vulnerabilities in the modern world. Particularly, we analyze geographic variation in demographic traits and body size among terrapin populations. The latter is important because reproductive traits in turtles are influenced by body size (e.g., Ryan and Lindeman 2007). Ernst and Lovich (2009) provide a summary of terrapin life history traits. In this chapter we analyze a comprehensive data set for the effects of geographic variation, specifically latitude, on terrapin life history traits.

The terrapin is a model organism for the study of life history traits for two reasons. First, it is one of the best-studied turtle species (Lovich and Ennen 2013). Abundant, albeit not always complete, data exist to examine various aspects of its life history in a comparative framework. A substantial terrapin life history literature emerged during the early 1900s (e.g., Coker 1906), aimed primarily at facilitating captive propagation because of the terrapin's economic importance as a gourmet food item (Hildebrand 1932; McCauley 1945;

Carr 1952). More recent research on terrapins focuses on causal factors contributing to population declines throughout their range (Seigel and Gibbons 1995; Gibbons et al. 2001; Dorcas et al. 2007). Second, terrapins are model organisms for studying life history variation because they have one of the widest latitudinal ranges of any nonmarine, temperate turtle species in the world (Buhlmann et al. 2009). Terrapins range from Cape Cod, Massachusetts, to Corpus Christi, Texas (Ernst and Lovich 2009), with another population of questionable provenance in Bermuda (Chapters 3 and 4). Their wide distribution exposes northern terrapin populations to the effects of the Labrador Current, while populations in the south are exposed to the warm embrace of tropical mangrove habitats (Hartsell and Ernst 2004). Under these circumstances, with widely different selective pressures, life history traits are expected to vary widely in response. The high site fidelity displayed by terrapins (Gibbons et al. 2001) suggests that populations are adapted to local environmental conditions.

Previous analyses of 169 populations of 146 turtle species demonstrated a negative relationship between latitude and egg size and a positive relationship between latitude and clutch size (Iverson et al. 1993). Those authors offered at least seven hypotheses (long day length, spring productivity, juvenile competition, size-selective predation, nest loss, climatic uncertainty or seasonality, and summer length hypotheses) that attempted to explain these relationships. The same relationships among latitude, clutch size, and egg size were supported for various combinations of terrapin populations by some researchers (Seigel 1980; Zimmerman 1992; Allman et al. 2012) but not others (Horn 2012). These and other studies provide testable hypotheses for our range-wide analysis, with data now available for a larger number of terrapin populations. We hypothesize that terrapins achieve larger mean body sizes at higher latitudes as demonstrated for other turtle species with extensive north-south ranges (Litzgus and Brooks 1998; Ashton and Feldman 2003; Werner et al. 2016), putatively in conformance with predictions from Bergmann's Rule (Blackburn et al. 1999), although this is not yet unequivocally

documented for turtles. Some authors suggest that Bergmann's Rule, founded on predictions for homeotherms, is poorly supported for turtles (Angielczyk et al. 2015). We also hypothesize that terrapins produce larger clutches of smaller eggs (even after standardization for body size) at higher latitudes relative to populations at lower latitudes. In contrast to all other US turtles, latitudinal comparisons are possible for terrapins along a near-linear gradient with minimal confounding influence from longitude, at least along the Atlantic Coast.

Our Approach

We systematically searched the scientific literature ($N > 150$ publications) used by Lovich and Ennen (2013) or posted on the Diamondback Terrapin Working Group site's (http://www.dtwg.org/) bibliography link for published data on population-specific traits for adult male and female body size; hatchling body size; age and body size at maturity; egg size; clutch size; clutch frequency; and survivorship. Information was based on terrapin publications and data available up to early 2016.

We used a meta-analysis approach (Arnqvist and Wooster 1995), defined as the quantitative summary of research domains—in this case, all terrapin data we could find in the literature for the variables listed above. Meta-analysis is a useful technique for synthesizing research findings, but it has potential shortcomings. One criticism is that it mixes "apples and oranges" because there is often a lack of uniformity among the studies. In the extreme case, meta-analysis might mix both "good" and "bad" studies. Despite these criticisms, Arnqvist and Wooster (1995) concluded that meta-analysis is a powerful tool for generating quantitative summaries that lead to "higher order conclusions about general trends and patterns." Meta-analysis has been used recently to compare and contrast life history traits among populations of another turtle species, the red-eared slider (*Trachemys scripta elegans*) (Taniguchi et al. 2017).

To minimize the limitations of meta-analysis, not all available terrapin data were included; some were excluded because of trait measuring or reporting inconsistencies among studies. For example, body size data were reported as means, minimums, and maximums, and often using different measures: carapace length (CL), plastron length (PL), or unknown. Because most studies report mean female body size as PL, we used PL in our analyses but report data for CL in our narrative review because it provides additional information on body size. We also recognize variation in measuring techniques (e.g., egg dimensions measured from images on X-radiographs vs. actual eggs), multiple data recorders, and various publication biases (e.g., Rosenthal 1979) that cannot easily be controlled.

We examined geographic trends of mean life history traits (see Appendix Table 6.1) from about 25° to 42° north latitude (3000000 to 4600000 Universal Transverse Mercator [UTM] northing values), south to north. General UTM coordinates were estimated to obtain northing values (interchangeable with "latitude" in the text) using Google Earth Pro version 7.1. Our approach allows synthesis of variation in terrapin life history traits to identify general trends, unanswered questions, and new hypotheses for further study.

We used linear regression analyses to examine relationships between latitude and various terrapin traits. All analyses were conducted using SYSTAT 13 with an alpha of 0.05. Given the variation in female body size previously reported among studies, we regressed reproductive traits on body size measures and used the associated residuals as "size-free" variables when possible, as recommended by Iverson et al. (1993). However, in most of our analyses, sample sizes were inadequate for particular combinations of variables to obtain enough residual estimates, so we used raw values of variables to obtain the benefit of a larger sample size. When multiple values for particular research locations were available, we generally used the most recent data compilation to reduce autocorrelation, unless the values showed a large range of variation in the parameters of interest.

Body Size Variation

Mean adult female size is greater than mean adult male size in all populations, consistent with the well-established sexual size dimorphism of terrapins (Ernst and Lovich 2009; Underwood et al. 2013). The grand mean (the mean of all population means) female CL was 168.6 mm (SD = 20.3; $N = 31$), and maximum female CL was 216 mm for a specimen from Cape May, New Jersey (Table 6.1), reported by Wood (1997). However, Ernst and Lovich (2009) reported that female terrapins can attain a CL of 238 mm (locality unknown). Grand mean male CL was 120.9 mm (SD = 7.6; $N = 27$), and maximum male CL was 150.0 mm (Gibbons et al. 2001) for a male from Kiawah Island, South Carolina (Table 6.1).

We compared mean and maximum body size among populations when paired values were available (i.e., sample sizes in those graphs are less than those reported in Table 6.1). Mean and maximum PL of females varied greatly among populations. Maximum body size may be a better predictor of maturity than mean body size, as some estimates based on the latter appear to use subadult females. We did not detect a significant relationship between maximum body size and latitude for either males ($F_{1,14} = 1.7$; $P = 0.2$) or females ($F_{1,24} = 1.4$; $P = 0.2$). The lack of a significant relationship between PL and latitude suggests that removing the effect of body size from the reproductive variables is not necessary. Horn (2012) also did not find a relationship between female body size and latitude for terrapins.

Size and Timing of Maturity

Few studies unequivocally determine the size or age of maturity for terrapins. Absolute determination of female maturity requires detection of first production of eggs or oviposition in known-age animals. For males, mating activity and transfer of viable sperm is a prerequisite for determination of maturity, although this is not necessarily proof of breeding for turtles (Kaufmann 1992). It is possible to infer size and timing of maturity in turtles (e.g., Lovich et al. 1990, 1998), including terrapins (Hildebrand 1932; Cagle 1952), based on slowed growth and the appearance of secondary sexual characteristics, including enlarged heads in females (Underwood et al. 2013) and elongated, thickened tails in males.

Table 6.1. Minimum and maximum body size statistics compiled from various combinations of diamond-backed terrapin (*Malaclemys terrapin*) populations listed in Appendix Table 6.1

	Body size (mm)							
	Female CL		Male CL		Female PL		Male PL	
Statistic	Minimum	Maximum	Minimum	Maximum	Minimum	Maximum	Minimum	Maximum
N	22	24	21	22	31	33	20	21
Smallest population value reported	85.0	132.0	83.0	110.0	76.0	122.0	63.0	94.0
Largest population value reported	186.0	216.0	124.0	150.0	176.0	225.0	139.7	172.7
Mean	130.0	199.3	103.1	135.5	127.1	176.8	93.5	122.0
SD	30.1	28.5	10.4	11.2	24.2	17.7	15.5	16.9

Note: CL = carapace length; PL = plastron length. Sample sizes (*N*) are based on the numbers of studies or population samples available for each metric. Means and standard deviations (SD) are based on grand mean of population means reported. Minimums include immature specimens, based on knowledge of size at maturity in this species.

Hildebrand (1932) suggested that under artificial conditions in North Carolina, males mature in as few as 4 years at a body size (PL or CL not specified) of 80 to 90 mm. Female maturity was a function of the amount of food provided. Females as small as 137 mm and 4 years of age produced eggs if provided with food during the winter instead of hibernating. However, some females fed during the winter reproduced for the first time at 8 years. In contrast, female terrapins that hibernated through the winter reproduced at 6 years, but more commonly at 7 years.

Cagle (1952) examined growth and maturity of 96 terrapins from Louisiana. He suggested that males as small as a PL of 98.7 mm were mature before they completed their third year of growth. Females were mature (as suggested by enlarged follicles) at 6 years and PL of 176 mm. Eleven other females with a PL ranging from 133 to 160 mm had immature follicles from 3 to 7 mm in diameter. However, immature terrapins may have follicles in various stages of development, suggesting that allocation of resources to reproduction may begin before reproductive maturity is attained (W. Roosenburg, Ohio University, pers. comm.).

Terrapins in Florida appear to grow and mature faster than populations in either Louisiana or North Carolina (Seigel 1984). The smallest female with oviductal eggs, or corpora lutea, had a PL of 135 mm, and all females with a PL of >140 mm were

mature at 4 to 5 years. The smallest male with enlarged testes and secondary sexual characteristics had a PL of 91 mm, and all males were mature at 2 to 3 years and a PL of >95 mm.

At Kiawah Island, South Carolina, the smallest male suspected as mature had a PL of 91 mm and was 3 years of age (Lovich and Gibbons 1990). The smallest gravid female determined by X-radiography had a PL of 138 mm and was 7 years of age. X-radiography revealed a gravid female with a PL of only 135 mm, and gravid females from nearby Edisto Island, South Carolina, ranged from 148 to 177 mm (Kern et al. 2016). In Maryland and Chesapeake Bay, males mature in 4 to 7 years at a PL of 100 mm, whereas females mature in 8 to 13 years at a PL as small as 175 mm (Roosenburg 1991). At Wellfleet on Cape Cod, Massachusetts, males mature at a PL of 80 mm, while females mature at 140 mm (Brennessel 2006).

Reproductive Output

Clutch Size and Clutch Frequency

Terrapin clutch size ranges from 1 to 23 eggs, and grand mean clutch size for 34 studies/populations was 9.6 eggs (Ernst and Lovich 2009). We examined the relationship between clutch size (using residuals of mean first clutch size vs. PL) and latitude from 11 studies/populations. Latitude explained 68% of the variation in body-size-adjusted clutch size (Fig. 6.1; $F_{1,9} = 13.28$; $P = 0.005$). Addi-

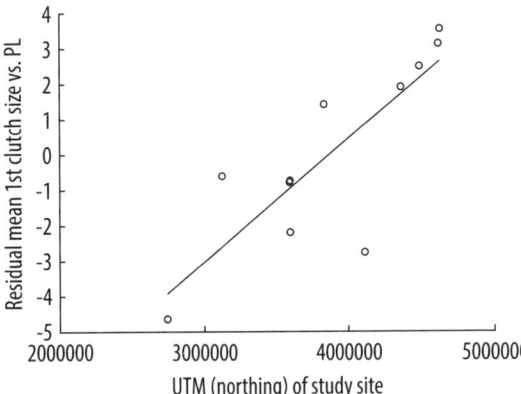

Fig. 6.1. Linear relationship between the residuals of mean first clutch size vs. mean female plastron length (PL) of a terrapin population and latitude. Latitude increases with UTM on the *x*-axis, here and in Figs. 6.3 to 6.6. The relationship is statistically significant (see text for details).

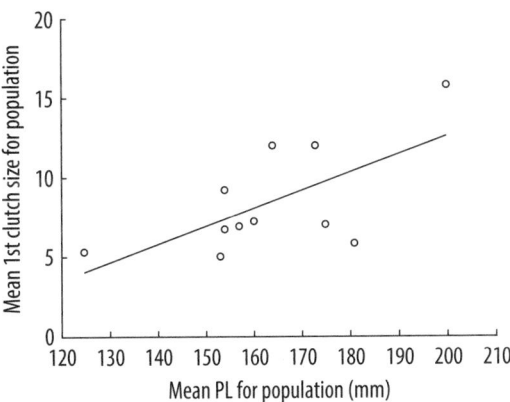

Fig. 6.2. Linear relationship between mean population first clutch size and mean population plastron length (PL). The relationship is statistically significant (see text for details).

tionally, populations with larger females, regardless of their geographic location, produce larger mean clutches (Fig. 6.2; $F_{1,9} = 6.3$; $P = 0.03$), with body size explaining 41% of the variation in mean clutch size.

Although captive terrapins can oviposit up to five clutches per year (Hildebrand and Hatsel 1926), most free-living terrapins have been documented to produce only one to two. Triple clutches per year have been suggested or reported from Florida (Seigel 1980, 1984; Butler et al. 2004) and Maryland (Roosenburg and Dunham 1997). The grand mean for 11 population means was 1.5 clutches per year. Clutch frequency did not change with latitude

($F_{1,10} = 0.2$; $P = 0.7$) in our data set. However, determining clutch frequency is difficult in free-living terrapins due to the open nature of their habitat, with multiple nesting locations.

Egg Size

Mean population egg width for terrapins ranged from 19.8 to 24.4 mm, with a grand mean of 22.1 mm (SD = 1.3; $N = 18$). Mean population egg width decreased with increasing latitude (Fig. 6.3; $F_{1,11} = 10.74$; $P = 0.007$; $N = 13$), with latitude explaining >49% of the variation in mean egg width among populations. Only seven studies that listed both the population mean PL and egg width necessary to generate size-free residuals could be georeferenced accurately. Size-free egg width did not change significantly with latitude ($F_{1,5} = 0.87$; $P = 0.39$; $N = 7$), although the sample size was small. Mean population egg length ranged from 27.1 to 39.7 mm, with a grand mean of 34.5 mm (SD = 3.5; $N = 18$). Mean population egg length decreased with increasing latitude (Fig. 6.4; $F_{1,10} = 64.33$; $P < 0.001$), with latitude explaining >86.5% of the variation in mean egg length among populations. Mean population egg mass ranged from 7.5 to 12.5 g, with a grand mean of 9.7 g (SD = 1.6; $N = 14$). The patterns of egg width and egg length decreasing with increasing latitude were paralleled in egg mass (Fig. 6.5; $F_{1,11} = 45.4$; $P < 0.001$; $N = 13$), with latitude explaining 80.5% of the variation in mean egg mass among populations.

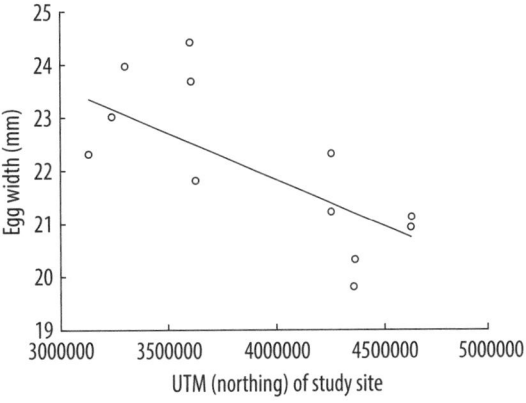

Fig. 6.3. Linear relationship between mean terrapin population egg width and latitude. The relationship is statistically significant (see text for details).

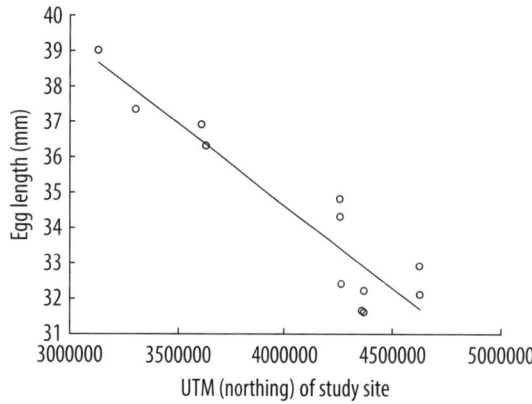

Fig. 6.4. Linear relationship between mean terrapin population egg length and latitude. The relationship is statistically significant (see text for details).

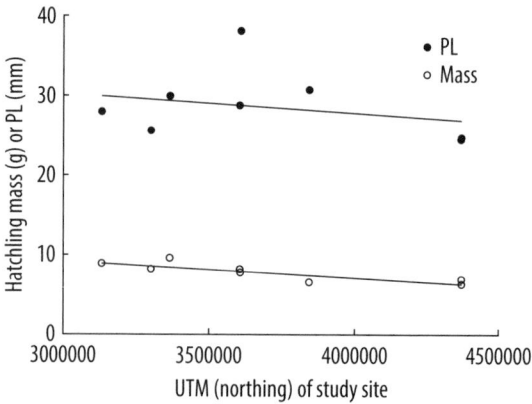

Fig. 6.6. Linear relationship between mean terrapin population hatchling mass and plastron length (PL) and latitude. The relationship is statistically significant only for mass (see text for details).

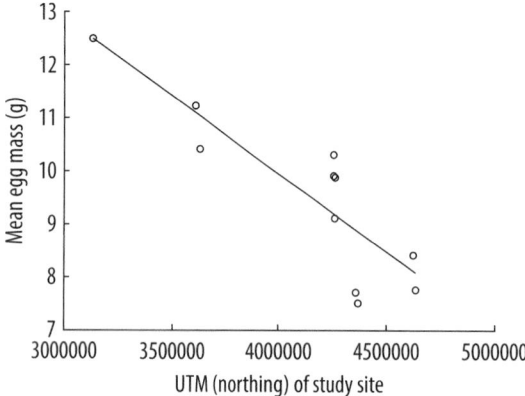

Fig. 6.5. Linear relationship between mean terrapin population egg mass and latitude. The relationship is statistically significant (see text for details).

Egg volumes ranged from 6.43 to 9.14 cc (Montevecchi and Burger 1975; Roosenburg and Dennis 2005).

Hatchling Size

Mean population hatchling PL ranged from 24.4 to 38.0 mm, with a grand mean of 28.0 mm (SD = 3.6; N = 13). Mean population hatchling mass ranged from 5.4 to 9.5 g, with a mean of 7.7 g (SD = 1.3; N = 13). Given the strong relationship between latitude and egg size and mass, a similar relationship was expected for hatchling size. The use of data

from locations that provided both mean hatchling PL and mean hatchling mass yielded mixed results. We were unable to detect a relationship between hatchling PL and latitude (Fig. 6.6; $F_{1,7}$ = 0.59; P = 0.47; N = 9). However, mean population hatchling mass decreased with latitude (Fig. 6.6; $F_{1,7}$ = 20.72; P = 0.003; N = 9), explaining 75% of the variation in hatchling mass.

Survivorship

Limited studies have estimated survivorship of terrapin populations. Sadly, terrapin populations do not appear to be doing well overall, according to a survey by Butler et al. (2006a). In responses from 54 individuals from all 16 states in the terrapin's range, 30% said that populations were declining in their state; 15%, that populations were stable; and the remainder (about 55%), that status was unknown. The third category underscores the need for more population studies.

One of the biggest threats to terrapin populations is drowning in recreational and commercial crab traps (Chapter 16). The magnitude of this problem is significant. For example, Grosse et al. (2009) found 133 terrapin carcasses in one crab trap in Georgia, more than double the estimated remainder of the population. Similarly, Roosenburg (1991) found a single unattended crab trap in Maryland that contained the carcasses of 49 terrapins, or about 1.6% to 2.8% of his study population. He later estimated that 15% to 78% of the local population could be trapped in a single year (Roosenburg et al. 1997). Population declines attributed to crab pots were also reported at the Kennedy Space Center in Florida between 1979 and 1993 (Seigel 1993; Seigel et al. 2002).

At Kiawah Island, South Carolina, Gibbons et al. (2001) documented the virtual extirpation of a terrapin population in a tidal creek from 1983 to 1996. In the same metapopulation, annual survivorship of females ranged from 0.748 to 0.971, depending on the tidal creek in which they lived (Tucker et al. 2001). Estimating mean life span from instantaneous mortality rates suggested that the average female terrapin did not survive to maturity in the South Carolina population (7 years). Given the high site fidelity observed in the population, migration rates were insufficient for recolonizing a creek that experienced extirpation. Later studies (2003-2013) of the same South Carolina population estimated annual survivorship of terrapins in five tidal creeks to range from 0.61 to 0.82, with no differences between the sexes (Witczak et al. 2014)—substantially lower than the survival estimates of Tucker et al. (2001). Mean annual survival of a terrapin population at North Inlet-Winyah Bay, South Carolina, was 0.78 (King and Ludlam 2014).

Increased rates of terrapin injuries and population declines have been reported. Sometimes-lethal boat propeller impacts increased from 1% to 2% in 1973 to 12% to 17% in 1990 among nesting females in New Jersey, coincident with increased human use of nesting beaches and increased boat traffic (Burger and Garber 1995; Chapter 15). Road mortality of nesting females was suspected as a cause of decreased mean body size and reduced frequency of capture of adult females elsewhere in New Jersey (Avissar 2006; Chapter 14).

Detailed demographic data are available from a Rhode Island mark-recapture study conducted between 1990 and 2001. The population growth parameter was estimated to be increasing ($\lambda = 1.034$; 95% confidence interval = 1.012-1.056; Mitro 2003).

Adult apparent survival was high with an estimated 188 breeding females but declined by 0.14% per year, from 0.959 in 1990 to 0.944 in 2000. Increased survival was deemed more important to population growth rate than recruitment of breeding females. Juvenile survival was estimated to be 0.565 at the estimated λ of 1.034, but it decreased to 0.446 for a stable population ($\lambda = 1$). In Maryland, Roosenburg and Niewiarowski (1998) estimated that adult female survival values of >0.87 were needed for a population to remain stable or to increase, depending on clutch size, nest survival, and age at maturity.

As in other turtle species, nest and hatchling survival varies greatly within and among years and populations (Ernst and Lovich 2009). In a Maryland site, hatchling recruitment decreased as nest success declined from 3.6% in 1987 to 2.4% in 1990, while hatchlings per successful nest declined from 10.75 to 3.4 (Roosenburg 1992). Hatchlings in northern regions overwinter within and outside nests, exposing them to increased thermal extremes and predation (see Chapter 8).

Discussion

The extensive latitudinal range of terrapins along the Atlantic Coast of the United States is reflected in variation of several life history traits. The lack of a clear relationship between body size and latitude further obfuscates the conformance or nonconformance of turtles to Bergmann's Rule (Ashton and Feldman 2003; Angielczyk et al. 2015; Werner et al. 2016). However, latitude, regardless of mean female body size of terrapin populations, exerts a strong influence on clutch size, egg width, egg length, egg mass, and hatchling mass. Overall, female terrapins produce larger clutches of smaller eggs at higher latitudes. Counterintuitively, they also apportion more lipid content to larger eggs at lower latitudes than at higher latitudes (Allman et al. 2012). Thus, females invest in more eggs with less energetic resources per offspring at higher latitudes and in fewer eggs with more lipids per offspring in the south. It is possible that this trade-off balances the number of surviving offspring needed to maintain populations in such widely variable environments.

Iverson et al. (1993) reviewed seven possible non-mutually exclusive hypotheses to explain why clutch size and egg size vary with latitude in a variety of organisms. Five of their hypotheses are applicable to turtles; the reader is referred to Iverson et al. (1993) for details and citations supporting each hypothesis. First is the spring productivity hypothesis, which posits that a rapid increase in productivity during the spring at temperate latitudes results in unlimited resources for reproduction, whereas the tropics are resource-limited. Spring productivity could cause increases in egg size or clutch mass, but the data do not support that for turtles (Iverson et al. 1993). Furthermore, the resources/energy used to produce eggs in many temperate turtles are harvested in the preceding year (Henen 2004; Rollinson et al. 2012), before hibernation. Seasonal allocation of energy to reproduction in terrapins has not been studied.

Second is the juvenile competition hypothesis. Because of the burst in productivity in the spring, it may be advantageous to produce more young at a time when resources are more abundant. However, most terrapins emerge from the nest in the summer, about 50 to 120 days after oviposition (Ernst and Lovich 2009). Third is the size-selective predation hypothesis. This is premised on the theory that predator diversity and specialization are reduced in temperate regions relative to lower latitudes. If true, advantages of producing many small eggs may prevail if hatchlings from smaller eggs have the same fitness and survival advantage as hatchlings produced from larger eggs. Fourth is the climatic uncertainty or seasonality hypothesis, which incorporates elements of the preceding hypotheses. Larger clutches are favored in unpredictable and seasonal climates. Thus, eggs would have to be smaller, because clutch mass is limited by a turtle's shell if body size does not increase in a predictable fashion with latitude as we show. However, the concept of larger clutches in unpredictable environments is in conflict with predictions under a bet-hedging reproductive strategy (Stearns 1976; Ennen et al. 2017).

The final hypothesis is based on summer length and predicts that small egg size at high latitudes may be a response to limited incubation times and cooler temperatures. However, Roosenburg (1996) speculated that female terrapins' nest site selection within a population depends on the size of the female's eggs and how this relates to environmental sex determination (Roosenburg and Kelley 1996). He predicted that females producing small eggs should oviposit in cool places where eggs will develop into males, because differences in egg size would not affect accelerated growth and age at maturity as they do in females. Females with larger eggs would be expected to oviposit in a warm place because of the premium of earlier maturity from larger eggs. Neither the seasonality nor summer length hypothesis was supported for terrapin egg mass or size analyses by Horn (2012), and more research is needed to determine which of these hypotheses, if any, best explains the variation observed in terrapins.

All of the hypotheses are complicated by optimal egg size, pelvic constraints on egg size, and various predictions of bet-hedging theory regarding propagule size (Congdon and Gibbons 1987; Lovich et al. 2012). More research is especially needed to understand the relationships among apportionment of yolk and lipids to eggs (Allman et al. 2012), clutch size, egg size, and latitude in terrapins. For example, our results suggest that hatchling PL is not strongly related to latitude, but hatchling mass is. This difference could be due to variation in post-hatching yolk sizes that vary predictably along the latitudinal gradient (larger in the south; Allman et al. 2012).

Future Research Needs

Even though the terrapin is well studied (Lovich and Ennen 2013), wide gaps persist in our understanding of its life history. Truly long-term demographic studies with peer-reviewed publications are available for only three locations throughout the terrapin's range: Kiawah Island, South Carolina (Dorcas and Gibbons 2013); Cape May Peninsula, New Jersey (e.g., Montevecchi and Burger 1975;

Wood 1997); and Mechanicsville, Maryland (e.g., Roosenburg and Dunham 1997), all on the Atlantic Coast. Another important Atlantic study spanning multiple decades is at the John F. Kennedy Space Center in Florida (Seigel et al. 2002). If ongoing research on mangrove terrapins continues (Hart and McIvor 2008), another important long-term study will be added. Numerous other efforts are underway to study terrapin demography and life history, including those by Burke and students in New York (e.g., Muldoon and Burke 2012; Burke and Calichio 2014) and studies at Barnegat Bay, New Jersey (Sheridan et al. 2010). Studies spanning two or more decades of intensive mark-recapture efforts are rare but are sorely needed to assess interannual variation in life history traits and its effects on demographic parameters (Hoyle and Gibbons 2000; Gibbons et al. 2001; Tucker et al. 2001). Since Cagle's (1952) early studies in Louisiana, there has been a scarcity of data on terrapin life history from the Gulf Coast, although Selman has made important recent contributions for Louisiana (Selman et al. 2014; Petre et al. 2015). Understanding the geographic variation in life history traits and how it affects population demographics is critical for the successful implementation of conservation and management strategies.

Other research questions remain unanswered. For example, how will climate change, particularly sea level rise, affect terrapin populations? In one study, terrapins ranked high in their sensitivity to rising sea levels because they nest on or near ocean beaches in many places (Hunter et al. 2015). Loss of nesting habitat and foraging areas is also expected under sea level rise (Woodland et al. 2017). However, tidal marsh stability is a complex interaction between sediment inputs, human modification of marsh habitat, sea level rise, and productivity of marsh plants (Kirwan et al. 2011; Kirwan and Megonigal 2013). Research that integrates and measures the effects of these factors (Isdell et al. 2015) will become increasingly important in efforts to conserve terrapins and their salt marsh habitats.

Another question needing further research is, how does variation in tidal amplitude affect terra-

pin populations? For example, along the Atlantic Coast from just below Cape Hatteras to the tip of Florida, tidal variation ranges from 0.2 to 2.1 m daily (Dame et al. 2000). Tidal range affects species composition and abundance in coastal environments (Dame et al. 2000), and this may affect terrapins as potential top-down predators in salt marsh ecosystems (Silliman and Bertness 2002), especially where they maintain high population densities (Hurd et al. 1979). Additional research should assess the effects of both natural and anthropogenic factors on life history traits. We know that mortality in commercial and recreational crab traps affects terrapin demography (Dorcas et al. 2007; Wolak et al. 2010). However, we know little about the demographic effects of environmental contaminants on terrapins (Blanvillain et al. 2007; Basile et al. 2011; Chapter 12) or how human modification of the marsh environment affects populations (Isdell et al. 2015).

How terrapin life history traits respond to increased mortality from crab trapping and other factors is a critical concern that needs continuing evaluation. Research by Wolak et al. (2010) suggested that in response to size-selected mortality in crab traps, terrapins in Chesapeake Bay shifted to a dramatically younger male age structure, decreased time to terminal female carapace size, a 15% increase in female carapace width, and increased sexual dimorphism. Such compensatory responses have only recently been suggested for turtles (Fordham et al. 2008a, b) and need more study to determine how populations respond, if at all, to increased mortality with changes in life history traits. Additional research is also needed on the survival of overwintering hatchlings (Muldoon and Burke 2012).

Environmental factors such as tidal amplitude, seasonal temperature variation, sea level rise, and other effects of climate change may all affect demography throughout terrapins' extensive range. Further research and initiation of new long-term studies at strategic locations in terrapin habitat will allow a better understanding of terrapins' ability to respond to environmental change and anthropogenic perturbations.

Appendix Table 6.1. Data reported for various morphological and life history attributes of diamond-backed terrapins throughout the range of the species

| Location | UTM northing | Mean adult CL (mm) | | Mean adult PL (mm) | | Mean clutch values | |
		Female	Male	Female	Male	1st clutch size	Clutch frequency
Ashley and Wando Rivers, Charleston, SC	3636299	—	—	121	100	—	—
Beaufort, NC	3843056	—	—	—	—	8	—
Dulac, LA	3253239	—	—	176.5	—	—	—
Merritt Island, Brevard County, FL	3132483	—	—	154	104	6.7	2
Little Beach, Brigantine, NJ	4362482	—	—	154.4	—	9.8	—
Comfort Island, Saint Bernard Parish, LA	3301538	195	—	—	—	8.5	1
Yorktown, VA	4122427	190	—	175	—	7	—
Nokum Hill, RI	4625081	—	—	—	—	16.1	—
Patuxent River, MD	4257035	—	—	—	—	12.2	—
Grice Island, SC	3627771	—	—	—	—	6	—
Kiawah Island, SC	3608227	160	121	143	103	5	—
Botany Island, SC	3603550	173.3	124	157.3	103.9	6.8	—
Patuxent River, MD	4256784	—	—	—	—	12.9	—
Kiawah Island, SC	3608227	191	123	—	—	—	—
Beaufort, NC	3843056	—	—	—	—	—	—
Schooner Creek, NJ	4380851	—	—	—	—	—	—
Ashley River, Charleston, SC	3636072	121	100	—	—	—	—
Little Creek, Fort George Island, FL	3366047	177.3	117.6	162.2	102.5	—	—
Cedar Point Marsh, Heron Bay, AL	3358625	—	—	168	—	—	—
Cedar Point Marsh, Heron Bay, AL	3358625	146.9	125	133.5	106.3	—	—
Jamaica Bay Wildlife Refuge, NY	4496875	—	—	172.9	—	12	2
Jamaica Bay Wildlife Refuge, NY	4496875	—	—	—	—	14.5	1
Kiawah Island, SC	3605428	—	—	157	—	6.9	—
Kiawah Island, SC	3608227	—	—	148	102	—	—
Kiawah Island, SC	3608227	160.2	120.5	144.2	102.6	—	—
Tangier Sound and St. Jerome Creek, Chesapeake Bay	4205217	—	—	178.2	115.6	—	—
Big Sable Creek Complex, FL	2795925	181	126	160	105	—	—
Richard Stockton College, NJ	4371459	—	—	—	—	—	—
Patuxent River, MD	4256695	—	—	—	—	—	—
Patuxent River, MD	4256695	—	—	—	—	—	—
Jamaica Bay Wildlife Refuge, NY	4496875	—	—	—	—	10.9	2
Patuxent River, MD	4262888	—	—	—	—	12.9	—
Patuxent River, MD	4262888	—	—	—	—	—	—
Little Beach, Brigantine, NJ	4370616	—	—	154	—	9.2	1
Jarrett Bay, NC	3849749	116.6	108.9	—	—	—	—
Beaufort, NC	3843056	—	—	—	—	14.5	—
Beaufort, NC	3843056	—	—	154	—	—	—
Beaufort, NC	3843056	—	—	124.7	151.6	5.3	—
Jamaica Bay Wildlife Refuge, NY	4497650	—	—	—	—	11.8	—
Monroe Co., FL		168.1	118.8	—	—	4	—
Monroe Co., FL		180.6	124.9	—	—	5.75	—
MS		168	121.4	—	—	—	—
Patuxent River, MD	4262888	—	—	—	—	13	2

Mean egg values			Mean hatchling values		
Length (mm)	Width (mm)	Mass (g)	PL (mm)	Mass (g)	Reference(s)
—	—	—	—	—	Bishop 1983
—	—	—	—	—	Hildebrand 1932
—	—	—	—	—	Cagle 1952
39	22.3	12.5	27.9	8.8	Seigel 1980, 1984
31.7	19.8	7.7	—	—	Burger and Montevecchi 1975; Montevecchi and Burger 1975
37.3	24.0	—	25.5	8.1	Burns and Williams 1972
—		—	27	—	Reid 1955
32.9	20.9	8.4	—	—	Allman et al. 2012
34.3	21.2	9.9	—	—	Allman et al. 2012
36.3	21.8	10.4	—	—	Allman et al. 2012
—	—	—	38	7.7	Unpublished
—	—	—	—	—	Unpublished
—	—	9.9	—	—	Roosenburg 1992
—	—	—	—	—	Akins et al. 2014
—	—	—	30.7	6.5	Allen and Littleford 1955
—	—	—	—	—	Avissar 2006
—	—	—	—	—	Bishop 1983
—	—	—	29.9	9.5	Butler 2002
—	—	—	—	—	Coleman et al. 2014
—	—	—	—	—	Coleman et al. 2011
—	—	—	—	—	Feinberg 2004
—	—	—	—	—	Cook 1980
36.9	22.2	11.2	28.7	8	Zimmerman 1992
—	—	—	—	—	Lovich and Gibbons 1990
—	—	—	—	—	Gibbons et al. 2001
—	—	—	—	—	Haramis et al. 2011
—	—	—	—	—	Hart and McIvor 2008
32.2	—	7.5	24.7	6.3	Herlands et al. 2004
—	—	—	—	7.3	Jeyasuria et al. 1994
34.8	22.3	10.3	—	—	Roosenburg and Dennis 2005
—	—	—	—	—	Feinberg and Burke 2003
—	—	9.9	—	—	Roosenburg and Dunham 1997
32.4	21.6	9.1	—	—	Roosenburg and Kelley 1996
31.6	20.3	—	24.4	6.8	Burger 1977
—	—	—	—	—	Hart and Crowder 2011
—	—	—	—	—	Hildebrand 1929
—	—	—	—	—	Barney 1922
—	—	—	25.4	—	Coker 1906
—	—	—	—	—	Giambanco 2002
—	—	—	—	—	Butler et al. 2006b
—	—	—	—	—	Butler et al. 2006b
—	—	—	—	—	Butler et al. 2006b
—	—	—	—	—	Roosenburg 1991

(continued)

Location	UTM northing	Mean adult CL (mm)		Mean adult PL (mm)		Mean clutch values	
		Female	Male	Female	Male	1st clutch size	Clutch frequency
Nockum Hill Wildlife Sanctuary, Barrington, RI	4626532	—	—	200	—	15.8	1
Duval Co., FL	3347007	—	—	—	—	6.7	—
Cape Cod, MA		—	—	—	—	—	1
MD		—	—	—	—	—	—
VA		179.3	124.8	160	103.2	11.1	—
Wellfleet, Cape Cod, MA	4637140	—	—	164	97	12	—
TX		—	—	161	126	—	—
Kiawah Island, SC	3608227	—	—	153.1	—	5	—
Botany Island, SC	3603550	—	—	160.1	—	7.2	—
Treatment site LK, Lower Florida Keys		166.6	119.8	149.2	101.4	—	—
Key West National Wildlife Refuge, FL	2718387	172.3	112.3	154.8	95.5	—	—
Everglades National Park, Middle Keys and Lower Keys of Florida Bay	2750061	—	—	181	125	5.8	—
Barnegat Bay, NJ	4384912	143	—	—	—	—	—
Little Beach, Brigantine, NJ	4370616	—	—	154	—	—	—
Intracoastal Waterway, Duval Co., FL	3374334	—	—	161.9	102.2	—	—
FL		173.2	123.5	—	—	—	—
Beaufort, NC	3843056	—	—	152.4	101.6	—	—
Saint Martins Marsh Aquatic Preserve, FL	3191280	—	—	175.4	118.2	—	—
Merritt Island, Brevard Co., FL	3132483	—	—	158	—	6.71	—
Tarpon Key in Pinellas Nation Wildlife Refuge, FL	3061381	—	—	165	110	—	—
Goodwin Islands, VA	4120069	201.1	130.2	—	—	—	—
Housatonic River, CT	4560110	172.5	124.4	—	—	—	—
Cheasapeake Bay (historical archives)	4120069	167.8	132.3	—	—	—	—
Sandy Neck, MA	4621039	—	—	—	—	15	1
Jarrett Bay, NC	3849749	123.3	105.3	—	—	—	—
Big Sable Creek Complex, FL	2795925	181	126	160	105	—	—
MA		—	—	170	100	15	—
LA		—	—	—	—	—	2
South Deer Island, Galveston Bay, TX	3239803	—	—	—	—	7	—
Kiawah Island, SC	3608227	—	—	118	99.6	—	—
North Inet-Winyah Bay, SC	3684955	158.4	117.7	—	—	—	—
Patuxent River, MD	4262888	—	—	157.9	110.9	—	—
Charleston Co., SC		170	120.1	153.2	102.4	—	—
Brevard Co., FL		180.4	119.4	160.5	99.4	—	—
Monroe Co., FL		178.7	128	160.7	110.7	—	—
Monroe Co., FL		172	117	155	97.7	—	—
Piedmont Marsh, Hudson River, NY	4542824	184	133	166.5	163.3	—	—
Jacques Cousteau National Estuarine Research Preserve, NJ	4376134	175	—	158	—	—	—
Stono River, James Island, SC	3622418	—	—	142	103	—	—
Kiawah Island, SC	3608227	—	—	—	—	—	—
Sea Isle City, NJ		—	—	—	—	—	—

Note: See text for abbreviations. Values are not always reported as means in each citation. In some cases, we used median or "typical" values based on examination of the source information. We included or excluded studies from our analyses as described in the text. For example, when multiple values for particular research locations were available, we generally used the most recent data compilation to reduce autocorrelation, unless the values showed a large range of variation in the parameters of interest. Also, various combinations of variables resulted in different sample sizes because not all studies reported all variables. Other approaches are likely to result in different interpretations, and we present this data summary to stimulate additional research and analysis.

| Mean egg values | | | Mean hatchling values | | |
Length (mm)	Width (mm)	Mass (g)	PL (mm)	Mass (g)	Reference(s)
32.1	21.1	—	26.2	—	Goodwin 1994
—	—	—	—	9.5	Butler et al. 2004
27.1	—	—	—	—	Auger and Giovannone 1979 (listed in Goodwin 1994)
31.1	21.2	—	—	—	Roosenburg 1992 (listed in Goodwin 1994)
34.4	21.2	8.5	27.5	7.9	Listed in Brennessel 2006
—	—	7.8	—	—	Brennessel 2006
—	—	—	—	—	Listed in Brennessel 2006
—	23.7	—	—	—	Kern et al. 2016
—	24.4	—	—	—	Kern et al. 2016
—	—	—	—	—	Mealey et al. 2014
—	—	—	—	—	Mealey et al. 2014
—	—	—	—	—	Baldwin et al. 2005
—	—	—	—	—	Burger 2002
—	—	—	—	—	Burger 1989
—	—	—	—	—	Butler et al. 2012
—	—	—	—	—	Listed in Butler et al. 2006b
—	—	—	—	—	Coker 1920
—	—	—	—	—	Boykin 2011
39	22.3	12.5	27.9	8.8	Seigel 1979
—	—	—	—	—	Boykin 1999 (listed in Boykin 2011)
—	—	—	—	—	Wolak et al. 2010
—	—	—	—	—	Wolak et al. 2010
—	—	—	—	—	Wolak et al. 2010
—	—	—	—	—	Hart 1999
—	—	—	—	—	Hart 2005
—	—	—	—	—	Hart 2005
—	—	—	—	—	Auger 1989 (listed in Hart 2005)
39.7	24.3	—	—	—	Dundee and Rossman 1989 (listed in Hart 2005)
39	23	—	—	—	Hogan 2003
—	—	—	—	—	Hoyle and Gibbons 2000
—	—	—	—	—	King and Ludlam 2014
—	—	—	—	—	Roosenburg et al. 1997
—	—	—	—	—	Schwartz 1955
—	—	—	—	—	Schwartz 1955
—	—	—	—	—	Schwartz 1955
—	—	—	—	—	Schwartz 1955
—	—	—	—	—	Simoes and Chambers 1999
—	—	—	—	—	Szerlag and McRobert 2006
—	—	—	—	—	Underwood et al. 2013
—	—	—	—	—	Witczak et al. 2014
—	—	—	—	5.4	Dimond 1987

ACKNOWLEDGMENTS

An earlier version of this chapter benefited greatly from comments provided by M. Agha, J. Iverson, and Shellie Puffer. Any use of trade, product, or firm names is for descriptive purposes only and does not imply endorsement by the US Government.

REFERENCES

Akins, C., C.D. Ruder, S.J. Price, L.A. Harden, J.W. Gibbons and M.E. Dorcas. 2014. Factors affecting temperature variation and habitat use in free-ranging diamondback terrapins. Journal of Thermal Biology 44:63-69.

Allen, J.F. and R.A. Littleford. 1955. Observations on feeding habits and growth of immature diamondback terrapins. Herpetologica 11:77-80.

Allman, P.E., A.R. Place and W.M. Roosenburg. 2012. Geographic variation in egg size and lipid provisioning in the diamondback terrapin *Malaclemys terrapin*. Physiological and Biochemical Zoology: Ecological and Evolutionary Approaches 85:442-449.

Angielczyk, K.D., R.W. Burroughs and C.R. Feldman. 2015. Do turtles follow the rules? Latitudinal gradients in species richness, body size, and geographic range area of the world's turtles. Journal of Experimental Zoology Part B: Molecular and Developmental Evolution 324:270-294.

Arnqvist, G. and D. Wooster. 1995. Meta-analysis: synthesizing research findings in ecology and evolution. Trends in Ecology and Evolution 10:236-240.

Ashton, K.G. and C.R. Feldman. 2003. Bergmann's Rule in nonavian reptiles: turtles follow it, lizards and snakes reverse it. Evolution 57:1151-1163.

Auger, P.J. 1989. Sex ratio and nesting behavior in a population of *Malaclemys terrapin* displaying temperature-dependent sex-determination. PhD dissertation, Tufts University, Medford, MA. 348 pages.

Auger, P.J. and P. Giovannone. 1979. On the fringe of existence: diamondback terrapins at Sandy Neck. Cape Naturalist 8:44-58.

Avissar, N.G. 2006. Changes in population structure of diamondback terrapins (*Malaclemys terrapin terrapin*) in a previously surveyed creek in southern New Jersey. Chelonian Conservation and Biology 5:154-159.

Baldwin, J.D., L.A. Latino, B.K. Mealey, G.M. Parks and M.R.J. Forstner. 2005. The diamondback terrapin in Florida Bay and the Florida Keys: insights into turtle conservation and ecology. Pages 180-186 in W.E. Meshaka, Jr. and K.J. Babbitt (eds.), Amphibians and Reptiles: Status and Conservation in Florida. Krieger Publishing Company, Malabar, FL.

Barney, R.L. 1922. Further notes on the natural history and artificial propagation of the diamond-back terrapin. Bulletin of the US Bureau of Fisheries 38:91-111.

Basile, E.R., H.W. Avery, W.F. Bien and J.M. Keller. 2011. Diamondback terrapins as indicator species of persistent organic pollutants: using Barnegat Bay, New Jersey, as a case study. Chemosphere 82:137-144.

Bishop, J.M. 1983. Incidental capture of diamondback terrapin by crab pots. Estuaries 6:426-430.

Blackburn, T.M., K.J. Gaston and N. Loder. 1999. Geographic gradients in body size: a clarification of Bergmann's Rule. Diversity and Distributions 5:165-174.

Blanvillain, G., J.A. Schwenter, R.D. Day, D. Point, S. J. Christopher, W.A. Roumillat and D.W. Owens. 2007. Diamondback terrapins, *Malaclemys terrapin*, as a sentinel species for monitoring mercury pollution of estuarine systems in South Carolina and Georgia, USA. Environmental Toxicology and Chemistry 26:1441-1450.

Boykin, C.S. 1999. The status of the ornate diamondback terrapin at Tarpon Key, in the Pinellas National Wildlife Refuge. Senior research project, Eckerd College, St. Petersburg, FL. 27 pages.

Boykin, C. 2011. The status and demography of the ornate diamondback terrapin (*Malaclemys terrapin macrospilota*) within the Saint Martins Marsh Aquatic Preserve. Report of the Department of Environmental Protection Florida. Tallahassee, FL. 47 pages.

Brennessel, B. 2006. Diamonds in the Marsh: A Natural History of the Diamondback Terrapin. University Press of New England, Hanover, NH, and London. 236 pages.

Buhlmann, K.A., T.S.B. Akre, J.B. Iverson, D. Karapatakis, R.A. Mittermeier, A. Georges, A.G. J. Rhodin, P.P. van Dijk and J.W. Gibbons. 2009. A global analysis of tortoise and freshwater turtle distributions with identification of priority conservation areas. Chelonian Conservation and Biology 8:116-149.

Burger, J. 1977. Determinants of hatching success in the diamondback terrapin, *Malaclemys terrapin*. American Midland Naturalist 97:444-464.

Burger, J. 1989. Diamondback terrapin protection. Plastron Papers 19:35-40.

Burger, J. 2002. Metals in tissues of diamondback terrapin from New Jersey. Environmental Monitoring and Assessment 77:255-263.

Burger, J. and S.D. Garber. 1995. Risk assessment, life history strategies, and turtles: could declines be prevented or predicted? Journal of Toxicology and Environmental Health 46:483-500.

Burger, J. and W.A. Montevecchi. 1975. Nest site selection in the terrapin *Malaclemys terrapin*. Copeia 1975:113-119.

Burke, R.L. and A.M. Calichio. 2014. Temperature-dependent sex determination in the diamond-backed terrapin (*Malaclemys terrapin*). Journal of Herpetology 48:466-470.

Burns, T.A. and K.L. Williams. 1972. Notes on the reproductive habits of *Malaclemys terrapin pileata*. Journal of Herpetology 6:237-238.

Butler, J.A. 2002. Population ecology, home range, and seasonal movements of the Carolina diamondback terrapin, *Malaclemys terrapin centrata*, in northeastern Florida. Report of the Florida Fish and Wildlife Conservation Commission. Tallahassee, FL. 65 pages.

Butler, J.A., C. Broadhurst, M. Green and Z. Mullin. 2004. Nesting, nest predation and hatchling emergence of the Carolina diamondback terrapin, *Malaclemys terrapin centrata*, in northeastern Florida. American Midland Naturalist 152:145-155.

Butler, J.A., G.L. Heinrich and M.L. Mitchell. 2012. Diet of the Carolina diamondback terrapin (*Malaclemys terrapin centrata*) in northeastern Florida. Chelonian Conservation and Biology 11:124-128.

Butler, J.A., G.L. Heinrich and R.A. Seigel. 2006a. Third workshop on the ecology, status, and conservation of diamondback terrapins (*Malaclemys terrapin*). Chelonian Conservation and Biology 5:331-334.

Butler, J.A., R.A. Seigel and B.K. Mealey. 2006b. *Malaclemys terrapin*—diamondback terrapin. Pages 279-295 in P.A. Meylan (ed.), Biology and Conservation of Florida Turtles. Chelonian Research Monographs 3. Chelonian Research Foundation, Lunenberg, MA.

Cagle, F.R. 1952. A Louisiana terrapin population. Copeia 1952:74-76.

Carr, A. 1952. Handbook of Turtles: The Turtles of the United States, Canada, and Baja California. Cornell University Press, Ithaca, NY. 542 pages.

Coker, R.E. 1906. The natural history and cultivation of the diamond-back terrapin with notes on other forms of turtles. North Carolina Geological Survey Bulletin 14:1-67.

Coker, R.E. 1920. The diamond-back terrapin: past, present, and future. Scientific Monthly 11:171-186.

Coleman, A.T., T. Roberge, T. Wibbels, K. Marion, D. Nelson and J. Dindo. 2014. Size-based mortality of adult female diamond-backed terrapins (*Malaclemys terrapin*) in blue crab traps in a Gulf of Mexico population. Chelonian Conservation and Biology 13:140-145.

Coleman, A.T., T. Wibbels, K. Marion, D. Nelson and J. Dindo. 2011. Effect of by-catch reduction devices (BRDS) on the capture of diamondback terrapins (*Malaclemys terrapin*) in crab pots in an Alabama salt marsh. Journal of the Alabama Academy of Sciences 82:145-157.

Congdon, J.D. and J.W. Gibbons. 1987. Morphological constraint on egg size: a challenge to optimal egg size theory? Proceedings of the National Academy of Sciences 84:4145-4147.

Cook, B. 1980. A natural history of the diamondback terrapin. Underwater Naturalist 18:25-30.

Dame, R., M. Alber, D. Allen, M. Mallin, C. Montague, A. Lewitus, A. Chalmers, R. Gardner, C. Gilman, B. Kjerfve, J. Pinckney and N. Smith. 2000. Estuaries of the South Atlantic coast of North America: their geographical signatures. Estuaries 23:793-819.

Dimond, M.T. 1987. The effects of incubation temperature on hatching time, sex, and growth of hatchlings of the diamondback terrapin, *Malaclemys terrapin* (Schoepff). Pranikee—Journal of Zoological Society of Orissa 8:1-5.

Dorcas, M.E. and J.W. Gibbons. 2013. Long-term ecological research on America's only estuarine turtle: the diamondback terrapin. Pages 447-461 in W.I. Lutterschmidt (ed.), Reptiles in Research: Investigations of Ecology, Physiology, and Behavior from Desert to Sea. Nova Biomedical, New York, NY.

Dorcas, M.E., J.D. Willson and J.W. Gibbons. 2007. Crab trapping causes population decline and demographic changes in diamondback terrapins over two decades. Biological Conservation 137:334-340.

Dundee, H.A. and D.A. Rossman. 1989. The Amphibians and Reptiles of Louisiana. Louisiana State University Press, Baton Rouge, LA. 300 pages.

Ennen, J.R., J.E. Lovich, R.C. Averill-Murray, C.B. Yackulic, M. Agha, C. Loughran, L. Tennant and B. Sinervo. 2017. The evolution of different maternal investment strategies in two closely related desert vertebrates. Ecology and Evolution 7:3177-3189. DOI 10.1002/ece3.2838.

Ernst, C.H. and J.E. Lovich. 2009. Turtles of the United States and Canada. 2nd edition. Johns Hopkins University Press, Baltimore, MD. 840 pages.

Feinberg, J.A. 2004. Nest predation and ecology of terrapin, *Malaclemys terrapin*, at the Jamaica Bay Wildlife Refuge. Pages 5-12 in C. Swarth, W.M. Roosenburg and E. Kiviat (eds.), Conservation and Ecology of Turtles of the Mid-Atlantic Region: A Symposium. Bibliomania, Salt Lake City, UT.

Feinberg, J.A. and R.L. Burke. 2003. Nesting ecology and predation of diamondback terrapins, *Malaclemys terrapin*, at Gateway National Recreation Area, New York. Journal of Herpetology 37:517-526.

Fordham, D.A., A. Georges and B.W. Brook. 2008a. Experimental evidence for density-dependent responses to mortality of snake-necked turtles. Oecologia 159:271-281.

Fordham, D.A., A. Georges and B.W. Brook. 2008b. Indigenous harvest, exotic pig predation and local persistence of a long-lived vertebrate: managing a tropical freshwater turtle for sustainability and conservation. Journal of Applied Ecology 45:52-62.

Giambanco, M.R. 2002. Comparison of viability rates, hatchling survivorship, and sex ratios of laboratory- and field-incubated nests of the estuarine, emydid turtle *Malaclemys terrapin*. MS thesis, Hofstra University, Hempstead, NY. 83 pages.

Gibbons, J.W., J.E. Lovich, A.D. Tucker, N.N. Fitzsimmons and J. L. Greene. 2001. Demographic and ecological factors affecting conservation and management of the diamondback terrapin (*Malaclemys terrapin*) in South Carolina. Chelonian Conservation and Biology 4:66-74.

Goodwin, C.C. 1994. Aspects of nesting ecology of the diamondback terrapin (*Malaclemys terrapin*) in Rhode Island. MS thesis, University of Rhode Island, Kingston, RI. 84 pages.

Grosse, A.M., J.D. van Dijk, K.L. Holcomb and J.C. Maerz. 2009. Diamondback terrapin mortality in crab pots in a Georgia tidal marsh. Chelonian Conservation and Biology 8:98-100.

Haramis, G.M., P.F. Henry and D.D. Day. 2011. Using scrape fishing to document terrapins in hibernacula in Chesapeake Bay. Herpetological Review 42:170-177.

Hart, K.M. 1999. Declines in diamondbacks: terrapin population modeling and implications for management. MS thesis, Duke University, Durham, NC. 59 pages.

Hart, K.M. 2005. Population biology of diamondback terrapins (*Malaclemys terrapin*): defining and reducing threats across their range. PhD dissertation, Duke University, Durham, NC. 259 pages.

Hart, K.M. and L.B. Crowder. 2011. Mitigating by-catch of diamondback terrapins in crab pots. Journal of Wildlife Management 75:264-272.

Hart, K.M. and C.C. McIvor. 2008. Demography and ecology of mangrove diamondback terrapins in a wilderness area of Everglades National Park, Florida, USA. Copeia 2008:200-208.

Hartsell, T.D. and C.H. Ernst. 2004. A review of environmental conditions along the coastal range of the diamondback terrapin, *Malaclemys terrapin*. Herpetological Bulletin 89:12-20.

Henen, B.T. 2004. Capital and income breeding in two species of desert tortoises. Transactions of the Royal Society of South Africa 59:65-71.

Herlands, R., R. Wood, J. Pritchard, H. Clapp and N. Le Furge. 2004. Diamondback terrapin (*Malaclemys terrapin*) head-starting project in southern New Jersey. Pages 13-18 in C. Swarth, W.M. Roosenburg and E. Kiviat (eds.), Conservation and Ecology of Turtles of the Mid-Atlantic Region: A Symposium. Bibliomania, Salt Lake City, UT.

Hildebrand, S.F. 1929. Review of experiments on artificial culture of diamond-back terrapin. Bulletin of the US Bureau of Fisheries 45:25-70.

Hildebrand, S.F. 1932. Growth of diamond-back terrapins: size attained, sex ratio and longevity. Zoologica 9:551-563.

Hildebrand, S.F. and C. Hatsel. 1926. Diamond-bask terrapin culture at Beaufort, N.C. US Bureau of Fisheries, Economic Circular 60:1-20.

Hogan, J.L. 2003. Occurrence of the diamondback terrapin (*Malaclemys terrapin littoralis*) at South Deer Island in Galveston Bay, Texas, April 2001-May 2002. US Geological Survey Open File Report 03-022, page 30. https://pubs.usgs.gov/of/2003/ofr03-022.

Horn, E.E. 2012. Life history variation in the diamondback terrapin (*Malaclemys terrapin*). MS thesis, Hofstra University, Long Island, NY. 145 pages.

Hoyle, M.E. and J.W. Gibbons. 2000. Use of a marked population of diamondback terrapins (*Malaclemys terrapin*) to determine impacts of recreational crab pots. Chelonian Conservation and Biology 3:735-737.

Hunter, E.A., N.P. Nibbelink, C.R. Alexander, K. Barrett, L.F. Mengak, R.K. Guy, C.T. Moore and R.J. Cooper. 2015. Coastal vertebrate exposure to predicted habitat changes due to sea level rise. Environmental Management 56:1528-1537.

Hurd, L.E., G.W. Smedes and T.A. Dean. 1979. An ecological study of a natural population of diamondback terrapins (*Malaclemys t. terrapin*) in a Delaware salt marsh. Estuaries 2:28-33.

Isdell, R.E., R.M. Chambers, D.M. Bilkovic and M. Leu. 2015. Effects of terrestrial-aquatic connectivity on an estuarine turtle. Diversity and Distributions 21:643-653.

Iverson, J.B., C.P. Balgooyen, K.K. Byrd and K.K. Lyddan. 1993. Latitudinal variation in egg and clutch size in turtles. Canadian Journal of Zoology 71:2448-2461.

Jeyasuria, P., W.M. Roosenburg and A.R. Place. 1994. Role of P450 aromatase in sex determination of the diamondback terrapin, *Malaclemys terrapin*. Journal of Experimental Zoology 270:95-111.

Kaufmann, J.H. 1992. The social behavior of wood turtles, *Clemmys insculpta*, in central Pennsylvania. Herpetological Monographs 6:1-25.

Kern, M.M., J.C. Guzy, J.E. Lovich, J.W. Gibbons and M.E. Dorcas. 2016. Relationships of maternal body size and morphology with egg and clutch size in the diamondback terrapin, *Malaclemys terrapin* (Testudines: Emydidae). Biological Journal of the Linnean Society 117:295-304.

King, P. and J.P. Ludlam. 2014. Status of diamondback terrapins (*Malaclemys terrapin*) in North Inlet-Winyah Bay, South Carolina. Chelonian Conservation and Biology 13:119-124.

Kirwan, M.L. and J.P. Megonigal. 2013. Tidal wetland stability in the face of human impacts and sea-level rise. Nature 504:53-60.

Kirwan, M.L., A.B. Murray, J.P. Donnelly and D.R. Corbett. 2011. Rapid wetland expansion during European settlement and its implication for marsh survival under modern sediment delivery rates. Geology 39:507-510.

Litzgus, J.D. and R.J. Brooks. 1998. Growth in a cold environment: body size and sexual maturity in a northern population of spotted turtles, *Clemmys guttata*. Canadian Journal of Zoology 76:773-782.

Lovich, J.E. and J.R. Ennen. 2013. A quantitative analysis of the state of knowledge of turtles of the United States and Canada. Amphibia-Reptilia 34:11-23.

Lovich, J.E., C.H. Ernst and J.F. McBreen. 1990. Growth, maturity, and sexual dimorphism in the wood turtle, *Clemmys insculpta*. Canadian Journal of Zoology 68:672-677.

Lovich, J.E., C.H. Ernst, R.T. Zappalorti and D.W. Herman. 1998. Geographic variation in growth and sexual size dimorphism of bog turtles (*Clemmys muhlenbergii*). American Midland Naturalist 139:69-78.

Lovich, J.E. and J.W. Gibbons. 1990. Age at maturity influences adult sex ratio in the turtle *Malaclemys terrapin*. Oikos 59:126-134.

Lovich, J.E., S.V. Madrak, C.A. Drost, A.J. Monatesti, D. Casper and M. Znari. 2012. Optimal egg size in a suboptimal environment: reproductive ecology of female Sonora mud turtles (*Kinosternon sonoriense*) in central Arizona, USA. Amphibia-Reptilia 33:161-170.

McCauley, R.H. 1945. The Reptiles of Maryland and the District of Columbia. Privately printed, Hagerstown, MD. 194 pages.

Mealey, B.K., J.D. Baldwin, G.G. Parks-Mealey, G.D. Bossart and M.R.J. Forstner. 2014. Characteristics of mangrove diamondback terrapins (*Malaclemys terrapin rhizophorarum*) inhabiting altered and natural mangrove islands. Journal of North American Herpetology 1:76-80.

Mitro, M.G. 2003. Demography and viability analyses of a diamondback terrapin population. Canadian Journal of Zoology 81:716-726.

Montevecchi, W.A. and J. Burger. 1975. Aspects of the reproductive biology of the northern diamondback terrapin *Malaclemys terrapin terrapin*. American Midland Naturalist 94:166-178.

Muldoon, K.A. and R.L. Burke. 2012. Movements, overwintering, and mortality of hatchling diamondbacked terrapins (*Malaclemys terrapin*) at Jamaica Bay, New York. Canadian Journal of Zoology 90:651-662.

Petre, C., W. Selman, B. Kreiser, S.H. Pearson and J.J. Wiebe. 2015. Population genetics of the diamondback terrapin, *Malaclemys terrapin*, in Louisiana. Conservation Genetics 16:1243-1252.

Reid, G.K., Jr. 1955. Reproduction and development in the northern diamondback terrapin, *Malaclemys terrapin terrapin*. Copeia 1955:310-311.

Rollinson, N., R.G. Farmer and R.J. Brooks. 2012. Widespread reproductive variation in North American turtles: temperature, egg size and optimality. Zoology 115:160-169.

Roosenburg, W.M. 1991. The diamondback terrapin: population dynamics, habitat requirements, and opportunities for conservation. Pages 227-234 in A. Chaney and J.A. Mihursky (eds.), New Perspectives in the Chesapeake System: A Research and Management Perspective. Proceedings of a Conference. Chesapeake Research Consortium Pub. no. 137. Chesapeake Research Consortium, Solomons, MD.

Roosenburg, W.M. 1992. Life history consequences of nest site choice by the diamondback terrapin, *Malaclemys terrapin*. PhD dissertation, University of Pennsylvania, Philadelphia, PA. 206 pages.

Roosenburg, W.M. 1996. Maternal condition and nest site choice: an alternative for the maintenance of environmental sex determination. American Zoologist 36:157-168.

Roosenburg, W.M., W. Cresko, M. Modesitte and M.B. Robbins. 1997. Diamondback terrapin (*Malaclemys terrapin*) mortality in crab pots. Conservation Biology 11:1166-1172.

Roosenburg, W.M. and T. Dennis. 2005. Egg component comparisons within and among clutches of the diamondback terrapin, *Malaclemys terrapin*. Copeia 2005:417-423.

Roosenburg, W.M. and A.E. Dunham. 1997. Allocation of reproductive output: egg- and clutch-size variation in the diamondback terrapin. Copeia 1997:290-297.

Roosenburg, W.M. and P.J. Kelley. 1996. The effect of egg size and incubation temperature on growth in the turtle, *Malaclemys terrapin*. Journal of Herpetology 30:198-204.

Roosenburg, W.M. and P. Niewiarowski. 1998. Maternal effects and the maintenance of environmental sex determination. Pages 307-322 in T.A. Mousseau and C.W. Fox (eds.), Maternal Effects as Adaptations. Oxford University Press, New York, NY.

Rosenthal, R. 1979. The file drawer problem and tolerance for null results. Psychological Bulletin 86:638-641.

Ryan, K.M. and P.V. Lindeman. 2007. Reproductive allometry in the common map turtle, *Graptemys geographica*. American Midland Naturalist 158:49-59.

Schwartz, A. 1955. The diamondback terrapins (*Malaclemys terrapin*) of peninsular Florida. Proceedings of the Biological Society of Washington 68:157-164.

Seigel, R.A. 1979. The reproductive biology of the diamondback terrapin, *Malaclemys terrapin tequesta*. MS thesis, University of Central Florida, Orlando, FL. 40 pages.

Seigel, R.A. 1980. Nesting habits of diamondback terrapins (*Malaclemys terrapin*) on the Atlantic Coast of Florida. Transactions of the Kansas Academy of Science 83:239-246.

Seigel, R.A. 1984. Parameters of two populations of diamondback terrapins (*Malaclemys terrapin*) on the Atlantic Coast of Florida. Pages 77-87 in R.A. Seigel, J.L. Knight, L. Malaret and N.L. Zuschlag (eds.), Vertebrate Ecology and Systematics: A Tribute to Henry S. Fitch. University of Kansas, Lawrence, KS.

Seigel, R.A. 1993. Apparent long-term decline in diamond-back terrapin populations at the Kennedy Space Center, Florida. Herpetological Review 24:102-103.

Seigel, R.A. and J.W. Gibbons. 1995. Workshop on the ecology, status, and management of the diamondback terrapin (*Malaclemys terrapin*), Savannah River Ecology Laboratory, 2 August 1994: final results and recommendations. Chelonian Conservation and Biology 1:240-243.

Seigel, R.A., R.B. Smith, J. Demuth, L.M. Ehrhart and F.F. Snelson, Jr. 2002. Amphibians and reptiles of the John F. Kennedy Space Center, Florida: a long-term assessment of a large protected habitat (1975-2000). Florida Scientist 65:1-12.

Selman, W., B. Baccigalopi and C. Baccigalopi. 2014. Distribution and abundance of diamondback terrapins (*Malaclemys terrapin*) in southwestern Louisiana. Chelonian Conservation and Biology 13:131-139.

Sheridan, C.M., J.R. Spotila, W.F. Bien and H.W. Avery. 2010. Sex-biased dispersal and natal philopatry in the diamondback terrapin, *Malaclemys terrapin*. Molecular Ecology 19:5497-5510.

Silliman, B.R. and M.D. Bertness. 2002. A trophic cascade regulates salt marsh primary production. Proceedings of the National Academy of Sciences 99:10,500-10,505.

Simoes, J.C. and R.M. Chambers. 1999. The diamondback terrapins of Piermont Marsh, Hudson River, New York. Northeastern Naturalist 6:241-248.

Stearns, S.C. 1976. Life-history tactics: a review of the ideas. Quarterly Review of Biology 51:3-47.

Szerlag, S. and S.P. McRobert. 2006. Road occurrence and mortality of the northern diamondback terrapin. Applied Herpetology 3:27-37.

Taniguchi, M., J.E. Lovich, K. Mine, S. Ueno and N. Kamezaki. 2017. Unusual population attributes of invasive red-eared slider turtles (*Trachemys scripta elegans*) in Japan: do they have a performance advantage? Aquatic Invasions 12:97-108.

Tucker, A.D., J.W. Gibbons and J.L. Greene. 2001. Estimates of adult survival and migration for diamondback terrapins: conservation insight from local extirpation within a metapopulation. Canadian Journal of Zoology 79:2199-2209.

Underwood, E.B., S. Bowers, J.C. Guzy, J.E. Lovich, C.A. Taylor, J.W. Gibbons and M.E. Dorcas. 2013. Sexual dimorphism and feeding ecology of diamond-backed terrapins (*Malaclemys terrapin*). Herpetologica 69:397-404.

Werner, Y.L., N. Korolker, G. Sion and B. Göçmen. 2016. Bergmann's and Rensch's rules and the spur-thighed tortoise (*Testudo graeca*). Biological Journal of the Linnean Society 117:796-811.

Witczak, L.R., J.C. Guzy, S.J. Price, J.W. Gibbons and M.E. Dorcas. 2014. Temporal and spatial variation in survivorship of diamondback terrapins (*Malaclemys terrapin*). Chelonian Conservation and Biology 13:146-151.

Wolak, M.E., G.W. Gilchrist, V.A. Ruzicka, D.M. Nally and R.M. Chambers. 2010. A contemporary, sex-limited change in body size of an estuarine turtle in response to commercial fishing. Conservation Biology 24:1268-1277.

Wood, R.C. 1997. The impact of commercial crab traps on northern diamondback terrapins, *Malaclemys terrapin terrapin*. Pages 21-27 in J.E. Van Abbema (ed.), Proceedings: Conservation, Restoration, and Management of Tortoises and Turtles—An International Conference. New York Turtle and Tortoise Society, State University of New York, Purchase, NY.

Woodland, R.J., C.L. Rowe and P.F.P. Henry. 2017. Changes in habitat availability for multiple life stages of diamondback terrapins (*Malaclemys terrapin*) in Chesapeake Bay in response to sea level rise. Estuaries and Coasts 40:1502-1515. DOI 10.1007/s12237-017-0209-2.

Zimmerman, T.D. 1992. Latitudinal reproductive variation in the salt marsh turtle, the diamondback terrapin (*Malaclemys terrapin*). MS thesis, University of Charleston, Charleston, SC. 54 pages.

7

Reproductive Behavior and Ecology

JOSEPH A. BUTLER
RUSSELL L. BURKE
WILLEM M. ROOSENBURG

Most aquatic turtles leave their aquatic habitats only to nest. Although many turtle species may perch on debris or river banks to bask, the gravid females of nearly all species must deposit their eggs in terrestrial habitats free of prolonged inundation. For the persistent researcher, nesting females are the most evident and most easily studied portion of the turtle life cycle. Further, egg shells left by nest predators facilitate the discovery of nest areas and can reveal details of nesting activity. Consequently, reproductive behavior and nesting ecology together are the best known stage of the life cycle in most turtle species. Diamond-backed terrapins are no exception. In the spring they form conspicuous courtship and mating aggregations in discrete areas for several days to weeks. Beginning in late spring and continuing into the summer, female terrapins enter critical terrestrial habitats in search of suitable oviposition sites above the mean high tide. Finally, in late summer and early fall and the following spring, hatchlings emerge from their nests.

In the early twentieth century, captive breeding populations in North Carolina (Coker 1906; Hay 1917; Hildebrand 1932) provided the earliest information on terrapin reproductive biology. Studies of natural terrapin nesting were pioneered in New Jersey (Burger and Montevecchi 1975; Montevecchi and Burger 1975; Burger 1976a, b, 1977) and central Florida (Seigel 1980a, b, c, 1984). Since then, numerous investigations of nesting behavior have been reported. Due to the terrapin's extensive geographic range, considerable variation exists in both the timing of nesting and the microhabitats where terrapins nest. In this chapter we review the literature on terrapin courtship through hatchling emergence and add our own observations collected over a cumulative 70 years of studying terrapin nesting biology.

Breeding Aggregations

Numerous researchers have observed large aggregations of terrapins in the spring, before the nesting season. Seigel (1980a) reported terrapin aggregations of 75 to 250 individuals in a 200 m² area in central Florida in late March to early April and described courtship and mating events. David Owens (University of Charleston, pers. comm.) noted mating congregations distant from hibernacula in South Carolina in late April. In Maryland and New York, similar aggregations have been observed in May that could be mating aggregations, but could also be large numbers of terrapins emerging from communal hibernacula (Haramis et al. 2011; Roosenburg, pers. obs.). Many of these turtles were basking and thus not actively mating. Similar large aggregations have been observed in May and June in Cape Cod, Massachusetts (Brennessel 2006), and Jamaica Bay, New York (D. Riepe, American Littoral Society, pers. comm.). Mating activity is shifted to later in the spring in colder northern locations. In Florida, the aggregations and mating were found to occur when water temperatures were between 24.8°C and 27.0°C and air temperatures between 22.8°C and 27.0°C (Seigel 1980a). The Maryland aggregations occurred in cooler water than those reported in Florida, but they typically happen in sheltered locations where water temperatures are warmer than in nearby open areas (Roosenburg, pers. obs.). It remains unclear whether these aggregations are purely for mating or whether terrapins are seeking warmer waters for physiological reasons and mating is simply more easily observed because of the aggregation. In Maryland, paired heads (females accompanied by males) can be seen throughout the entire active period between April and October, frequently as isolated pairs.

Little is known about the mating behavior of terrapins because the estuarine waters where they live are frequently turbid, such that the terrapins are difficult to observe in their aquatic environment. Frequently, the presence of a smaller head immediately behind the larger head of a female is interpreted as mating, although it might be mate guarding (preventing access to a female by other males), potentially important because terrapins can store sperm (Barney 1922). Seigel (1980a) described the following mating behavior in Florida: "while the female floated at the surface, the male approached from the rear and nudged her cloacal region. If she remained in place, the male mounted and copulation was completed at the water surface within 2 min." The basis for mate choice is unclear, but Dominy (2015) measured the visual acuity of terrapins in shallow water and showed that females seem to prefer males with higher color contrast between the shell and the skin, as well as greater hue saturation of the shell. Head bobbing, touching of heads, and stroking of the female's head with the male's claws also have been noted in captivity, but with much less vigor (Sachesse 1984) than in other emydid turtles. If the female swam away before copulation, the male pursued her for some time. Male terrapins lack the elongate foreclaws characteristic of most turtle species that exhibit the stroking behavior. Terrapins in central Florida were observed to mate only during daylight hours (10:40–16:10; Seigel 1980a), but Hay (1904) reported nocturnal copulation in pen-reared terrapins in North Carolina. It is unclear whether the intraspecific behavioral differences are real or are noted because comparisons of captive versus wild populations are limited by a lack of observed mating behavior in most wild populations.

Nesting Behavior

Oviposition begins in the spring and continues into August throughout the terrapin's range. Like mating, nesting commences earlier, in late April to early May, in the south (Seigel 1980b; Zimmerman 1992; Mann 1995; Butler et al. 2004); in the north it does not begin until June (Burger 1977; Feinberg and Burke 2003; Brennessel 2006), with June 10 being the latest first date reported (Lazell and Auger 1981; Goodwin 1994). By late July, oviposition is mostly finished throughout the range, but August nests have been reported in New York (Feinberg and Burke 2003), Maryland (Roosenburg, pers. obs.), and Florida (Butler et al. 2004). Burns and Williams (1972) documented captive individuals nesting into September in Louisiana, but these data may not reflect nesting times in natural populations. Onset of nesting also can vary locally; in the Patuxent River, Maryland, the first days of nesting ranged from May 18 to June 3 between 1987 and 1994, and on Poplar Island, Maryland, from May 23 to June 7 between 2004 and 2015. Although their effects are untested, temperatures during late April and early May may be important predictors of the start of the nesting season (Roosenburg, pers. obs.).

Female terrapins frequently aggregate offshore in large groups of 50 or more before nesting (Roosenburg 1992; Goodwin 1994; Feinberg 2004; Burke, unpub. data). Terrapins are wary and, singly or in groups, assess beaches thoroughly, extending their heads well out of the water, intently focused on the area where they may come ashore. This behavior frequently occurs in the shallow surf, the estuary bottom providing a platform to stand on to obtain better visibility of the nesting area. We suggest that terrapins are looking for predators such as raccoons, which kill adult females (Feinberg and Burke 2003; Roosenburg, pers. obs.), or are searching for favored nesting spots. The decision to come ashore may take several hours and frequently involves several false attempts, investigating the shoreline and the beach before committing to nesting. Throughout the terrapin's range, time spent ashore, searching and traveling (<10–1,000 m) to a nest site and completing nesting, ranges from 12 min to 48 h (Burger 1977; Auger and Giovannone 1979; Roosenburg 1992; Goodwin 1994; Feinberg and Burke 2003). During the search, terrapins sometimes touch the sand with their snouts, "ground-nuzzling" (Lazell and Auger 1981) and "throwing sand" with their forelimbs. beginning as soon as they emerge from the water. In Jamaica Bay, New York, they may travel for more than an hour before nesting (Scholz 2006). In Maryland, terrapins are easily disturbed while nesting and, unlike sea turtles, which continue with nesting once egg laying commences, terrapins will abandon nesting if disturbed while laying (Roosenburg, pers. obs.); in Rhode Island, searching females are far more tolerant of disturbance (P. Allman, Florida Gulf Coast University, pers. comm.).

One factor potentially contributing to the vigilance and observation of nesting areas before nesting is nest site philopatry. In Maryland, a detailed mark-recapture study of two nesting beaches had by the third year marked 95% of females nesting at these sites. Of the more than 200 females caught on the two beaches, which were less than 1 km apart, only four used both beaches as nesting sites over 5 years (Roosenburg, unpub. data). Similarly, in Rhode Island, all females used a known nesting area, and no evidence of these females nesting elsewhere has been reported (Goodwin 1994). Despite the apparently high level of philopatry, it remains unknown whether the individuals that constitute a nesting population originated from that beach or immigrated from another site. For example, Sheridan et al. (2010) documented that females move considerable distances, possibly migrating between nesting and foraging areas.

Diurnal nesting is common for diamond-backed terrapins (Burger and Montevecchi 1975; Feinberg and Burke 2003; Butler et al. 2004), but nocturnal nesting has been observed (Roosenburg 1992; Auger and Giovannone 1979; Bauer 2004; Burke, unpub. data). Roosenburg (1992) reported peak nesting times between 10:00 and 14:00, with minor peaks coinciding with dusk and dawn; nocturnal nesting occurred later in the summer when air and water temperatures were warmer. Where tidal amplitude significantly affects travel distances to nest sites, nesting occurs during and around high tide (Butler 2002; Feinberg and Burke 2003; Bauer 2004), especially during high spring tides (Bauer 2004). This may reduce the female's exposure to predators and shorten the long trek from the water's edge to the nesting site (Burger and Montevecchi 1975; Roosenburg 1992; Goodwin 1994; Butler 2002). However, where tidal amplitude is less than 1 m, nesting is not synchronized with high tide. Most nesting in these areas occurs within 10 m of the mean high water mark, and thus the distance traveled and additional exposure to predators is minimal. Synchronizing nesting with high tides also reduces the probability that females choose nest sites below the high tide line. Nesting too close to the water's edge can result in embryonic mortality due to drowning or wash-out of nests by spring high tides or storms (Butler et al. 2004; Roosenburg et al. 2014). Terrapins prefer to nest on hot, sunny to partly cloudy days, and nesting may be even more intense on sunny days after rain (Burger and Montevecchi 1975; Goodwin 1994). The only report of terrapins nesting in the rain is by Feinberg and Burke (2003).

Terrapins typically nest in loose, sandy soil with grain size large enough to preclude the clogging of egg pores, thus ensuring adequate gas exchange between the embryo and the environment (Roosenburg and Place 1995). Many Gulf Coast populations nest in shell hash, soils with unusually large particle sizes made up of broken mollusk shells and small amounts of sand (T. Wibbels, University of Alabama, Birmingham, pers. comm.). Terrapins also nest in agricultural fields, lawns, mulched beds, and gravel roads and walkways (Feinberg and Burke 2003; Roosenburg, pers. obs.).

Vegetation is another component of nesting beaches that affects nest site choice differently throughout the range. Burger and Montevecchi (1975) found that shrub cover of 25% was ideal for nesting in New Jersey, but more wooded areas harbored additional predators leading to greater nest predation. Scholz (2006) suggested that females in New York chose nest sites with more bare ground, less overhead cover, and higher solar exposure than randomly chosen sites. In a two-year study, percentage grass cover and nest depth were the most significant indicators of emergence success (percentage of eggs that hatched and emerged), but the role of nest depth switched: in 2004, warmer, shallower nests were more successful, whereas in 2005, deeper, cooler nests were more successful. Further, terrapins favored nesting in 5% to 25% grass cover, and areas more open than this led to higher avian nest predation (Scholz 2006). Roosenburg (1996) demonstrated that Maryland terrapins preferred to nest in open areas with stem densities of less than 50/0.25 m². Additionally, an experiment that created open areas on a nesting beach that had become heavily vegetated increased nesting in open plots relative to plots that remained heavily vegetated (Clowes 2013). In Charleston, South Carolina, terrapins frequently nest in dense vegetation, possibly as a mechanism to shade nests and reduce incubation temperatures (Roosenburg, pers. obs.). Because increasing vegetation decreases nest tem-

peratures (Jeyasuria et al. 1994), placing nests in different vegetation densities may be a behavioral option for females to manipulate temperature (Roosenburg 1996) but also to respond to latitudinal variation in temperature throughout the nesting season.

Shrubs are often common on nesting areas and may serve as cues to females to identify habitat above mean high tide. Marsh elder (*Iva frutescens*; Burger and Montevecchi 1975; Roosenburg 1992) and Christmas berry (*Lycium carolinianum*; Butler and Heinrich 2013) grow above mean high tide and often are indicative of terrapin nesting areas. Another important component for a terrapin nesting site in most locales is the presence of adjacent salt marsh that will serve as habitat for hatchlings and juveniles (Mann 1995; Feinberg and Burke 2003).

Once the nest site is selected, terrapins sometimes begin nesting by clearing surface material by "throwing sand" with their forelimbs. Later they switch to alternating their hindlimbs and using the webbing of their feet to scoop soil (Burger 1977), efficiently creating a flask-shaped hole into which eggs are deposited (Montevecchi and Burger 1975; Roosenburg 1992; Butler 2002). The female covers the nest by using her hindlimbs alternately to scoop sand into the flask and then pats down the area in a ritualistic dance. Terrapins spend nearly the same amount of time covering as they do digging the nest (Burger 1977). Once complete, the depth to the top of the uppermost egg averages about 10 cm (Montevecchi and Burger 1975; Butler 2000); the mean depth to the bottom of the nest varies from 15.5 to 16.5 cm from year to year (Scholz 2006; Roosenburg, unpub. data).

Turtle incubation periods span from oviposition until eggs pip; then it may take 1 to 4 days for hatchlings to completely free themselves from the shell (Ewert 1979). Terrapin hatchlings remain in the nest for several days or may overwinter in the nest (Chapter 8) before emerging. The incubation periods of natural nests are challenging to record, requiring the researcher to dig into the nest each day as hatching becomes more imminent. Such treatment can alert predators or alter incubation and emergence times (Burger 1977). Natural nests can be covered with nest protectors/boxes to trap emerging hatchlings (Auger 1989; Chapter 2), potentially providing data on emergence period (time from oviposition to appearance of hatchlings at the surface). Records of incubation and emergence periods can differ by up to 12 days or more. In New Jersey, the mean incubation period for in situ nests was found to be 76.1 days, and the emergence period was 78.6 days (Burger 1976a, 1977); this is similar to the emergence period in Rhode Island of 73.7 days (Goodwin 1994). In Florida, hatchlings emerged from natural nests in 68.9 days, and incubation in laboratory nests took 65.6 days (Seigel 1980b; Butler et al. 2004).

As is typical for turtles, warmer incubation temperatures lead to shorter incubation periods (Ewert 1979). Burger (1976b) established that northern diamond-backed terrapin nests in New Jersey had shorter mean incubation periods when the eggs were deposited on warmer, south-facing slopes (71 days) than on cooler, north-facing slopes (79 days). Roosenburg and Kelley (1996) demonstrated that terrapin eggs artificially incubated at 32°C hatched much earlier (47 days) than those incubated at 26°C (81 days) and that the warmer incubation temperature positively affected future hatchling growth rate. Giambanco (2002) similarly found that on average, each additional 1.0°C in incubation temperature reduced incubation duration by 1.92 days.

Burger (1976a) reported that most hatchlings emerged during daylight hours, and all sought cover in vegetation. Hatchlings display both synchronous and asynchronous nest emergence, which can take up to 10 days, or even longer when individuals from the same nest emerge in both fall and spring (Wojakowski et al. 2005; Roosenburg et al. 2014).

There is both anecdotal and experimental evidence that, unlike nearly all other aquatic turtles, many fall-emerging terrapin hatchlings do not move to water. Burger (1976a) challenged hatchlings on an artificial incline and found that they would seek vegetative cover even if it meant traveling uphill; however, if no vegetation was available they moved downhill. When hatchlings in South Carolina were released offshore near their nesting beach, they returned to shore and found shelter under tidal wrack (Lovich et al. 1991). Pitler (1985) located 12 hatchlings or juveniles over a 3-year period under tidal wrack, rocks, or boards. In Maryland, Roosenburg (1991) described hatchlings seeking cover in adjacent salt marshes instead of entering the water, and in Florida, 93% of hatchling crawls headed up the beach toward the salt marsh and vegetation instead of to the water (Butler et al. 2004). Burger (1976a) proposed that this hatchling cover-seeking behavior may be a response to potential diurnal predation by birds.

Draud et al. (2004) reported that some hatchlings overwintered in individual hibernacula near the shoreline. Muldoon and Burke (2012) showed that, in New York, the majority of fall-emerging hatchlings moved away from water, and they suggested that terrapin hatchlings commonly overwinter on land. They found that some hatchlings remained on land for as long as 9 months. Duncan and Burke (2016) tracked hatchlings upland as far as 50 m to their overwintering refugia, where the hatchlings buried themselves 5 to 10 cm deep until late the following spring. Little is known about hatchling overwintering in temperate aquatic turtles, but terrapins may be unique in using terrestrial environments during their first year.

Nests may be considered successful if one or more eggs hatch and hatchlings emerge, but the rate at which this occurs varies among sites and from season to season. In New Jersey, Burger (1977) observed that in one year, 84% of nests hatched, but the next year only 25% hatched. In Maryland, 3.3% of nests survived over a five-year period, but survival varied among nesting areas and across years (Roosenburg 1992). Goodwin (1994) reported that 9.7% of terrapin nests in Rhode Island were successful. At a site in Maryland where nest predators are absent, nest survival averaged 73% (range 59%–85% across years; Roosenburg et al. 2014). In New York, Feinberg (2004) found that in one year, all nests were depredated, and only 5.2% of nests hatched in the following year. In northeastern Florida, in one year, 22.8% of nests that were discovered intact hatched; in another year, 33.8% were successful (Butler et al. 2004). Finally, nests deposited earlier in the season were more likely to survive, probably because predators are not keyed into nesting activities at that time (Burger 1977; Roosenburg 1992; Goodwin 1994).

Another measure of nest success is the number of eggs that hatch within an emerging nest. For terrapins, this was found to range from 18% to 39% (successive years; Burger 1977) in New Jersey and was 54% in New York (Feinberg 2004) and 85.5% in Rhode Island (Goodwin 1994). In Maryland, on a site where no nest predators were present, within-nest survival rates averaged 63% (range 0%–100%) among nests; nests with 0% survival contained intact eggs and no sign of predation (Roosenburg et al. 2014). Of 721 eggs in 63 New York nests incubated in situ, 639 (88.6%) commenced embryo development and were therefore fertile, 635 (88.1%) survived to hatch, and 624 (86.5%) successfully emerged from the nest (Giambanco 2002). In a head-starting project in New Jersey, the successful hatching percentage of artificially incubated eggs ranged from 32% to 58% over a nine-year period, but many of these eggs were harvested from road-kill terrapins and thus may have been exposed to lethal temperatures (Herlands et al. 2004). Other studies in which eggs from natural nests were artificially incubated had nearly 100% survival rates (Roosenburg and Kelley 1996; Giambanco 2002). These cumulative findings suggest that most terrapin eggs are viable and that within-nest mortality most likely occurs as a function of the incubation environment, except during unusually dry or hot summers (Scholz 2006).

The major deterrent to the success of terrapin nests is predation (Roosenburg 1991). Nest predation of northern diamond-backed terrapins ranges from 24% to 100% (Burger 1977; Auger and Giovannone 1979; Roosenburg 1991; Feinberg and Burke 2003), and for Carolina diamond-backed terrapins in Florida it is 82% to 87% (Butler et al. 2004). Most nest predation occurs within 24 to 48 h after oviposition, when nests are most easily detected by predators (Roosenburg 1992; Goodwin 1994; Feinberg and Burke 2003; Butler et al. 2004). Nest detection by raccoons is primarily scent based, probably associated with odors emanating from soils excavated during the nesting process (Burke et al. 2005; Buzuleciu et al. 2016). Throughout incubation, nest predation is less likely, but a second peak occurs at hatching/emergence (Burger 1977; Auger and Giovannone 1979; Roosenburg 1992; Butler et al. 2004). Once a hatchling leaves the nest, the exit hole allows scents from hatched eggs and remaining hatchlings to attract predators (Burke, pers. obs.). Goodwin (1994) did not find a second peak, however, and attributed this to hatchlings emerging within 48 h, before predators became aware of nests.

Nest, Hatchling, and Adult Female Mortality

Numerous predators destroy terrapin nests and devour hatchlings, but raccoons (*Procyon lotor*) are generally the most efficient and will also depredate nesting females. Munscher et al. (2012) found that by removing 29 raccoons from a terrapin nesting beach in Florida, overall nest predation dropped from 82% and 87% in two previous years to 17.2%. Raccoons also eat hatchlings (Muldoon and Burke 2012; Rulison et al. 2012). Other midsized mammalian predators include foxes (*Vulpes fulva* and *Urocyon cinereoargenteus*), otters (*Lutra canadensis*), striped skunks (*Mephitis mephitis*—most important terrapin nest predator in Cape Cod), and armadillos (*Dasypus novemcinctus*) (Burger 1977; Auger and Giovannone 1979; Roosenburg 1992; Goodwin 1994; Butler et al. 2004; Feinberg 2004). In Florida Bay, black rats (*Rattus rattus*) are suspected nest predators (B. Mealey, Institute of Wildlife Sciences, unpub. data). In New York, Norway rats (*R. norvegicus*) depredate hatchlings at emergence; however, they do not take eggs (Draud et al. 2004; Muldoon and Burke 2012). In Maryland, white-footed deer mice (*Peromyscus leucopus*) will partially depredate nests throughout incubation and eat newly emerged hatchlings (Roosenburg et al. 2014)

Burger (1977) reported that American crows (*Corvus brachyrhynchos*) and laughing gulls (*Leucophaeus atricilla*) are diurnal nest predators that watched terrapins nesting from afar, then approached, causing the terrapins to abandon their nest before it was covered. The birds dug into the nest with their bills and flew off with one or more eggs from the open nests, sometimes feeding them to their own young. Similar observations exist for American and fish crows (*C. ossifragus*) in Florida and Maryland (Butler et al. 2004; Roosenburg et al. 2014), Boat-tailed grackles (*Quiscalus major*) in northern Florida (Butler et al. 2004), and ruddy turnstones (*Arenaria interpres*) on the Florida Gulf Coast (Butler, unpub. data). Willets (*Tringa semipalmata*) depredate terrapin nests by probing the nest cavity and puncturing the eggs (Roosenburg et al. 2014). Because birds use visual cues to locate turtles and nests, they are effective predators only during the laying period. Later in the season, both laughing gulls and black-crowned night herons (*Nycticorax nycticorax*) depredate emergent hatchlings if they are found on the beach (Burger 1976a). Yellow-crowned night herons (*N. violacia*) are likely predators of juveniles as well (Draud et al. 2004). Finally, in Maryland, eastern king snakes (*Lampropeltis getula*) and eastern rat snakes (*Pantherophis alleghaniensis*) depredate terrapin nests (Roosenburg et al. 2014).

A number of invertebrates destroy eggs and hatchlings. Ghost crabs (*Ocypode quadrata*) depredated emerged terrapin hatchlings in New Jersey (Arndt 1991, 1994). In northeast Florida, Munscher et al. (2012) found four hatched nests attacked, presumably by ghost crabs, and one or more dead hatchlings near the mouths of nearby crab holes. In addition to the late-season hatchling deaths, in May and June they found three nests, close to ghost crab holes, where one or all eggs had been dragged or pushed to the surface. The egg shells had not been penetrated, but surface heat had killed them. Praying mantids (species unknown) are also known to eat terrapin hatchlings (D. Riepe, American Littoral Society, pers. comm.). Roosenburg (1992) reported an unidentified species of ant mining calcium from eggs and causing nest failure, and ants accounted for some egg loss in New Jersey (Burger 1977). Fire ants (*Solenopsis invicta* and *Conomyrma* sp.) were discovered within nests and on hatchling terrapin carcasses in Florida, but these probably entered after other predators had exposed the nest (Butler et al. 2004; Munscher et al. 2012). Roosenburg (1992) found fly maggots living in live hatchlings and suggested that they entered through the yolk sac. Auger and Giovannone (1979) reported that 12 of 17 eggs in one clutch

were destroyed by maggots of the flesh fly family, Sarcophagidae.

Nests deposited near vegetation may be subject to mortality caused by plant roots. Rhizomes of beach grass (*Ammophila breviligulata*) can destroy some terrapin eggs by penetrating them; in other cases, root balls prevented hatching/emergence as clutches became enmeshed (Auger and Giovannone 1979; Lazell and Auger 1981). Eggs affected in this manner were shriveled and desiccated. Stegmann et al. (1988) demonstrated that even when *A. breviligulata* roots do not penetrate eggs, they are capable of absorbing nutrients that are incorporated into their roots and stems. As this occurs during the growing season, the roots become more prolific and surround the nest. Stegmann et al. (1988) cautioned against stabilizing beach erosion by planting such grasses in terrapin habitat. Other plants reported to cause egg/hatchling mortality in this manner are cordgrass (*Spartina* sp.), saltwort (*Batis maratima*), seashore saltgrass (*Distichlis spicata*), tall wormwood (*Artemisia campestris caudata*), and bayberry (*Myrica pensylvanica*) (Roosenburg 1992; Feinberg and Burke 2003; Butler et al. 2004).

Nests deposited too close to the tide line are subject to inundation or wash-out by storm surges or spring tides (Roosenburg 1991, 1992), which makes the value of nesting at high tide and using water height as an indicator of safe nesting habitat (see above) particularly relevant. The effects of such events vary annually; in Florida, Butler et al. (2004) recorded that 21.9% of nests were washed out in one year and only 9% in another. In a subsequent study at the same Florida site, Munscher et al. (2012) found a 23.7% loss in one year and only a 2.2% loss in another. In New York, 2.6% of nests were lost to flooding (Feinberg 2004). Auger and Giovannone (1979) found that 12% of nests in Cape Cod suffered wind erosion that caused desiccation and mortality in both eggs and hatchlings. Similarly, on Poplar Island in Maryland, a small fraction of nests were washed out and lost during spring high tides (Roosenburg et al. 2014). Muldoon and Burke (2012) suggested that the risk of nests being destroyed during the fall hurricane season might be the reason that hatchlings left nests and moved to upland refugia (Duncan and Burke 2016) rather than over-wintering in their nests.

Nesting is also the period when females are the most vulnerable to terrestrial predators and anthropogenic sources of mortality. Females are readily killed and eviscerated by raccoons for the eggs (Seigel 1980; Feinberg and Burke 2003; Roosenburg, pers. obs.). This appears to be a learned behavior and can significantly reduce the female population on a nesting area; after 7 years of no such predation of females on a nesting area, >10% were killed in the following year (Roosenburg, pers. obs.). Similar episodic predation by raccoons on nesting terrapins has been observed in New York (Burke, pers. obs.). Another peril faced by nesting females is automobiles (Chapter 14). As encroaching development limits and bisects native nesting habitat, females are more attracted to elevated roads and causeways that traverse salt marshes. Road mortality occurs on these elevated sites in New Jersey (Wood and Herlands 1997; Szerlag and McRobert 2006), Georgia (Crawford et al. 2014a, b), and Delaware (H. Niederriter, Delaware Department of Natural Resources and Environmental Control, pers. comm.). Simple conservation measures such as barrier fences, vegetation control, and supplemented nesting sites can collectively reduce this mortality (Crawford et al. 2014a, b, 2015; Chapter 14). Similar problems confront terrapins attempting to nest between the runways at John F. Kennedy Airport (Burke and Francoeur 2014). These extrinsic sources of mortality during the nesting season pose one of the largest threats to terrapin populations.

A detailed discussion of the life history traits of egg and clutch size associated with terrapin reproduction is provided in Chapter 6.

Future Studies

Terrapins' nesting ecology is unique because of their nearshore tidal habitat and broad geographic, especially latitudinal, range. In combination with the nesting area philopatry of many females, these characteristics suggest that nesting habitats are critical to the preservation of populations. Because the behavior and ecology of nesting vary throughout the terrapin's range, precise knowledge of nesting behavior specific to locations may prove essential to developing successful management techniques. Although fairly well described for the eastern seaboard, our understanding of nesting remains limited for Gulf Coast populations, and detailed studies of both behavior and life history traits associated with terrapin nesting are sorely needed for this region. Far too little is known about movements and survivorship of hatchlings, which appear to be different from those in other turtles. Detailed behavioral studies integrated with life history trait assessment are still absent for many populations. Lastly, genetic studies could further evaluate philopatry and annual clutch frequency through comparisons of samples taken solely from nests.

Terrapins are highly affected by anthropogenic activity during the nesting portion of their life cycle. Nesting areas are increasingly popular recreational sites, and vehicular traffic in these areas increases adult and hatchling mortality. Additionally, development and shoreline hardening (Chapters 14 and 18) frequently destroy or eliminate access to nesting areas (Roosenburg et al. 2014; Winters et al. 2015). The loss of nesting sites and increased vulnerability of terrapins on their nesting sites demand special consideration in plans for shoreline development and terrapin conservation. While sound conservation strategies seem relatively straightforward, the political and economic nuances associated with shoreline preservation are far more challenging. However, predator management and restoration of nesting areas (Chapter 18) can dramatically alter population dynamics and result in positive population growth.

REFERENCES

Arndt, R.G. 1991. Predation on hatchling diamondback terrapin, *Malaclemys terrapin* (Schoepff), by the ghost crab, *Ocypode quadrata* (Fabricius). Florida Scientist 54:215-217.

Arndt, R.G. 1994. Predation on hatchling diamondback terrapin, *Malaclemys terrapin* (Schoepff), by the ghost crab, *Ocypode quadrata* (Fabricius). II. Florida Scientist 57:1-5.

Auger, P.J. 1989. Sex ratio and mating behavior in a population of *Malaclemys terrapin* displaying temperature-dependent sex determination. PhD dissertation, Tufts University, Boston, MA. 174 pages.

Auger, P.J. and P. Giovannone. 1979. On the fringe of existence: diamondback terrapins at Sandy Neck. Cape Naturalist 8:44-58.

Barney, R.L. 1922. Further notes on the natural history and artificial propagation of the diamond-back terrapin. Bulletin of the US Bureau of Fisheries 38:91-111.

Bauer, B.A. 2004. Nesting ecology of the northern diamondback terrapin (Order Testudines; *Malaclemys terrapin terrapin*). MS thesis, C.W. Post Long Island University, Brookville, NY. 45 pages.

Brennessel, B. 2006. Diamonds in the Marsh: A Natural History of the Diamondback Terrapin. University Press of New England, Hanover, NH, and London. 236 pages.

Burger, J. 1976a. Behavior of hatchling diamondback terrapins (*Malaclemys terrapin*) in the field. Copeia 1976:742-748.

Burger, J. 1976b. Temperature relationships in nests of the northern diamondback terrapin, *Malaclemys terrapin terrapin*. Herpetologica 32:412-418.

Burger, J. 1977. Determinants of hatching success in the diamondback terrapin, *Malaclemys terrapin*. American Midland Naturalist 97:444-464.

Burger, J. and W.A. Montevecchi. 1975. Nest site selection in the terrapin *Malaclemys terrapin*. Copeia 1975:113-119.

Burke, R.L. and L. Francoeur. 2014. Cleared for takeoff. The Tortoise 1:111-117.

Burke, R.L., C. Schneider and M.T. Dolinger. 2005. Cues used by raccoons to find turtle nests: effects of flags, human scent, and diamond-backed terrapin sign. Journal of Herpetology 39:312-315.

Burns, T.A. and K.L. Williams. 1972. Notes on the reproductive habits of *Malaclemys terrapin pileata*. Journal of Herpetology 6:237-238.

Butler, J.A. 2002. Population ecology, home range, and seasonal movements of the Carolina diamondback terrapin, *Malaclemys terrapin centrata*, in northeastern Florida. Report of the Florida Fish and Wildlife Conservation Commission. Tallahassee, FL. 65 pages.

Butler, J.A., C. Brodhurst, M. Green and Z. Mullin. 2004. Nesting, nest predation, and hatchling emergence of the Carolina diamondback terrapin, *Malaclemys terrapin centrata*, in northeastern Florida. American Midland Naturalist 152:145-155.

Butler, J.A. and G.L. Heinrich. 2013. Distribution of the ornate diamondback terrapin (*Malaclemys terrapin macrospilota*) in the Big Bend region of Florida. Southeastern Naturalist 12:552-567.

Buzuleciu, D.P., D.P. Crane and S.L. Parker. 2016. Scent of disinterred soil as an olfactory cue used by raccoons to locate nests of diamond-backed terrapins (*Malaclemys terrapin*). Herpetological Conservation and Biology 11:539-551.

Clowes, E.L. 2013. Influence of vegetation on northern diamondback terrapin (*Malaclemys terrapin terrapin*) nesting. Honors tutorial thesis, Ohio University, Athens, OH. 86 pages.

Coker, R.E. 1906. The cultivation of the diamond-back terrapin with notes on other forms of turtles. North Carolina Geological Survey Bulletin 14:1-67.

Crawford, B.A., J.C. Maerz, N.P. Nibbelink, K.A. Buhlmann and T.M. Norton. 2014a. Estimating the consequences of multiple threats and management strategies for semi-aquatic turtles. Journal of Applied Ecology 51:359-366.

Crawford, B.A., J.C. Maerz, N.P. Nibbelink, K.A. Buhlmann, T.M. Norton and S.E. Albeke. 2014b. Hot spots and hot moments of diamondback terrapin road-crossing activity. Journal of Applied Ecology 51:367-375.

Crawford, B.A., N.C. Poudyal and J.C. Maerz. 2015. When drivers and terrapins collide: assessing stakeholder attitudes toward wildlife management on the Jekyll Island Causeway. Human Dimensions of Wildlife 20:1-14.

Dominy, A.E. 2015. Modeling underwater visual ability and varied color expression in the diamondback terrapin (*Malaclemys terrapin*) in relation to potential mate preference. PhD dissertation, Drexel University, Philadelphia, PA. 104 pages.

Draud, M., M. Bossert and S. Zimnavoda. 2004. Predation on hatchling and juvenile diamondback terrapins (*Malaclemys terrapin*) by the Norway rat (*Rattus norvegicus*). Journal of Herpetology 38:467-470.

Duncan, N.P. and R.L. Burke. 2016. Dispersal of newly-emerged diamond-backed terrapin (*Malaclemys terrapin*) hatchlings at Jamaica Bay, NY. Chelonian Conservation and Biology 15:249-256.

Ewert, M.A. 1979. The embryo and its egg: development and natural history. Pages 333-413 in M. Harless and H. Morlock (eds.), Turtles: Perspectives and Research. John Wiley and Sons, New York, NY.

Feinberg, J.A. 2004. Nest predation and ecology of terrapins, *Malaclemys terrapin terrapin*, at the Jamaica Bay Wildlife Refuge. Pages 5-12 in C.W. Swarth, W.M. Roosenburg and E. Kiviat (eds.), Conservation and Ecology of Turtles of the Mid-Atlantic Region: A Symposium. Bibliomania, Salt Lake City, UT.

Feinberg, J.A. and R.L. Burke. 2003. Nesting ecology and predation of diamondback terrapins, *Malaclemys terrapin*, at Gateway National Recreation Area, NY. Journal of Herpetology 37:517-526.

Giambanco, M. 2002. Comparison of viability rates, hatchling survivorship, and sex ratio of laboratory- and field-incubated nests of the estuarine, emydid turtle *Malaclemys terrapin*. MS thesis, Hofstra University, Hempstead, NY. 82 pages.

Goodwin, C.C. 1994. Aspects of nesting ecology of the diamondback terrapin (*Malaclemys terrapin*) in Rhode

Island. MS thesis, University of Rhode Island, Kingston, RI. 84 pages.

Haramis, G.M., P.F. Henry and D.D. Day. 2011. Using scrape fishing to document terrapins in hibernacula in Chesapeake Bay. Herpetological Review 42:170-177.

Hay, W.P. 1904. A revision of *Malaclemys*, a genus of turtles. Bulletin of the US Bureau of Fisheries 5:1-21.

Hay, W.P. 1917. Artificial propagation of the diamond-back terrapin. Bulletin of the US Bureau of Fisheries 24:1-20.

Herlands, R., R. Wood, J. Pritchard, H. Clapp and N. Le Furge. 2004. Diamondback terrapins (*Malaclemys terrapin*) head-starting project in southern New Jersey. Pages 13-18 in C.W. Swarth, W.M. Roosenburg and E. Kiviat (eds.), Conservation and Ecology of Turtles of the Mid-Atlantic Region: A Symposium. Bibliomania, Salt Lake City, UT.

Hildebrand, S.F. 1932. Growth of diamond-back terrapins: size attained, sex ratios and longevity. Zoologica 9:551-563.

Jeyasuria, P., W.M. Roosenburg and A.R. Place. 1994. Role of P450 aromatase in sex determination of the diamondback terrapin, *Malaclemys terrapin*. Journal of Experimental Zoology 270:95-111.

Lazell, J.D., Jr. and P.J. Auger. 1981. Predation on diamondback terrapin (*Malaclemys terrapin*) eggs by dunegrass (*Ammophila breviligulata*). Copeia 1981:723-724.

Lovich, J.E., A.D. Tucker, D.E. Kling and J.W. Gibbons. 1991. Behavior of hatchling diamondback terrapins (*Malaclemys terrapin*) released in a South Carolina salt marsh. Herpetological Review 22:81-83.

Mann, T.M. 1995. Population surveys for diamondback terrapins (*Malaclemys terrapin*) and Gulf salt marsh snakes (*Nerodia clarkii clarkii*) in Mississippi. Mississippi Museum of Natural Sciences Technical Report no. 37. Jackson, MS. 75 pages.

Montevecchi, W.A. and J. Burger. 1975. Aspects of the reproductive biology of the northern diamondback terrapin, *Malaclemys terrapin terrapin*. American Midland Naturalist 94:166-175.

Muldoon, K.A. and R.A. Burke. 2012. Movements, overwintering, and mortality of hatchling diamond-backed terrapins (*Malaclemys terrapin*) at Jamaica Bay, New York. Canadian Journal of Zoology 90:651-662.

Munscher, E.C., E.H. Kuhns, C.A. Cox and J.A. Butler. 2012. Decreased nest mortality for the Carolina diamondback terrapin (*Malaclemys terrapin centrata*) following removal of raccoons (*Procyon lotor*) from a nesting beach in northeastern Florida. Herpetological Conservation and Biology 7:167-184.

Pitler, R. 1985. *Malaclemys terrapin terrapin* (northern diamondback terrapin) behavior. Herpetological Review 16:82.

Roosenburg, W.M. 1991. The diamondback terrapin: habitat requirements, population dynamics, and opportunities for conservation. Pages 237-234 in J.A. Mihursky and A. Chaney (eds.), New Perspectives in the Chesapeake System: A Research and Management and Partnership. Proceedings of a Conference. Chesapeake Research Consortium Pub. no. 137. Chesapeake Research Consortium, Solomons, MD.

Roosenburg, W.M. 1992. Life history consequences of nest site choice by the diamondback terrapin, *Malaclemys terrapin*. PhD dissertation, University of Pennsylvania, Philadelphia, PA. 206 pages.

Roosenburg, W.M. 1996. Maternal condition and nest site choice: an alternate for the maintenance of environmental sex determination. American Zoologist 36:157-168.

Roosenburg, W.M. and K.C. Kelley. 1996. The effect of egg size and incubation temperature on growth in the turtle *Malaclemys terrapin*. Journal of Herpetology 30:198-204.

Roosenburg, W.M. and A.R. Place. 1995. Nest predation and hatchling sex ratio in the diamondback terrapin. Pages 65-70 in P. Hill and S. Nelson (eds.), Implications for Management and Conservation: Towards a Sustainable Coastal Watershed: The Chesapeake Experiment. Proceedings of a Conference. Chesapeake Research Consortium Pub. no. 149. Chesapeake Research Consortium, Solomons, MD.

Roosenburg, W.M., D.M. Spontak, S.P. Sullivan, E.L. Matthews, M.L. Heckman, R.J. Trimbath, R.P. Dunn, E.A. Dustman, L. Smith and L.J. Graham. 2014. Nesting habitat creation enhances recruitment in a predator free environment: *Malaclemys* nesting at the Paul S. Sarbanes ecosystem restoration project. Restoration Ecology 22:815-823.

Rulison, E.L., L. Luiselli and R.L. Burke. 2012. Habitat outweighs geography: the feeding strategies of a generalist carnivore, *Procyon lotor*, in salt marsh ecosystems. American Midland Naturalist 168:231-246.

Sachesse, W. 1984. Long term studies of the reproduction of *Malaclemys terrapin centrata*. Acta Zoologica et Pathologica Antverpiensia 78:297-308.

Scholz, A.L. 2006. Impacts of nest site choice and nest characteristics on hatchling success in the diamond-back terrapins of Jamaica Bay, NY. MS thesis, Hofstra University, Hempstead, NY. 97 pages.

Seigel, R.A. 1980a. Courtship and mating behavior of the diamondback terrapin, *Malaclemys terrapin tequesta*. Journal of Herpetology 14:420-421.

Seigel, R.A. 1980b. Nesting habits of diamondback terrapins (*Malaclemys terrapin*) on the Atlantic coast of Florida. Transactions of the Kansas Academy of Science 83:239-246.

Seigel, R.A. 1980c. Predation by raccoons on diamond-back terrapins, *Malaclemys terrapin tequesta*. Journal of Herpetology 14:87-89.

Seigel, R.A. 1984. Parameters of two populations of diamondback terrapins (*Malaclemys terrapin*) on the Atlantic coast of Florida. Pages 77-87 in R.A. Seigel, L. Hunt, J.L. Knight, L. Malaret and N. L. Zuschlag (eds.), Vertebrate Ecology and Systematics: A Tribute to Henry S. Fitch. Museum of Natural History, University of Kansas, Special Pub. no. 10. University of Kansas, Lawrence, KS.

Sheridan, C.M., J.R. Spotila, W.F. Bien and H.W. Avery. 2010. Sex-biased dispersal and natal philopatry in the diamondback terrapin, *Malaclemys terrapin*. Molecular Ecology 19:5497-5510.

Stegmann, E.W., R.B. Primark and G.S. Ellmore. 1988. Absorption of nutrient exudates from terrapin eggs by roots of *Ammophila breviligulata*. Canadian Journal of Botany 66:714-718.

Szerlag, S. and S.P. McRobert. 2006. Road occurrence and mortality of the northern diamondback terrapin. Applied Herpetology 3:27-37.

Winters, J.M., H.W. Avery, E.A. Standora and J.R. Spotila. 2015. Between the bay and a hard place: altered diamondback terrapin nesting movements demonstrate the effects of coastal barriers upon estuarine wildlife. Journal of Wildlife Management 79:682-688.

Wojakowski, M.M., Widrig, A.L. and R.L. Burke. 2005. Emergence of the diamondback terrapin (*Malaclemys terrapin*) in Jamaica Bay Wildlife Refuge, New York City. In Abstracts, Annual Meetings of the American Society of Ichthyologists and Herpetologists, Tampa, FL. American Society of Ichthyologists and Herpetologists, Lawrence, KS.

Wood, R.C. 1997. The impact of commercial crab traps on northern diamondback terrapins, *Malaclemys terrapin terrapin*. Pages 21-27 in J.E. Van Abbema (ed.), Proceedings: Conservation, Restoration, and Management of Tortoises and Turtles—An International Conference. New York Turtle and Tortoise Society, State University of New York, Purchase, NY.

Wood, R.C. and R. Herlands. 1997. Turtles and tires: the impact of roadkills on northern diamondback terrapin, *Malaclemys terrapin terrapin*, populations on the Cape May Peninsula, southern New Jersey, USA. Pages 46-53 in J.V. Abbema (ed.), Proceedings: Conservation, Restoration, and Management of Tortoises and Turtles—An International Conference. New York Turtle and Tortoise Society, State University of New York, Purchase, NY.

Zimmerman, T.D. 1992. Latitudinal reproductive variation of the salt marsh turtle, the diamondback terrapin (*Malaclemys terrapin*). MS thesis. University of Charleston, Charleston, SC. 54 pages.

Hatchling Behavior and Overwintering

PATRICK J. BAKER
RALPH E. J. BOERNER
ROGER C. WOOD

Few aquatic turtle species have as detailed observations across their entire life cycle as those of the diamond-backed terrapin, *Malaclemys terrapin*. Information about the ecology of its early life stages (post-hatching to sexual maturity) is sparse, however, due to the cryptic nature of juvenile terrapins. The earliest published accounts of the behavior, ecology, and growth of hatchlings and juveniles are from captive breeding facilities operated by the US Bureau of Fisheries in Beaufort, North Carolina (Coker 1906; Hay 1917; Hildebrand 1932). While it is reasonable to suspect that these findings may have been artefacts of captivity, many of the careful observations at Beaufort have since been validated by studies conducted under natural conditions.

Knowledge of the life history of the terrapin should always be put into geographic and ecological context. Terrapins have a long coastal distribution that ranges from the temperate north to the subtropical south, encompassing estuaries, tidal salt marshes, and mangrove islands. Differences in climatic conditions across the terrapin's range contribute to latitudinal variation in life history traits among populations. In the north, hatchling terrapins contend with a shortened growing season and prolonged exposure to subfreezing temperatures; in the south, hatchlings that seek food and freshwater during an extended activity season are at increased risk of predation.

We summarize here the first 3 years in the life cycle of the diamond-backed terrapin, with emphasis on emergence from the nest, post-emergence movements, and hibernation. The chapter includes available information for hatchlings (up to 1 year old; carapace length [CL] < 35 mm) and juvenile terrapins (up to 3 years old; CL < 80 mm) because both groups appear to be distinct in their behavior, ecology, and physiology from sub-adults and adults. The ontogenetic shifts in habitat use and behavior that occur after three winters are likely to be related to changes in anatomy and physiology that improve salinity tolerance.

Nest Emergence

Nesting season varies across the terrapin's range (Chapter 7) but is generally restricted to spring and early summer, with peak activity in mid-June. Terrapin eggs typically hatch in August and September (Coker 1906; Burger 1977; Feinberg and Burke 2003; Butler et al. 2004). Shallow nests are, on average, warmer than deep nests, and the eggs in shallow nests hatch earlier (Burger 1976b). At the time of hatching, a terrapin's mobility is impaired until the residual yolk sac, which can be substantial, is internalized through the umbilicus (Burger 1976a). Emergence from the nest can begin 1 to 10 days after hatching (Burger 1977).

Although direct observations are limited, hatchling terrapins appear to emerge from the nest primarily during the daytime. On Little Beach Island, New Jersey, hatchling emergence was observed from 07:00 to 19:00, with peak activity between 12:00 and 19:00 (Burger 1976a). Diurnal emergence was also observed at this location by Arndt (1994). This pattern is similar to that observed for other northerly distributed turtles (e.g., Tucker 1999; Tuttle and Carroll 2005). In northeastern Florida, daytime surveys of nesting beaches found evidence of emergence activity (i.e., hatchling tracks) but no surface-active hatchlings (Butler et al. 2004). Surface temperatures are higher in the south, and the hatchlings of some freshwater turtle species (e.g., *Apalone mutica*) emerge from the nest after sunset (Plummer 2007). In Alabama, hatchling terrapins have succumbed to heat stress on coastal beaches (Coleman et al. 2014). Sea turtles, which also nest on beaches of the Atlantic and Gulf coasts, primarily emerge at night (Drake and Spotila 2002; Glen et al. 2005). A propensity for nocturnal emergence by hatchling sea turtles results from an inhibition of emergence behavior by high temperatures at the surface (Drake and Spotila 2002). Given that the surface above terrapin nests may present similar thermal challenges,

hatchlings in southern locales may shift to nighttime emergence to avoid potentially lethal surface temperatures. The use of cameras (e.g., Plummer 2007) or time-specific sampling (e.g., dawn, midday, dusk, midnight) in studies that monitor natural emergence from the nest can improve our knowledge of emergence timing.

Synchronous emergence, when all hatchlings emerge from the nest within one day, is thought to be advantageous to turtles because cooperative digging enables the hatchlings to reach the surface more easily and reduces per capita predation risk (Carr and Hirth 1961; Burger 1976a; Tucker et al. 2008; Spencer and Janzen 2011; Peterson et al. 2013). However, hatchling terrapins do not always emerge synchronously, and emergence activity can span days, weeks, or months (Burger 1976a, 1977; Baker et al. 2006). Unlike sea turtles, which may rely on cooperative digging by many siblings to reach the surface from deep nests (Carr and Hirth 1961), hatchling terrapins are exceptional climbers (Hay 1917), and individuals can reach the surface from their relatively shallow nests without the assistance of siblings. Accordingly, Burger (1976a) found no difference in emergence success between nests with fewer than three hatchlings and those with more.

Asynchronous emergence has been observed for terrapin nests in southern New Jersey (Burger 1976a, 1977). On Little Beach Island, emergence duration for a single nest ranged from 1 to 11 days (Burger 1976a, 1977). In two years of study (2002–2003), staff and volunteers observed 60 nests on the grounds of the Wetlands Institute in Stone Harbor, New Jersey, from which at least three hatchlings emerged. The hatchlings from the majority of nests (60%) emerged synchronously, but 40% of the nests exhibited asynchronous emergence. In these latter nests, emergence duration ranged from 1 to 32 days (mean = 9 days).

Split emergence is an extreme form of asynchrony in which some hatchlings depart the nest in fall, but their siblings overwinter in the nest chamber and emerge in the following spring. Split emergence is typically rare. Baker et al. (2013a) observed split emergence from 4% of painted turtle nests observed over six years in northern Indiana. On average, ~24% of all terrapin nests observed at the Wetlands Institute over nine years exhibited split emergence (Fig. 8.1). The degree of split emergence observed in southern New Jersey is unexpected because hatchlings that exit the nest cavity may create cues (e.g., scent) that could be detected by predators, such that the siblings remaining in the nest may be at greater risk after the first has emerged (Burger 1977). Intranest asynchrony of emergence suggests a diversified bet-hedging strategy (e.g., Danforth 1999) in which variation in emergence duration and emergence period is modulated by hatchlings' condition and environmental cues.

Season of emergence varies among turtle species, and temperate species generally favor one season (fall vs. spring) strongly over the other (Gibbons and Nelson 1978; Costanzo et al. 2008; Baker et al. 2013a; Gibbons 2013). For example, in a study of seven species in northern Indiana, the hatchlings of four species emerged from the nest exclusively in the fall, and the hatchlings of three species emerged primarily (89%-100% of nests observed) in the spring (Baker et al. 2013a). However, terrapins exhibit considerable plasticity within and among populations in their pattern of emergence from the nest (Roosenburg 1994; Baker et al. 2006). In southern New Jersey, in 53% of successful nests the hatchlings emerged exclusively in the fall, and in 24% of successful nests they emerged exclusively in the spring (Fig. 8.1). The remaining nests produced hatchlings that emerged in both fall and spring. A similar proportion of fall and spring emergence has been observed on Poplar Island, Maryland (Roosenburg et al. 2010); however, spring emergence is rarely observed on Little Beach Island, New Jersey (Burger 1977) and Jamaica Bay, New York (Scholz 2007).

Emergence period, the number of days from the first hatchling to emerge from a nest until the last, varies across the terrapin's range. Roosenburg et al. (2003) collected 565 hatchlings on Poplar Island, Maryland, during a 72 day fall emergence period. In southern New Jersey, Baker et al. (2006) observed a 90 day fall emergence period (August 5–November 3) and a 33 day spring emergence period (April 18–May 21). Based on observations by Burger

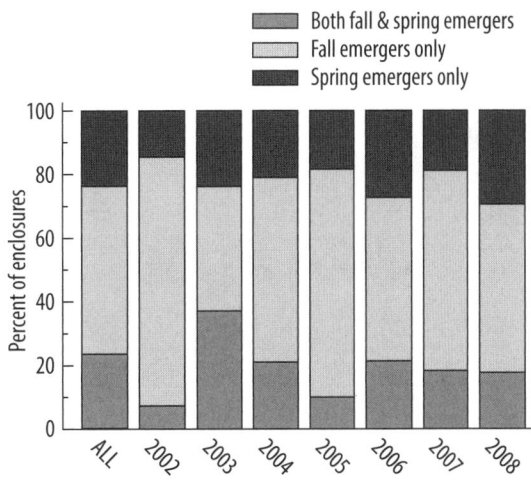

Fig. 8.1. Percentage of all terrapin nest enclosures from which hatchlings emerged in fall only, both fall and spring, and spring only, from 2001 to 2009 at the Wetlands Institute, Stone Harbor, New Jersey.

(1977), we calculated an emergence period of 53 days (August 20-October 12) for Little Beach Island. Arndt's (1994) subsequent observations of nest emergence on October 23 and hatchling activity on October 31 extend the emergence period at this site to 72 days.

Emergence by *M. t. centrata* in northeastern Florida begins in July, peaks in August and September, and ends in mid-October (Butler et al. 2004). The emergence period (July 3-October 11) is at least 100 days (Butler 2002; Butler et al. 2004). The earlier onset of emergence in Florida is a function of an earlier nesting season; however, emergence activity ends in mid-October, despite a higher ambient temperature than in the mid-Atlantic, where emergence from terrapin nests continues until the end of October or early November (Arndt 1994; Baker et al. 2006; Roosenburg et al. 2007). Butler et al. (2004) suggested that earlier nesting would lead to earlier hatching, which would in turn give hatchlings time to forage and grow in their initial season after emerging. Although the expected relationship between date of oviposition and date of hatching is supported by observations that eggs laid in June have shorter incubation times than those laid in July (Burger 1976b), it is not clear that emergence date is related to date of oviposition. For terrapin populations where spring emergence has been

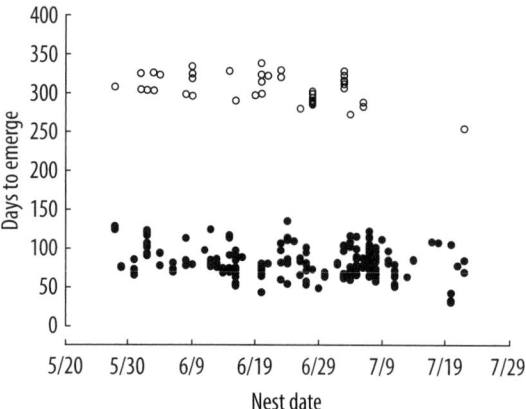

Fig. 8.2. Days to emergence as a function of nest (oviposition) date for 919 hatchling terrapins observed from 2001 to 2009 at the Wetlands Institute, Stone Harbor, New Jersey. Solid circles represent individual hatchlings that emerged in fall; open circles, those that emerged in spring.

documented, no relationship exists between the date of oviposition (nest date) and the day that resulting hatchlings emerge from the nest (Graham 2009) (Fig. 8.2).

Emergence Cues

Direct observation of emergence from the terrapin nest is rare; thus we know little about the environmental cues that trigger emergence. Rainfall does not appear to stimulate emergence (Burger 1976a), and hatchlings are more likely to be active on days without precipitation (Muldoon and Burke 2012). Burger (1976a) reported a mean air temperature of 25°C ± 1°C on days when hatchlings emerged synchronously, but lower air temperature (19°C ± 2°C) on days when hatchlings emerged asynchronously. Asynchrony of emergence increased with declining air temperature from August to September (Burger 1976a). Air temperature alone is insufficient to explain emergence activity, however, as it is significantly higher during the fall activity period than during the spring activity period (Muldoon and Burke 2012).

Hatchling turtles may respond to changes in soil temperature (Bleakney 1963; Tucker 1999; Drake and Spotila 2002; Glen et al. 2005). Like air temperature, soil temperature at nest depth is higher during the fall emergence period than during the

spring emergence period (Bleakney 1963; Tucker 1999; Baker et al. 2013a). In southern New Jersey, mean soil temperatures at nest depth are between 25°C and 30°C at the onset of the fall emergence period and between 5°C and 10°C when emergence activity resumes in the spring (Baker, Boerner, and Wood, unpub. data).

Rather than responding to an absolute temperature cue, hatchlings that overwinter terrestrially may emerge in response to a change in the soil depth–temperature gradient (Bleakney 1963; Gibbons and Nelson 1978; Tucker 1999; Baker et al. 2013a). Soil cools from the top down during the fall. For this reason, a negative temperature gradient is established in late October, with the top of the nest colder than the bottom (Bleakney 1963; Tucker 1999; Baker et al. 2013a). Hatchling terrapins may sense this gradient as a cue to delay emergence until after the winter. The temperature gradient reverses in spring as rising air temperature warms the surface. A sustained positive temperature gradient, with warmer soil at the top of the nest than at the bottom, may be a reliable cue that forecasts the onset of the growing season.

Ecology of Hatchling and Juvenile Terrapins

Post-emergence Movements

While it is often assumed that hatchling turtles will move directly to aquatic habitats after emerging from the nest (Gibbons 2013), hatchling terrapins apparently do not. Field observations and laboratory experiments demonstrate that recently hatched terrapins prefer terrestrial habitats and seek refuge in or below nearby vegetation (Hay 1917; Burger 1976a; Lovich et al. 1991; Hoden and Able 2003; Butler et al. 2004; Roosenburg et al. 2007; Coleman et al. 2014). Tracks of hatchling terrapins that emerged naturally from their nests indicate movement toward upland vegetation and away from open water (Burger 1976a; Arndt 1994; Butler et al. 2004).

In the laboratory, hatching terrapins ($N = 8$) that were given a choice between aquatic and terrestrial habitats spent 92% of the time out of the water (Hoden and Able 2003). Although initially released in water, all eight terrapins swam directly to a piece of vegetated marsh peat in the center of the arena. Similarly, laboratory-hatched terrapins avoid open water when released in their native salt marsh (Lovich et al. 1991; Coleman et al. 2014). Whether released in water or on land, hatchlings consistently move toward vegetation (Lovich et al. 1991). Post-emergent movements of hatchling terrapins are not influenced by compass direction (Burger 1976a; Coleman et al. 2014). Instead, terrapins rely on visual cues to orient themselves toward marsh vegetation (Burger 1976a; Coleman et al. 2014). Coleman et al. (2014) observed that hatchlings, post-hatchlings, and yearlings perform "orientation circles" with their heads extended before moving toward marsh habitat.

Young terrapins frequently bury themselves shortly after emergence or, in the case of laboratory-hatched terrapins, after release (Lovich et al. 1991; Coleman et al. 2014). This behavior may reduce their exposure to diurnal predators (Burger 1976a). Burying may also reduce thermal and hydric stress if hatchlings enter an environment with a lower ambient temperature and higher relative humidity (Lovich et al. 1991; Coleman et al. 2014).

The Missing Years

Little is known about terrapins after emergence from the nest until they reach a CL >80 mm or complete three growing seasons. Few studies are able to effectively sample juvenile turtles in aquatic habitats (Ream and Ream 1966; Gibbons 1970). For this reason, early age classes may be underrepresented or missing in population studies across the terrapin's range (Coker 1906; Hurd et al. 1979; Seigel 1984; Auger 1989; Morreale 1992; Wood 1997; Gibbons et al. 2001; Butler 2002; Mitro 2003; Hart and McIvor 2008). Sampling bias cannot be ruled out, especially for studies that primarily use crab pots or seines; however, Hoden and Able (2003) did not encounter any terrapins with a CL <80 mm in aquatic habitats, despite sampling a variety of habitat types and employing multiple capture methods. Of 725 individuals captured by hand (snorkel and SCUBA), seine, or trammel net in aquatic habitats at several locations on Long Island, New York, only 2 individuals (67 mm and 73 mm) were smaller

than the smallest discernible male (Morreale 1992); as estimated from growth annuli, the smallest terrapins in these populations were 3 years old. In contrast, Hoden and Able (2003) collected 395 hatchlings and 31 juveniles in upland habitats along a causeway adjacent to a salt marsh. The majority of these encounters, often the result of roadkill, took place from early March to mid-June, with peak activity in late April. Similarly, Mitchell and Denmon (2005) found a solitary hatchling terrapin along a beach access road in mid-March. Because hatchling and juvenile terrapins have little temporal overlap with nesting females, road surveys outside the nesting season may prove profitable in documenting the terrestrial activity, movements, and mortality rates of hatchlings and juveniles.

Drift fences with pitfall traps are another effective way to document terrestrial behavior in turtles (Gibbons 1970). Movements of hatchling terrapins in Jamaica Bay, New York, were estimated from capture-recapture at drift fences set above the shoreline (Muldoon and Burke 2012). In two years of trapping, 324 hatchling terrapins were captured in upland environments in both fall (August to late October) and spring (April–June). Within the fall emergence period, 111 of 185 hatchlings were recaptured while moving away from water, whereas 61 of 86 hatchlings recaptured in spring were moving toward water. The latter finding suggests that many of these hatchlings sought to remain in upland habitats through late spring. Incorporating extensive terrestrial monitoring, particularly the use of drift fences (e.g., Muldoon and Burke 2012), thread trailing (e.g., Burke and Capitano 2011), or radio-telemetry (e.g., Draud et al. 2004), into population studies across the range of terrapins would greatly improve our knowledge of hatchling and juvenile ecology.

The absence of hatchlings from open water habitats may be explained, in part, by the physiological challenges of prolonged submergence in a hypertonic environment. Hatchlings have a high surface area to volume ratio relative to adults and have not fully developed the ability to regulate sodium influxes in saline environments (Dunson 1985). Growth is optimal at salinities between 10 and 15, and hatchling terrapins do not grow without occasional access to freshwater (Dunson 1985; Holliday et al. 2009). Paradoxically, nest sites are often located in areas where the salinity is much higher. In southern New Jersey, low-salinity habitats are available where the salt marsh abuts the mainland and freshwater inlets mix with tidal waters; however, nesting terrapins prefer the barrier islands, where the salinity of adjacent waters is similar to that of the ocean. Alternatively, hatchlings may seek out microhabitats near the nest site where freshwater is periodically available. For example, terrapins inhabiting the high marsh can avoid the dehydrating effects of continuous immersion in seawater and more easily obtain freshwater after rains (Dunson 1985; Davenport and Macedo 1990). Salinity tolerance improves with increased body size, and juvenile terrapins (42–50 g) are capable of growth in 100% seawater (Dunson 1985). Other factors (e.g., aquatic predators) may be important, too, as this size range is still 50 to 100 g below that of the smallest adult males (CL > 90 mm) captured in aquatic habitats in southern New Jersey (Baker, Boerner, and Wood, unpub. data).

A population model developed by Mitro (2003) suggested that juveniles (1–3 years old) account for ~35% of all females in a stable population, yet juvenile terrapins are rare in population studies. Cryptic behavior of juvenile *M. t. terrapin* was documented by Pitler (1985) over three years of observation in Barnegat Bay, New Jersey. In this study, terrapins (*N* = 12) were found at low tide sheltered beneath surface debris, rocks, and matted marsh grass within 100 m of water. In coastal Mississippi, Mann (1995) encountered eight *M. t. pileata* juveniles (plastron length [PL] = 42–52 mm) actively foraging in high marsh habitat several hundred meters from open water. Unlike the young terrapins in Barnegat Bay, the juvenile terrapins in Mississippi were not associated with cover and were presumed to be basking or foraging (Mann 1995). Juveniles of *M. t. rhizophorarum* are occasionally found in upland habitats in the lower Florida Keys (Wood, unpub. data). Of 699 original captures in Key West National Wildlife Refuge, only 10 (1.4% of all captures) had a CL < 90 mm (Wood, unpub. data). These small terrapins were found exclusively on the interior of small islands, in areas dominated

by black mangrove (*Avicennia germinans*). However, in these populations, adult terrapins also exhibit a high degree of terrestrial behavior (Wood 1981). The only source of freshwater on these islands is rain, which falls primarily during the rainy season (May–October).

Predators

Predation risk may drive hatchling behavior (Burger 1976a). Predation rates on terrapin nests are highest within 24 h of laying and during emergence periods (Burger 1977; Butler et al. 2004). The predators of hatchling turtles are generalists that switch prey seasonally based on abundance (Janzen et al. 2000; Tucker et al. 2008; Spencer and Janzen 2011). Hatchling turtles that emerge early may experience lower predation pressure due to the lag time between the onset of emergence activity and the switch by predators to a new prey item (Tucker et al. 2008). Hatchlings that delay emergence after the first hatchling has left are presumed to be at increased risk from predators (e.g., red fox [*Vulpes vulpes*], raccoon [*Procyon lotor*]) that can detect the scent of an open nest chamber. White-footed deer mice (*Peromyscus leucopus*) are nest predators that opportunistically feed on hatchling terrapins (Roosenburg et al. 2014). Burger (1976a, 1977) observed that predators often incompletely depredate a nest, with surviving hatchlings emerging a few days later.

Emergence from the nest does not eliminate predation risk and may increase exposure to predators that would otherwise not have access to hatchlings. Raccoons, red fox, Norway rats (*Rattus norvegicus*), and ghost crabs (*Ocypode quadrata*) kill surface-active hatchling terrapins (Burger 1977; Arndt 1994; Butler 2002; Butler et al. 2004; Draud et al. 2004; Rulison et al. 2012). Birds eat terrapin eggs at the time of laying (Burger 1977; Butler et al. 2004), but they do not seem to prey on hatchling terrapins in the nest (Burger 1977). Birds are known to be important predators on surface-active hatchling turtles elsewhere (Janzen et al. 2000), and several bird species that inhabit coastal areas are likely predators of hatchling terrapins (Arndt 1994; see also Muldoon and Burke 2012). However, only laughing gulls (*Leucophaeus atricilla*) and black-crowned night herons (*Nycticorax nycticorax*) have been reported to prey on hatchling terrapins while moving on land (Burger 1976a). Draud et al. (2004) cited unpublished evidence that fish crows (*Corvus ossifragus*) and yellow-crowned night herons (*Nycticorax violacea*) also kill hatchlings. The remains of juvenile terrapins have been found beneath *N. violacea* nests (Riegner 1982). Although direct observations of the predators were lacking, 60% of the 43 carcasses recovered by Muldoon and Burke (2012) were missing limbs or heads or were punctured through the shell, suggesting that terrestrially active hatchlings in this population were killed by both mammalian and avian predators.

Predation risk is higher at night, when nocturnal predators (e.g., raccoons, rats, foxes, ghost crabs) are active. Sixteen of the 24 hatchlings radio-tracked by Draud et al. (2004) were killed, presumably by Norway rats, between 21:00 and 06:00. Predation on hatchlings is density dependent, with most mortality occurring during periods of peak hatchling activity (Burger 1977; Draud et al. 2004; Muldoon and Burke 2012). Experimental releases of hatchling red-eared sliders on Onslow Beach, North Carolina, demonstrated that group releases decreased per capita risk of predation by ghost crabs relative to hatchlings released individually (Peterson et al. 2013). While this result suggests a benefit for synchronous emergence from the nest, the group sizes (8, 20, 40) used in the study were intentionally large as they were intended to simulate emergence groups of hatchling sea turtles (Peterson et al. 2013). A similar study design with smaller group sizes (e.g., 5, 10, 15) would provide insight into the effect of synchronous emergence on predation risk in terrapins.

There is a long and diverse list of potential predators in aquatic habitats (e.g., Stancyk 1995), but it remains unknown whether any present a significant threat to hatchling and juvenile terrapins. An alternative hypothesis is that hatchling and juvenile terrapins remain in terrestrial habitats where predation risk is lower (Iverson 1991) until they attain a body size greater than the gape of some, but not all, aquatic predators. Bald eagles (*Haliaeetus leucocephalus*), presumably fishing in estuarine

habitats, exhibit a preference for young terrapins (Baldwin et al. 2005; Wood and L. Smith, unpub. data). However, the minimum size (CL = 92 mm) of terrapins found in and adjacent to eagles' nests in New Jersey (Wood and L. Smith, unpub. data) is consistent with the hypothesis that juvenile terrapins (CL < 80 mm) occupy different habitats from adults.

Overwintering Inside the Nest

Hatchling terrapins can facultatively delay emergence and remain in the natal nest chamber until the following spring (Hay 1917; Auger and Giovannone 1979; Baker et al. 2006; Scholz 2007; Roosenburg et al. 2009). Of the 102 successful nests (defined as having at least one hatchling emerge) observed by Baker et al. (2006), approximately half harbored hatchlings through the winter. Roosenburg et al. (2010) reported hatchlings overwintering in 222 (34%) of all 658 terrapin nests monitored on Poplar Island over four nesting seasons (2006–2009). However, if their analysis is restricted to the 511 nests from which at least one hatchling emerged, the percentage of nests that successfully overwintered increases to 38% (range 26.6%–50.7%). While these studies show relatively high rates of overwintering inside the nest, this behavior is rare at well-studied sites in Jamaica Bay, New York (Scholz 2007), Little Beach Island, New Jersey (Arndt 1994; Burger 1976b, 1977), and northeastern Florida (Butler 2002; Butler et al. 2004). Studies that excavate the nest in the fall to assess hatching success (e.g., Burger 1976b, 1977; Feinberg and Burke 2003; Butler et al. 2004) limit the possibility of observing overwintering inside the nest. However, studies that leave some nests intact (e.g., Scholz 2007) may capture the natural variation in emergence phenology at a particular site.

Nest soil characteristics may partially explain variation in nest emergence time (Costanzo et al. 2008). The "passive response" hypothesis, in which hatchlings do not emerge in the fall because overlying soil must first be softened by spring rains, has little support for most turtles (Tucker 1999; Costanzo et al. 2008; Gibbons 2013) and no physical basis for terrapins that nest in sand (Burger 1976a). However, the bulk density of soil adjacent to overwintering nests on Poplar Island, a US Army Corp of Engineers restoration site, was higher than that for nests with early emergence (Graham 2009). It is not known whether more compacted soils prevent hatchlings from emerging or are more likely to maintain a nest chamber with greater structural integrity that promotes overwintering (Costanzo et al. 2008). Our understanding of the relationship between soil characteristics and overwintering in the nest would be greatly enhanced by additional studies across the terrapin's range.

Terrestrial Hibernation Outside the Nest

Hatchlings that vacate the nest during the fall remain active in terrestrial habitats until mid-October, at which point they bury in friable soils and overwinter above the high-tide line (Draud et al. 2004; M. Draud and Baker, unpub. data). Muldoon and Burke (2012) recaptured hatchlings ($N = 18$) in spring in the same terrestrial habitats where they were captured the previous fall. While hibernacula were not located, it was assumed that these hatchlings remained in upland habitats through the winter. In late December 2003, Draud and Baker (unpub. data) located 24 hatchling terrapins while searching upland habitats adjacent to a salt marsh in Bayville, New York. Hatchlings were found individually, buried (depth < 10 cm) in moist sandy soils (mean water content = 5.2%; range 3.0%-9.8%; $N = 24$) near vegetation (e.g., *Solidago* sp.). Given the activity patterns of hatchling terrapins and the winter climate on Long Island (Draud et al. 2004; Muldoon and Burke 2012), these terrapins would probably have remained dormant until early April.

Winter ecology of juvenile terrapins is poorly known and is limited to a single report from Virginia, where a young terrapin (CL = 54 mm) was found buried in a grassy area 8 m above the high-tide line (Lawler and Musick 1972). Initially discovered buried in moist sand to a depth of 30 cm, this terrapin made small (2-8 cm) vertical and horizontal movements during the winter. In early April, the terrapin began to move toward the surface and was 3 to 5 cm from the surface on April 23, when it most likely emerged (Lawler and Musick 1972). This finding accords well with the earliest appearance of

juvenile terrapins along Great Bay Boulevard in Tuckerton, New Jersey (Hoden and Able 2003).

Aquatic Hibernation

Adult terrapins hibernate in a variety of estuarine habitats (Yearicks et al. 1981; Haramis et al. 2011). In southern New Jersey, 311 terrapins (CL = 97-186 mm) were found in the winter resting on the bottom of tidal creeks, buried in creek banks near the high-tide limit, and in undercut banks in the intertidal zone (Yearicks et al. 1981). Haramis et al. (2011) did not report a minimum body size for terrapins ($N = 1,175$) collected from hibernacula (water depth = 1.5-3.5 m; $N = 7$) in Chesapeake Bay; however, they concluded that terrapins younger than 6 years were rarely caught in their winter surveys of aquatic hibernacula. In North Carolina, where winters are relatively mild, 10 female terrapins were visually located 58 times in aquatic hibernacula between October 2 and February 28 (Williard and Harden 2011). During these observations, 8% of the terrapins were swimming, 2% were resting on top of the mud, and 90% were buried in the mud of the intertidal zone. The hibernacula of juvenile terrapins in this population are unknown (L. A. Harden, Loyola University Chicago, pers. comm.). Hibernacula of adult male (CL = 111-133 mm) and female (CL = 186-202 mm) terrapins in South Carolina were inferred by comparing hourly temperature readings from data loggers attached to the carapace to temperatures recorded in available microhabitats (Akins et al. 2014). An apparent seasonal shift in microhabitat occurred in October, when the terrapins' body temperatures were more similar to those of shallow mud, and then reversed in April, when body temperatures were more similar to water temperature (Akins et al. 2014). A similar approach may prove profitable for discerning microhabitat use by juveniles without having to make direct observations.

Physiological Ecology of Overwintering

Remaining inside the nest through the winter is thought to benefit hatchling terrapins by reducing their exposure to predators at a time when resources necessary for rapid growth are in decline (Gibbons and Nelson 1978). However, at high latitudes, winter survival may depend on the hatchlings' ability to reduce metabolic rate, minimize water loss, and tolerate exposure to subfreezing temperatures (Costanzo et al. 2008).

Energetics

At oviposition, terrapin eggs contain more than enough nutrients to complete development (Allman et al. 2012). The proportion of polar lipid (average = 26%–29%) in terrapin eggs (Ricklefs and Burger 1977; Roosenburg and Dennis 2005) is high relative to that of some freshwater species but comparable to that of turtle species that overwinter within the nest (Congdon et al. 1983). Residual yolk in hatchling turtles, the portion not used during embryogenesis, can vary depending on incubation conditions, with greater yolk reserves found in hatchlings from dry nests (Packard et al. 1999). Allman et al. (2012) reported larger eggs with more lipids in terrapins from southern locales, potentially meeting the increased metabolic demands of hatchlings that experience higher ambient temperatures during a longer pre-hibernation period and a milder winter. In hatchling painted turtles (*Chrysemys picta*), the energetic reserves in yolk are consumed during the warm months immediately after hatching and converted to storage materials (Muir et al. 2013). For hatchlings that overwinter in the nest, glycogen stored in the liver may be the main source of energy (Muir et al. 2013). These finite energy reserves are critical for survival from hatching through emergence; however, they may be conserved through the winter as a consequence of the reduced metabolic rates resulting from low temperatures. Winter-acclimated hatchlings of *M. t. terrapin* from New Jersey had the most residual yolk (2.23% ± 0.91% carcass dry mass) among nine taxa studied by Costanzo et al. (2006). Baker et al. (2006) found that hatchling terrapins that overwintered in the nest were nominally, but not significantly, lighter than fall-emerging hatchlings. Similarly, lipid content did not differ between fall- and spring-emerging hatchlings on Poplar Island (Graham 2009). In southern New Jersey, hatchlings that overwintered in the nest were similar in mass to siblings that emerged from the nest in the fall, with the exception of years when the winter was warm and dry (Baker, Boerner, and Wood, unpub. data).

Hatchlings that exhibit early emergence have access to food and water before hibernation (Draud et al. 2004; Muldoon and Burke 2012). When kept in warm aquatic or terrestrial environments, hatchling terrapins will feed and grow during the first winter (Coker 1906; Hay 1917; Herlands et al. 2004; Kinneary 2008). While abundant prey may be available to hatchling terrapins in post-emergent refugia (e.g., Lovich et al. 1991), it is unclear whether naturally emerging hatchling terrapins eat before winter arrives (Coker 1906). Hatchlings that hibernated outside the nest in New York were smaller when recaptured the following spring (Muldoon and Burke 2012). For some individuals, CL decreased by 3.6% and PL decreased by 10%. While some, if not all, of this decrease in body size may be attributed to water loss (see below), early emergence does not appear to be driven by an energetic advantage to hatchlings that vacate the nest before winter in more northern populations.

Evaporative Water Loss

In cold, dry winters, dehydration may influence body mass more than does energy consumption. Hatchling terrapins that remain in terrestrial habitats are at risk of desiccation during the winter because they may not have access to freshwater to drink. In laboratory studies, terrestrially hibernating species are generally more resistant to evaporative water loss (EWL) than those that emerge in the fall (e.g., snapping turtle [*Chelydra serpentina*]) and, presumably, those that hibernate in water (Costanzo et al. 2001; Kinneary 2008). When tested under conditions that simulate dry nest conditions (5°C; 75%–80% relative humidity), the rate of EWL for winter-acclimated hatchling terrapins (2.7 ± 0.4 mg $\cdot$ g^{-1} $\cdot$ d^{-1}; $N = 7$) was comparable to that for hatchlings of other terrestrially hibernating species tested under identical conditions (Costanzo et al. 2001; Baker et al. 2003). The ability to resist EWL is largely morphological, being highest in species with a large plastron relative to carapace

(Costanzo et al. 2001). Terrapins maintain positive water balance and even grow when reared in terrestrial environments, whereas hatchling snapping turtles reared under identical conditions become dehydrated and die (Kinneary 2008). The greater resistance to EWL observed in *M. terrapin* relative to *C. serpentina* is associated with the terrapin's superior tolerance of marine environments (Dunson 1984). Terrapins are highly resistant to water loss when immersed in seawater, losing only 0.3% of body mass per day (Robinson and Dunson 1976). By contrast, a morphologically similar freshwater species, painted turtle (*C. picta*), loses 2.9% of body mass each day (Robinson and Dunson 1976). Because 80% to 85% of water exchange in seawater occurs across the skin in terrapins, the permeability of the integument is believed to be an important mediator of water loss (Robinson and Dunson 1976).

Despite their well-developed resistance to EWL, terrestrially hibernating hatchlings may become so dehydrated that they shrink during the winter. Muldoon and Burke (2012) found the PLs of 10 recaptured hatchling terrapins were 0.4% to 10.0% smaller in spring than in the previous fall. Similar decreases in body size in a desert tortoise were attributed to bone loss during extended dry periods (Loehr et al. 2007). Shells of hatchling terrapins have higher water content than those of adults because they are poorly ossified, and dehydration during a terrapin's first winter can lead to a decrease in body size under desiccating conditions. Juvenile terrapins are less susceptible to EWL than hatchlings because they have a lower surface area to volume ratio and a more completely ossified shell.

Flooding

Terrapin nests are often constructed in close proximity to the high-tide line and may be lost to extreme tides during incubation (Coker 1906; Feinberg and Burke 2003; Roosenburg et al. 2003; Butler et al. 2004). Embryos may survive brief submergence, but prolonged inundation of the nest chamber can be lethal (Roosenburg 1994). Although the likelihood of coastal flooding increases during August and September in the Atlantic hurricane season, little is known about the effect of flooding on hatchling survivorship. At Jamaica Bay, New York, 6 of 35 nests containing hatchling terrapins were destroyed by a 2011 hurricane (R. Burke, unpub. data cited in Muldoon and Burke 2012). In southern New Jersey, extreme lunar high tides (e.g., king tide) and storm surges from winter storms (e.g., nor'easters) inundate nesting sites that harbor hatchlings during fall and winter (Baker, pers. obs.). While it is unclear how flooding at the surface influences the nest microenvironment, live hatchlings have emerged after successfully overwintering in nests that were flooded with seawater multiple times for short durations throughout the winter. Temperature at the time of inundation may determine survival; survival time during submergence is significantly reduced at higher temperatures (Baker et al. 2013b). For this reason, late summer and early fall floods, when soil and water temperatures are still relatively high, are more likely to be lethal than winter floods. Indirect effects of flooding (i.e., increased soil moisture content) have been observed to increase winter mortality in several species (see Costanzo et al. 2008).

Anoxia

Hatchling terrapins may avoid aquatic habitats for hibernation because of the physiological challenges of prolonged submergence in a potentially anoxic environment. Adult terrapins rarely, if ever, surface during hibernation (Yearicks et al. 1981), and some may remain buried under a thin layer of mud for months (Yearicks et al. 1981; Mann 1995; Haramis et al. 2011; Williard and Harden 2011; Harden and Williard 2012). In related species, long-term submergence (3–6 months) during hibernation is achieved by significantly reducing metabolic rate and using skeletal elements (shell, long bones) to sequester lactate and buffer against a lethal lactacidosis (Jackson 2000). While the adults of some turtle species can survive extended periods of anoxic submergence at 3°C, this ability is much reduced in hatchlings (Reese et al. 2004). Further, hatchlings of *M. t. terrapin* are poorly tolerant of anoxia compared with other aquatic species (Dinkelacker et al. 2005). When kept under anoxic conditions at 4°C, hatchling terrapins survived a mean of 17 ± 1 days. Of the seven species tested by Dinkelacker et al. (2005), hatchling terrapins had the highest rate of lactate accumulation (6.2 ± 0.2 mmol · L^{-1} · d^{-1}). Young turtles may lack sufficient energy reserves to support prolonged anaerobic respiration or the capacity to sequester and buffer lactate that accumulates during anoxic submergence (Reese et al. 2004; Dinkelacker et al. 2005). Thus, hibernation in an anoxic habitat (e.g., mud) does not appear to be a viable option for hatchling terrapins. The anoxia tolerance of juvenile terrapins is unknown and deserves further study.

Subfreezing Temperatures

While the reason that some hatchlings emerge from the nest in fall and others emerge in spring remains unknown, both strategies can result in exposure to freezing temperatures. The highest temperature at which terrapins' tissues can be frozen is determined by the concentration of solutes in their body fluids. For winter-acclimated terrapins, tissues begin to freeze at ~−0.6°C (Baker et al. 2006; Costanzo et al. 2006). In southern New Jersey, overwintering hatchlings experienced up to 16 critical chilling episodes (CCEs), discrete periods when nest temperature fell below −0.6°C (Baker et al. 2006). However, the majority of CCEs at nest depth were mild (minimum temperature of −1.2°C) and relatively brief (Baker et al. 2006). Typical CCEs were less than 12 h, but subfreezing conditions occasionally persisted for more than 100 h. The lowest temperature recorded at nest depth was −4.7°C; all of the hatchlings from this nest had emerged in the fall (Baker et al. 2006).

Hatchling terrapins that emerge from the nest and excavate their own hibernacula may be at greater risk of freezing. Hatchlings hibernating outside the nest are superficially buried (depth < 10 cm) relative to hatchlings overwintering in nest chambers, which can be 15 to 20 cm deep (Baker et al. 2006; Draud and Baker, unpub. data). In the absence of snow and other insulating material, the thermal regimen near the surface

is characterized by higher daily highs and lower daily lows than at nest depth (Costanzo et al. 1995; Baker et al. 2013a). On Long Island, hatchling terrapins that hibernate outside the nest may experience thermal environments similar to those described for hatchling box turtles (*Terrapene carolina*). Burke and Capitano (2011) found that *T. carolina* hatchlings in shallow, self-excavated hibernacula were exposed to temperatures as low as −7°C, but most CCEs were relatively brief (2–3 h). Hatchlings hibernating near the surface may experience the most extreme temperatures, but those that survive are the first to receive environmental cues that stimulate activity at the onset of the next growing season (Baker et al. 2010, 2013a).

While it is indisputable that hatchling turtles in northern populations are serially exposed to subfreezing temperatures during hibernation, there is debate in the literature about the mechanisms that promote their survival (Costanzo et al. 2008). Some authors assert that hatchlings of various turtle species survive by avoiding freezing through supercooling (Packard and Packard 1993), while others contend that they survive by tolerating freezing (Storey et al. 1988). Although these strategies are often considered to be mutually exclusive, some organisms employ a combination of both freeze avoidance through supercooling and freeze tolerance, depending on prevailing microenvironmental conditions (Costanzo et al. 1995, 2008; Voituron et al. 2002).

For some turtle species, hatchlings exposed to subzero temperatures may avoid somatic freezing by becoming supercooled, a state in which their body tissues remain liquid below the equilibrium freezing point (−0.6°C). The probability that a supercooled hatchling will freeze spontaneously increases as body temperature decreases, until the temperature of crystallization (T_c) is reached and ice begins to form in body fluids (Packard and Packard 1993; Costanzo et al. 2003).

Hatchlings of *M. terrapin* have a well-developed capacity for supercooling when reared under sterile laboratory conditions (Baker et al. 2006). Pristine, cold-acclimated hatchlings of several species exhibit a similar capacity to supercool (from −8°C to −20°C), regardless of phylogenetic affinity and winter ecology (Costanzo et al. 2001, 2003, 2008; Baker et al. 2003). Although these results are sometimes used to suggest a freeze-avoidance strategy (e.g., Graham 2009), they simply demonstrate that winter-acclimated hatchling turtles lack endogenous ice-nucleating agents (INAs; e.g., ice-nucleating proteins). In experiments that exposed hatchling turtles to INAs, even briefly, the ability to supercool was greatly reduced (Baker et al. 2003, 2006; Costanzo et al. 2003).

Innate supercooling capacity is reduced in naturally incubated hatchlings because soil surrounding the nest chamber contains exogenous INAs (e.g., quartz crystals, ice-nucleating *Pseudomonas* bacteria) that facilitate the freezing process (Costanzo et al. 2000, 2003; Baker et al. 2003, 2006; Storey 2006; Graham 2009). Ice formation is lethal to most freeze-avoidant animals (Lee and Costanzo 1998), and thus a survival strategy based on supercooling in the presence of ice and INAs requires a cutaneous barrier that prevents ice and INAs from contacting body fluids (Packard and Packard 1993; Willard et al. 2000). Further, recently hatched turtles may ingest INA-laden soil and eggshell from the nest chamber (Costanzo et al. 2003). While the nutritional benefit of consuming nest materials remains unknown, the presence of INAs in their guts renders hatchlings incapable of supercooling to the same extent as those incubated on sterile vermiculite (Costanzo et al. 2003, 2008).

Hatchlings of *M. terrapin* are poorly resistant to inoculation by ice or INAs found in their nest microenvironment (Baker et al. 2006). When chilled in the presence of frozen nest soil, hatchlings from southern New Jersey froze at high subzero temperatures ($T_c = -1.1°C \pm 0.2°C$; $N = 8$). Hatchling terrapins collected from shallow, individually excavated hibernacula in Bayville, New York (Draud and Baker, unpub. data), were also highly susceptible to inoculation ($T_c = -1.3°C \pm 0.2°C$; $N = 8$) when chilled in the presence of frozen soil (soil moisture = 6.0% ± 2.3%; $N = 8$). Inoculation resistance may be determined by barrier properties of the integument (Willard et al. 2000). However, despite the intermediate levels of lipid in the skin of hatchling terrapins, inoculation resistance was poor relative to that of other terrestrial hibernators (J.

Costanzo and J. Jack, unpub. data; see also Costanzo et al. 2008, Fig. 12). Thus, whether hibernating inside or outside the nest chamber, *M. terrapin* hatchlings are probably inoculated at high subzero temperatures (~–1°C) by the ice or environmental INAs that surround them (Baker et al. 2006). Susceptibility to inoculation is a common feature of freeze-tolerant organisms because ice formation proceeds slowly at high subzero temperatures, reducing the incidence of injury (Storey and Storey 2004; Costanzo and Lee 2013).

Hatchling terrapins are considered naturally freeze tolerant because they survive somatic freezing under ecologically relevant conditions (Baker et al. 2006). In laboratory studies, hatchling terrapins survived freezing for 7 days to a minimum body temperature of –2.5°C (Dinkelacker et al. 2005; Baker et al. 2006). For organisms that tolerate somatic freezing, ice formation can minimize the temperature stress on their tissues because latent heat released during crystallization stabilizes body temperature near the freezing point until ice content approaches equilibrium (Storey and Storey 1996). The degree of freeze tolerance exhibited by *M. t. terrapin* is congruent with the winter conditions near the northern limit of its distribution (Baker et al. 2006). Typical chilling episodes observed by Baker et al. (2006) were brief (~12 h) and relatively mild (minimum temperature of ~–1.2°C).

Ice content stabilizes when a hatchling is held at a given temperature, and thus the mortality of terrapins observed by Baker et al. (2006) during extended freeze tolerance trials (11 days at –2.5°C) was more likely the result of a profound acidosis in oxygen-deprived tissues or a reperfusion injury on thawing. Extended freezing bouts are challenging because without a patent circulatory system, the limited buffering capacity of the hatchling's skeletal system is inaccessible, and an ischemic lactacidosis eventually develops (Costanzo et al. 2006). On thawing, circulatory and respiratory functions return, and tissues are reoxygenated. Resumption of aerobic metabolism after freezing can lead to the generation of free radicals (e.g., O_2^-) that damage lipids, proteins, and nucleic acids (Storey 2006; Costanzo et al. 2008). Freeze-tolerant turtles, including terrapins, possess high constituent antioxidant levels that manage radical formation and reduce injury to tissues (Dinkelacker et al. 2005; Baker et al. 2007).

Hatchling mortality observed at lower temperatures (≤−3.5°C) is likely to be related to the increasing amount of ice that forms as body temperature decreases (Baker et al. 2006). In freeze-tolerant vertebrates, ice crystals propagate throughout extracellular spaces during the freezing process; however, intracellular ice formation is lethal because ice crystals damage cell structure and function (Storey and Storey 1996; Lee and Costanzo 1998). The probability that ice crystals will form inside a cell increases at lower temperatures and when freezing proceeds rapidly. Bodily fluids become more concentrated as extracellular ice crystals grow and exclude solutes. Water is drawn out of cells to restore osmotic balance, causing a decrease in cell volume (Lee and Costanzo 1998). Despite the well-developed capacity for freezing in some species, cell membranes eventually collapse when freezing-induced dehydration extends below a critical cell volume (~65%). The ice content of *M. t. terrapin* frozen to an equilibrium body temperature of −3.0°C was estimated to be ~54% of body water (Costanzo et al. 2006). This value was the lowest among the seven species examined, primarily because of the terrapin's lower body water content and higher plasma osmolality (Costanzo et al. 2006).

Although the physiological basis for freeze tolerance in reptiles is poorly understood, the mechanisms that regulate cell volume during dehydration may promote survival during freezing (Storey and Storey 2004; Storey 2006; Costanzo and Lee 2013). Terrapins may be predisposed to tolerate freezing through their adaptations to hyperosmotic marine habitats (Baker et al. 2006; Costanzo et al. 2006). Terrapins exposed to seawater maintain high interstitial fluid volumes to prevent dehydration (Robinson and Dunson 1976). Similar partitioning of body water during terrestrial hibernation is likely to be beneficial during the freezing process.

Cryoprotectants, a class of solutes that confer protection during the freezing process, work colligatively (i.e., by increasing the solute-to-solvent ratio) to reduce the amount of body water that can freeze at a given temperature (Storey and Storey 1996; Lee and Costanzo 1998; Costanzo et al. 2006). Some low molecular weight molecules, termed penetrating cryoprotectants, are taken up by cells and reduce cellular dehydration (Storey and Storey 1996; Costanzo et al. 2006). In addition to their colligative effects, cryoprotectants may preserve macromolecules, membranes, and other cellular structures, or fuel metabolism in frozen or recently thawed tissues (Costanzo et al. 2008).

Urea, glucose, and lactate appear to be the most important cryoprotectants in terrapins (Costanzo et al. 2006). Whether hibernating on land or in water, terrapins accumulate high plasma concentrations of urea during the winter (Gilles-Baillien 1973; Costanzo et al. 2006; Muir et al. 2010; Harden et al. 2015). Increased osmotic pressure associated with uremia defends cell volume during dehydration or freezing (Gilles-Baillien 1973; Costanzo et al. 2006). In laboratory trials, significant increases in glucose and lactate in frozen or thawed terrapins jointly raised the plasma solute concentration by an average of 48 mmol/L. Without these three osmolytes, plasma osmolality would be ~30% lower and ice content at −3°C would approach the critical level (Costanzo et al. 2006). High urea concentrations have the additional benefit of suppressing metabolic rate and thus helping to conserve energy stores through the winter (Muir et al. 2010).

The survival strategy of terrapins that overwinter in terrestrial environments is unknown. Direct observation of overwintering hatchlings is difficult without disturbing the nest or hibernaculum. Laboratory experiments suggest that terrapins are capable of supercooling when ice or environmental INAs do not seed the freezing of body tissues (Baker et al. 2006). Hatchlings also are tolerant of somatic freezing so long as equilibrium body temperature does not drop below −3.5°C (Dinkelacker et al. 2005; Baker et al. 2006; Costanzo et al. 2006). Their dual overwintering ability to supercool or tolerate freezing may promote survival under dynamic environmental conditions (Baker et al. 2006). Such plasticity in response to freezing conditions is consistent with a bet-hedging strategy in which the fitness of individuals in unpredictable environments is optimized by not specializing in a particular mode of survival (Voituron et al. 2002; Costanzo and Lee 2013).

Beyond the First Winter

Although freezing exposure may be reduced for juvenile and adult terrapins, individuals hibernating in the intertidal zone remain at risk (e.g., Yearicks et al. 1981; Williard and Harden 2011). In North Carolina, adult terrapins were found to be occasionally exposed to temperatures as low as −3.4°C (Williard and Harden 2011). Adult turtles, owing to their larger size, do not supercool appreciably (Costanzo and Claussen 1990; Claussen and Zani 1991), and freezing is initiated near the equilibrium freezing point of their tissues (~−0.6°C). Although freezing survival may be enhanced in adults, given that the rate of ice formation is inversely related to body size (Claussen and Zani 1991), the survivable minimum temperature for adult painted turtles is closer to the equilibrium freezing point than that for hatchlings. According to Coker (1906), short-term survival of "freezing" by adult terrapins was well known to terrapin dealers; however, the duration and extent of freezing episodes were not described. To better understand the winter ecology and physiology of adult terrapins, the freezing risk, especially in northern populations, should be quantified with temperature data loggers (e.g., Williard and Harden 2011; Akins et al. 2014) and coupled with ecologically relevant tests of freeze tolerance (e.g., Baker et al. 2006).

Management Implications

Whether inside the nest, actively foraging, or buried in self-excavated refugia, hatchling and juvenile terrapins can be found in high marsh and upland habitats throughout the first 3 years of life. Depending on the season, construction projects that disturb these habitats may inadvertently crush or entomb eggs, hatchlings, or juvenile terrapins. Advanced planning for these projects should include fencing that excludes nesting females, as well as pre-construction surveys to remove any eggs, hatchlings, or juveniles that may have been enclosed by the fence. Eggs (May–August) or hatchlings (July–June of the following year) are potentially present in suitable nesting habitat in any month of the year (Draud et al. 2004; Baker et al. 2006; Muldoon and Burke 2012).

Conclusions

We reviewed more than 100 years of observations on diamond-backed terrapins, but there are still large gaps in our understanding of the ecology and physiology of early age classes, especially juveniles. Where we do have detailed information on a particular topic, we lack range-wide observations to capture variation across ecologically distinct populations. Future studies should examine the diel and seasonal timing of emergence from the nest, habitat use and movements, diet, mortality rates in the terrestrial life stages, characteristics of hibernacula in the first three winters, and ontogenetic changes in tolerance of low temperatures and hyperosmotic environments.

REFERENCES

Akins, C.D., C.D. Ruder, S.J. Price, L.A. Harden, J.W. Gibbons and M.E. Dorcas. 2014. Factors affecting temperature variation and habitat use in free-ranging diamondback terrapins. Journal of Thermal Biology 44:63-69.

Allman, P.E., A.R. Place and W.M. Roosenburg. 2012. Geographic variation in egg size and lipid provisioning in the diamondback terrapin *Malaclemys terrapin*. Physiological Zoology 85:442-449.

Arndt, R.G. 1994. Predation on hatchling diamondback terrapin, *Malaclemys terrapin* (Schoepff), by the ghost crab, *Ocypode quadrata* (Fabricius). II. Florida Scientist 57:1-5.

Auger, P.J. 1989. Sex ratio and nesting behavior in a population of *Malaclemys terrapin* displaying temperature-dependent sex determination. MS thesis, Tufts University, Medford, MA. 173 pages.

Auger, P.J. and P. Giovannone. 1979. On the fringe of existence: diamondback terrapins at Sandy Neck. Cape Naturalist 8:44-58.

Baker, P.J., J.P. Costanzo, R. Herlands, R.C. Wood and R.E. Lee. 2006. Inoculative freezing promotes winter survival in the diamondback terrapin, *Malaclemys terrapin*. Canadian Journal of Zoology 84:116-124.

Baker, P.J., J.P. Costanzo, J.B. Iverson and R.E. Lee. 2003. Adaptations to terrestrial overwintering of hatchling northern map turtles, *Graptemys geographica*. Journal of Comparative Physiology B 173:643-651.

Baker, P.J., J.P. Costanzo, J.B. Iverson and R.E. Lee. 2013a. Seasonality and interspecific and intraspecific asynchrony in emergence from the nest by hatchling freshwater turtles. Canadian Journal of Zoology 91:451-461.

Baker P.J., J.P. Costanzo and R.E. Lee. 2007. Oxidative stress and antioxidant capacity of a terrestrially-hibernating hatchling turtle. Journal of Comparative Physiology B 177:875-883.

Baker, P.J., J.B. Iverson, R.E. Lee and J.P. Costanzo. 2010. Winter severity and phenology of spring emergence from the nest in freshwater turtles. Naturwissenschaften 97:607-615.

Baker, P.J., A. Thompson, I. Vatnick and R C. Wood. 2013b. Estimating survival times for northern diamondback terrapins, *Malaclemys terrapin terrapin*, in submerged crab pots. Herpetological Conservation and Biology 8:667-680.

Baldwin, J.D., L.A. Latino, B.K. Mealey, G.M. Parks and M.R.J. Forstner. 2005. The diamondback terrapin in Florida Bay and the Florida Keys: insights into turtle conservation and ecology. Pages 180-186 in W.E. Meshaka, Jr. and K.J. Babbitt (eds.), Amphibians and Reptiles: Status and Conservation in Florida. Krieger Publishing Company, Malabar, FL.

Bleakney, J.S. 1963. Notes on the distribution and life histories of turtles in Nova Scotia. Canadian Field-Naturalist 77:67-76.

Burger, J. 1976a. Behavior of hatchling diamondback terrapin (*Malaclemys terrapin*) in the field. Copeia 1976:742-748.

Burger, J. 1976b. Temperature relationships in nests of the northern diamondback terrapin, *Malaclemys terrapin terrapin*. Herpetologica 32:412-418.

Burger, J. 1977. Determinants of hatching success in the diamondback terrapin, *Malaclemys terrapin*. American Midland Naturalist 97:444-464.

Burke, R.L. and W. Capitano. 2011. Eastern box turtle, *Terrapene carolina*, neonate overwintering ecology on Long Island, New York. Chelonian Conservation and Biology 10:256-259.

Butler, J.A. 2002. Population ecology, home range, and seasonal movements of the Carolina diamondback terrapin, *Malaclemys terrapin centrata*, in northeastern Florida. Florida Fish and Wildlife Conservation Commission Project Report NG96-007. Tallahassee, FL. 65 pages.

Butler, J.A., C. Broadhurst, M. Green and Z. Mullin. 2004. Nesting, nest predation and hatchling emergence of the Carolina diamondback terrapin, *Malaclemys terrapin centrata*, in northeastern Florida. American Midland Naturalist 152:145-155.

Carr, A.F. and H. Hirth. 1961. Social facilitation in green turtle siblings. Animal Behavior 9:68-70.

Claussen, D.L. and P.A. Zani. 1991. Allometry of cooling, supercooling, and freezing in the freeze-tolerant turtle *Chrysemys picta*. American Journal of Physiology 261:R626-R632.

Coker, R.E. 1906. The natural history and cultivation of the diamond-back terrapin with notes on other forms of turtles. North Carolina Geological Survey Bulletin 14:1-67.

Coleman, A.T., T. Wibbels, K. Marion, T. Roberge, D. Nelson and J. Dindo. 2014. Dispersal behavior of diamond-backed terrapin post-hatchlings. Southeastern Naturalist 13:572-586.

Congdon, J.D., D.W. Tinkle and P.C. Rosen. 1983. Egg components and utilization during development in aquatic turtles. Copeia 1983:264-268.

Costanzo, J.P., P.J. Baker, S.A. Dinkelacker and R.E. Lee. 2003. Endogenous and exogenous ice-nucleating agents constrain supercooling in the hatchling painted turtle. Journal of Experimental Biology 206:477-485.

Costanzo, J.P., P.J. Baker and R.E. Lee. 2006. Physiological responses to freezing in hatchlings of freeze-tolerant and -intolerant turtles. Journal of Comparative Physiology B 176:697-707.

Costanzo, J.P. and D.L. Claussen. 1990. Natural freeze tolerance in the terrestrial turtle, *Terrapene carolina*. Journal of Experimental Zoology 254:228-232.

Costanzo, J.P., J.B. Iverson, M.F. Wright and R.E. Lee. 1995. Cold hardiness and overwintering strategies of hatchlings in an assemblage of northern turtles. Ecology 76:1772-1785.

Costanzo, J.P. and R.E. Lee. 2013. Commentary: avoidance and tolerance of freezing in ectothermic vertebrates. Journal of Experimental Biology 216:1961-1967.

Costanzo, J.P., R.E. Lee and G.R. Ultsch. 2008. Physiological ecology of overwintering in hatchling turtles. Journal of Experimental Zoology 309A:297-379.

Costanzo, J.P., J.D. Litzgus, J.B. Iverson and R.E. Lee. 2000. Ice nuclei in soil compromise cold hardiness of hatchling painted turtles (*Chrysemys picta*). Ecology 81:346-360.

Costanzo, J.P., J.D. Litzgus, J.B. Iverson and R.E. Lee. 2001. Cold-hardiness and evaporative water loss in hatchling turtles. Physiological and Biochemical Zoology 74:510-519.

Danforth, B.N. 1999. Emergence dynamics and bet hedging in a desert bee, *Perdita portalis*. Proceedings of the Royal Society of London B 266:1985-1994.

Davenport, J. and E.A. Macedo. 1990. Behavioral osmotic control in the euryhaline diamondback terrapin *Malaclemys terrapin*: responses to low salinity and rainfall. Journal of Zoology 220:487-496.

Dinkelacker, S.A., J.P. Costanzo and R.E. Lee. 2005. Anoxia tolerance and freeze tolerance in hatchling turtles. Journal of Comparative Physiology B 175:209-217.

Drake, D.L. and J.R. Spotila. 2002. Thermal tolerances and the timing of sea turtle hatchling emergence. Journal of Thermal Biology 27:71-81.

Draud, M., M. Bossert and S. Zimnavoda. 2004. Predation on hatchling and juvenile diamondback terrapins (*Malaclemys terrapin*) by the Norway rat (*Rattus norvegicus*). Journal of Herpetology 38:467-470.

Dunson, W.A. 1984. The contrasting roles of the salt glands, the integument and behavior in osmoregulation of marine reptiles. Pages 107-129 in A. Pequeux, R. Gilles and L. Bolis (eds.), Osmoregulation in Estuarine and Marine Animals: Lecture Notes on Coastal and Estuarine Studies. Springer, Berlin.

Dunson, W.A. 1985. Effect of water salinity and food salt content on growth and sodium efflux of hatchling diamondback terrapins (*Malaclemys*). Physiological Zoology 58:736-747.

Feinberg, J.A. and R.L. Burke. 2003. Nesting ecology and predation of diamondback terrapins, *Malaclemys terrapin*, at Gateway National Recreation Area, New York. Journal of Herpetology 37:517-526.

Gibbons, J.W. 1970. Terrestrial activity and the population dynamics of aquatic turtles. American Midland Naturalist 83:404-414.

Gibbons, J.W. 2013. A long-term perspective of delayed emergence (aka overwintering) in hatchling turtles: some they do and some they don't, and some you just can't tell. Journal of Herpetology 47:203-214.

Gibbons, J.W., J.E. Lovich, A.D. Tucker, N.N. Fitzsimmons and J.L. Greene. 2001. Demographic and ecological factors affecting conservation and management of diamondback terrapins (*Malaclemys terrapin*) in South Carolina. Chelonian Conservation and Biology 4:66-74.

Gibbons, J.W. and D.H. Nelson. 1978. The evolutionary significance of delayed emergence from the nest by hatchling turtles. Evolution 32:297-303.

Gilles-Baillien, M. 1973. Hibernation and osmoregulation in the diamondback terrapin *Malaclemys centrata centrata* (Latreille). Journal of Experimental Biology 59:45-51.

Glen, F., A.C. Broderick, B.J. Godley and G.C. Hays. 2005. Patterns in the emergence of green (*Chelonia mydas*) and loggerhead (*Caretta caretta*) turtle hatchlings from their nests. Marine Biology 146:1039-1049.

Graham, L.J. 2009. Diamondback terrapin, *Malaclemys terrapin*, nesting and over-wintering ecology. MS thesis, Ohio University, Athens, OH. 69 pages.

Haramis, G.M., P.F. Henry and D.D. Day. 2011. Using scrape fishing to document terrapins in hibernacula in Chesapeake Bay. Herpetological Review 42:170-177.

Harden, L.A., S.R. Midway and A.S. Williard. 2015. The blood biochemistry of overwintering diamondback terrapins (*Malaclemys terrapin*). Journal of Experimental Marine Biology and Ecology 466:34-41.

Harden, L.A. and A.S. Williard. 2012. Using spatial and behavioral data to evaluate the seasonal bycatch risk of diamondback terrapins *Malaclemys terrapin* in crab pots. Marine Ecology Progress Series 467:207-217.

Hart, K.M. and C.C. McIvor. 2008. Demography and ecology of mangrove diamondback terrapins in a wilderness area of Everglades National Park, Florida, USA. Copeia 2008:200-208.

Hay, W.P. 1917. Artificial propagation of the diamondback terrapin. Bulletin of the US Bureau of Fisheries 24:1-20.

Herlands, R., R.C. Wood., J. Pritchard, H. Clapp and N. Le Furge. 2004. Diamondback terrapin, (*Malaclemys terrapin*) head-starting project in southern New Jersey. Pages 13-18 in C. Swarth, W.M. Roosenburg and E. Kiviat (eds.), Conservation and Ecology of Turtles of the Mid-Atlantic Region: A Symposium. Bibliomania, Salt Lake City, UT.

Hildebrand, S.F. 1932. Growth of diamond-back terrapins: size attained, sex ratio and longevity. Zoologica (New York) 9:551-563.

Hoden, R. and K.W. Able. 2003. Habitat use and road mortality of diamondback terrapins (*Malaclemys terrapin*) in the Jacques Cousteau National Estuarine Research Reserve at Mullica River-Great Bay in southern New Jersey. Jacques Cousteau NERR Technical Report 100-24. Tuckerton, NJ. 31 pages.

Holliday, D.K, A.A. Elskus and W.M. Roosenburg. 2009. Impacts of multiple stressors on growth and metabolic rate of *Malaclemys terrapin*. Environmental Toxicology and Chemistry 28:338-345.

Hurd, L.E., G.W. Smedes and T.A. Dean. 1979. An ecological study of a natural population of diamondback terrapins (*Malaclemys t. terrapin*) in a Delaware salt marsh. Estuaries 2:28-33.

Iverson, J.B. 1991. Patterns of survivorship in turtles (Order Testudines). Canadian Journal of Zoology 69:385-391.

Jackson, D.C. 2000. Living without oxygen: lessons from the freshwater turtle. Comparative Biochemistry and Physiology 125A:299-315.

Janzen, F.J., J.K. Tucker and G.L. Paukstis. 2000. Experimental analysis of an early life-history stage: avian predation selects for larger body size of hatchling turtles. Journal of Evolutionary Biology 13:947-954.

Kinneary, J.J. 2008. Observation on terrestrial behavior and growth in diamondback terrapin (*Malaclemys*) and snapping turtle (*Chelydra*) hatchlings. Chelonian Conservation and Biology 7:118-119.

Lawler, A.R. and J.A. Musick. 1972. Sand beach hibernation by a northern diamondback terrapin, *Malaclemys terrapin terrapin* (Schoepff). Copeia 1972:389-390.

Lee, R.E. and J.P. Costanzo. 1998. Biological ice nucleation and ice distribution in cold-hardy ectothermic animals. Annual Review of Physiology 60:55-72.

Loehr, V.J.T., M.D. Hofmeyr and B.T. Heinen. 2007. Growing and shrinking in the smallest tortoise, *Homopus signatus signatus*: the importance of rain. Oecologia 153:479-488.

Lovich, J.E., A.D. Tucker, D.E. Kling, J.W. Gibbons and T.D. Zimmerman. 1991. Hatchling behavior of diamondback terrapins (*Malaclemys terrapin*) released in a South Carolina salt marsh. Herpetological Review 22:81-83.

Mann, T.M. 1995. Population surveys for diamondback terrapins (*Malaclemys terrapin*) and Gulf salt marsh snakes (*Nerodia clarkii clarkii*) in Mississippi. Mississippi Museum Technical Report, no. 37, pages 1-75.

Mitchell, J.C. and P. Denmon. 2005. Early terrestrial emergence of a hatchling northern diamond-backed terrapin (*Malaclemys terrapin terrapin*) on the Eastern Shore of Virginia. Banisteria 26:19-20.

Mitro, M.G. 2003. Demography and viability analysis of a diamondback terrapin population. Canadian Journal of Zoology 81:716-726.

Morreale, S.J. 1992. The status and population ecology of the diamondback terrapin, *Malaclemys terrapin*, in New York. Final Report Submitted to New York Department of Environmental Conservation, Contract no. C002656. Hampton Bays, NY. 75 pages.

Muir, T.J., J.P. Costanzo and R.E. Lee. 2010. Evidence for urea-induced hypometabolism in isolated organs of dormant ectotherms. Journal of Experimental Zoology 313A:28-34.

Muir, T.J., B.D. Dishong, R.E. Lee and J.P. Costanzo. 2013. Energy use and management of energy reserves in hatchling turtles (*Chrysemys picta*) exposed to variable winter conditions. Journal of Thermal Biology 38:324-330.

Muldoon, K.A. and R.A. Burke. 2012. Movements, overwintering, and mortality of hatchling diamond-backed terrapins (*Malaclemys terrapin*) at Jamaica Bay, New York. Canadian Journal of Zoology 90:651-662.

Packard, G.C., K. Miller, M.J. Packard and G.F. Birchard. 1999. Environmentally induced variation in body size and condition in hatchling snapping turtles (*Chelydra serpentina*). Canadian Journal of Zoology 77:278-289.

Packard, G.C. and M.J. Packard. 1993. Delayed inoculative freezing is fatal to hatchling painted turtles (*Chrysemys picta*). Cryo-Letters 14:273-284.

Peterson, C.H., S.R. Fegley, C.M. Voss, S.R. Marschhauser and B.M. VanDusen. 2013. Conservation implications of density-dependent predation by ghost crabs on hatchling sea turtles running the gauntlet to the sea. Marine Biology 160:629-640.

Pitler, R. 1985. *Malaclemys terrapin terrapin* (northern diamondback terrapin) behavior. Herpetological Review 16:82.

Plummer, M.V. 2007. Nest emergence of smooth softshell turtle (*Apalone mutica*) hatchlings. Herpetological Conservation and Biology 2:61-64.

Ream, C. and R. Ream. 1966. The influence of sampling methods on the estimation of population structure in painted turtles. American Midland Naturalist 75:325-338.

Reese, S.A., G.R. Ultsch and D.C. Jackson. 2004. Lactate accumulation, glycogen depletion, and shell composition of hatchling turtles during simulated aquatic hibernation. Journal of Experimental Biology 207:2889-2895.

Ricklefs, R.E. and J. Burger. 1977. Composition of eggs of the diamondback terrapin. American Midland Naturalist 97:232-235.

Riegner, M.F. 1982. The diet of yellow-crowned night-herons in the eastern and southern United States. Colonial Waterbirds 5:173-176.

Robinson, G.D. and W.A. Dunson. 1976. Water and sodium balance in the estuarine diamondback terrapin (*Malaclemys*). Journal of Comparative Physiology 105:129-152.

Roosenburg, W.M. 1994. Nesting habitat requirements of the diamondback terrapin: a geographic comparison. Wetlands Journal 6:9-12.

Roosenburg, W.M., P.E. Allman and B.J. Fruh. 2003. Diamondback terrapin nesting on the Poplar Island Environmental Restoration Project. Proceedings of the 13th Biennial Coastal Zone Conference, Baltimore, July 13-17. NOAA/CS/20322-CD. CD-ROM. National Oceanic and Atmospheric Administration, Charleston, SC. 6 pages.

Roosenburg, W.M. and T. Dennis. 2005. Egg component comparisons within and among clutches of the diamondback terrapin, *Malaclemys terrapin*. Copeia 2005:417-423.

Roosenburg, W.M., R. Dunn, and N.L. Smeenk. 2010. Terrapin monitoring at Poplar Island, 2009. Final Report Submitted to the Army Corps of Engineers, Baltimore Office. Baltimore, MD. 23 pages.

Roosenburg, W.M., L.J. Graham and M.L. Heckman. 2009. Terrapin monitoring at Poplar Island, 2007. Final Report Submitted to the Army Corps of Engineers, Baltimore Office. Baltimore, MD. 69 pages.

Roosenburg, W.M., L.J. Graham and E. Matthews. 2007. Terrapin monitoring at Poplar Island, 2006. Final Report Submitted to the Army Corps of Engineers, Baltimore Office. Baltimore, MD. 48 pages.

Roosenburg, W.M., D.M. Spontak, S.P. Sullivan, E.L. Matthews, M.L. Heckman, R.J. Trimbath, R.P. Dunn, E.A. Dustman, L. Smith and L.J. Graham. 2014. Nesting habitat creation enhances recruitment in a predator-free environment: *Malaclemys* nesting at the Paul S. Sarbanes Ecosystem Restoration Project. Restoration Ecology 22:815-823.

Rulison, E.L., L. Luiselli and R.L. Burke. 2012. Relative impacts of habitat and geography on raccoon diets. American Midland Naturalist 168:231-246.

Scholz, A.L. 2007. Impacts of nest site choice and nest characteristics on hatching success in the diamondback terrapins of Jamaica Bay, New York. MS thesis, Hofstra University, Hempstead, NY. 97 pages

Seigel, R.A. 1984. Parameter of two populations of the diamondback terrapin (*Malaclemys terrapin*) on the Atlantic coast of Florida. Pages 77-87 in R.A. Seigel, L.E. Hunt, J.L. Knight, L. Malaret and N.L. Zushlag (eds.), Vertebrate Ecology and Systematics: A Tribute to Henry S. Fitch. Museum of Natural History, University of Kansas, Lawrence, KS.

Spencer, R.J. and F.J. Janzen. 2011. Hatching behavior in turtles. Integrative and Comparative Biology 51:100-110.

Stancyk, S.E. 1995. Non-human predators of sea turtles and their control. Pages 139-152 in K.A. Bjorndal (ed.), Biology and Conservation of Sea Turtles. Smithsonian Institution Press, Washington, DC.

Storey, K.B. 2006. Reptile freeze tolerance: metabolism and gene expression. Cryobiology 52:1-16.

Storey, K.B. and J.M. Storey. 1996. Natural freezing survival in animals. Annual Review of Ecology and Systematics 27:365-386.

Storey, K.B. and J.M. Storey. 2004. Physiology, biochemistry, and molecular biology of vertebrate freeze tolerance: the wood frog. Pages 243-274 in B.J. Fuller, N. Lane and E.E. Benson (eds.), Life in the Frozen State. CRC Press, Boca Raton, FL.

Storey, K.B., J.M. Storey, S.P.J. Brooks, T.A. Churchill and R. J. Brooks. 1988. Hatchling turtles survive freezing during winter hibernation. Proceedings of the National Academy of Sciences 85:8350-8354.

Tucker, J.K. 1999. Environmental correlates of hatchling emergence in the red-eared turtle, *Trachemys scripta elegans*, in Illinois. Chelonian Conservation and Biology 3:401-406.

Tucker, J.K., G.L. Paukstis and F J. Janzen. 2008. Does predator swamping promote synchronous emergence of turtle hatchlings among nests? Behavioral Ecology 19:35-40.

Tuttle, S.E. and D.M. Carroll. 2005. Movements and behavior of hatchling wood turtles (*Glyptemys insculpta*). Northeastern Naturalist 12:331-348.

Voituron, Y., N. Mouquet, C. de Mazancourt and J. Clobert. 2002. To freeze or not to freeze? An evolutionary perspective on the cold-hardiness strategies of overwintering ectotherms. American Naturalist 160:255-270.

Willard, R., G.C. Packard, M.J. Packard and J.K. Tucker. 2000. The role of the integument as a barrier to penetration of ice into overwintering hatchlings of the painted turtle (*Chrysemys picta*). Journal of Morphology 246:150-159.

Williard, A.S. and L.A. Harden. 2011. Seasonal changes in thermal environment and metabolic enzyme activity in the diamondback terrapin (*Malaclemys terrapin*). Comparative Biochemistry and Physiology 158A:477-484.

Wood, R.C. 1981. The mysterious mangrove terrapin. Florida Naturalist, July-September, pages 6-7.

Wood, R.C. 1997. The impact of commercial crab traps on northern diamondback terrapins, *Malaclemys terrapin terrapin*. Pages 21-27 in J.E. Van Abbema (ed.), Proceedings: Conservation, Restoration, and Management of Tortoises and Turtles—An International Conference. New York Turtle and Tortoise Society, State University of New York, Purchase, NY.

Yearicks, E.F., R.C. Wood and W.S. Johnson. 1981. Hibernation of the northern diamondback terrapin, *Malaclemys terrapin terrapin*. Estuaries 4:78-80.

Osmoregulation

LEIGH ANNE HARDEN
AMANDA SOUTHWOOD
WILLIARD

O f the extant species of reptiles, fewer than 2% live in estuarine or marine habitats (Heatwole 1999). While numerous biotic and abiotic factors influence habitat use by a given species, the distribution of reptiles in coastal and oceanic environments is constrained in large part by water salinity (Dunson and Mazzotti 1989; Kinneary 1993; Hirayama 1998; Brischoux et al. 2012). Blood and tissue osmolality of marine reptiles is typically one-third the osmolality of seawater. In other words, the body fluids of marine reptiles are more dilute than the water in which they live. In this desiccating environment, behavioral and physiological adjustments are necessary to prevent water loss and combat excessive influx of salts. Compensatory strategies are those that involve active (i.e., energy-requiring) uptake or extrusion of water or salts, whereas evasive strategies employ behavioral or physiological adjustments to reduce the passive exchange of water and salts between the organism and its environment (Kirschner 1970).

Regulation of the volume and composition of body fluids is critically important, as the aqueous solution bathing molecules and cells comprises the immediate environment in which biochemical reactions and the processes necessary for life take place. Dissolved solutes that contribute significantly to the osmolality of body fluids and are regulated to preserve physiological function are referred to as osmolytes (Hochachka and Somero 2002). Osmolytes include inorganic ions (Na^+, Cl^-, K^+, Ca^{2+}, Mg^{2+}) and organic molecules (amino acids, carbohydrates, urea, trimethylamine N-oxide [TMAO]). Certain types of osmolytes may interact with and affect the structure and function of macromolecules. Additionally, the total concentration of osmolytes in body fluids is an important determinant of water movement between an organism and its environment and the maintenance of cell volume. Osmotic and ionic regulation are integral to preserving physiological function at the molecular, cellular, and organismal levels.

The invasion of marine habitats occurred independently in several lineages of reptiles. Initial exploitation of brackish and marine habitats may have been made possible by morphological features inherent to reptiles in general, along with simple behavioral adjustments (Dunson and Mazzotti 1989). Passive exchange of water and salts between reptiles and their environment is typically low due to the reduced permeability of their integument (see "Strategies for Osmoregulation" below) (Dunson 1986; Lillywhite 2006). Reptiles that are tolerant of high salinity and exploit marine habitats have lower rates of water exchange across the integument compared with freshwater species (Bentley and Schmidt-Nielsen 1970; Robinson and Dunson 1976; Dunson and Seidel 1986); thus, alterations in integument permeability may have enhanced the ability of transitional forms to take advantage of resources in estuarine or coastal environments.

Preference for drinking water may also play a role in maintaining osmotic balance for transitional forms. A preference for drinking freshwater or dilute brackish water rather than seawater appears to be an important component of water balance for filesnakes (*Acrochordus granulatus*; Lillywhite and Ellis 1994) and sea kraits (*Laticauda colubrina, L. laticaudata, L. semifasciata*; Lillywhite et al. 2008), which use estuarine and marine environments. Estuarine-adapted species of crocodylians (*Crocodylus porosus, C. johnstoni*) alter drinking behavior in response to salinity fluctuations, such that drinking ceases at salinities of 18 to 20 (Kuchel and Franklin 1998; Taplin et al. 1999). Turtles that use brackish water (*Batagur borneoensis, B. baska, Malaclemys terrapin*) also show a preference for drinking water at lower salinities and do not drink full-strength seawater (Dunson and Moll 1980; Davenport and Wong 1986; Davenport and Macedo 1990).

The reptile kidney is not capable of generating concentrated urine (Hildebrandt 2001), but many species of estuarine and marine reptiles have modified cranial (cephalic) salt glands that facilitate excretion of excess salt and contribute to regulation of body fluid osmolality. The integral role of salt glands in permitting the expansion of reptiles into the marine environment is highlighted by the fossil record, which shows that the presence of cephalic salt glands preceded the appearance of highly modified limbs in the ancient marine turtles (Hirayama 1998). Extrarenal mechanisms for salt excretion have arisen multiple times in different reptile lineages (Schmidt-Nielsen and Fänge 1958). Excretion of excess salts is facilitated in marine iguanas (*Amblyrhynchus cristatus*) by nasal salt glands (Shoemaker and Nagy 1984), in marine and estuarine snakes (*Pelamis* spp., *Laticauda* spp., *A. granulatus*) by sublingual glands (Dunson 1979), in estuarine crocodiles (*C. porosus*) by lingual salt glands (Taplin and Grigg 1981), and in sea turtles (Cheloniidae spp., *Dermochelys coriacea*) and diamond-backed terrapins (*M. terrapin*) by modified lachrymal glands (Schmidt-Nielsen and Fänge 1958; Holmes and McBean 1964; Dunson 1970). The secretory capacity of the salt gland varies greatly among marine and estuarine reptile species, with species that are most tolerant of high salinities possessing the most effective salt glands (Dunson 1970; Marshall and Cooper 1988; Brischoux et al. 2012). In sea snakes, for example, species with higher maximum rates of salt gland secretion exploited habitats with higher maximum annual salinity and had broader geographic distributions (Brischoux et al. 2012). The powerful lachrymal salt glands of sea turtles permit secretion of large amounts of excess salts ingested as the result of foraging on invertebrate prey items and purposefully drinking seawater (Marshall and Cooper 1988; Reina et al. 2002). The estuarine diamond-backed terrapin, by contrast, does not drink seawater, and its salt glands appear to be used primarily for ridding the body of salt loads resulting from prey ingestion and subsequent long-term accumulation (Dunson and Mazzotti 1989).

The combination of morphological features, behaviors, and physiological adjustments made by a marine reptile to regulate water and salt balance depends on the evolutionary history of the species and the osmotic and ionic challenges presented by its immediate environment. In this chapter we describe in detail the osmoregulatory strategies of the estuarine terrapin and consider this information in the light of strategies employed by other turtles and other species of marine reptiles. This comparative approach permits us to assess how terrapins survive in the dynamic estuarine environment and provides insight as to why the terrapin is the only temperate species of turtle that lives all of its life in coastal estuaries.

The Estuarine Environment

The estuarine salt marsh is a dynamic ecotone between the oceanic and terrestrial and freshwater habitats. It is a physiologically rigorous abiotic regime that has resulted in low species richness and high productivity when compared with adjacent freshwater systems (Dunson and Travis 1994). It is not surprising, therefore, that this coastal ecosystem is used by organisms with unique behavioral, morphological, and physiological traits that permit them to tolerate and exploit this nutrient-rich environment (Chan et al. 2006). Estuarine-colonizing organisms can serve as models for gaining insights into the evolution of physiological specialization (Dunson and Travis 1994).

A limited number of aquatic turtles periodically use tidally influenced brackish water habitats with salinities less than 15 (Dunson and Moll 1980; Dunson and Seidel 1986; Kinneary 1993; Lee et al. 2006), but only a few species are known to spend the majority of their life in these habitats. Three of these species are found in tropical estuarine systems of Southeast Asia (*B. borneoensis*, *B. baska*, and *Orlitia borneensis*; Davenport and Wong 1986; Hart and Lee 2006). The diamond-backed terrapin is the only turtle endemic to temperate estuarine ecosystems with salinities ranging from 11 to full-strength seawater (Hart and Lee 2006). This North American emydid turtle occurs exclusively in tidally influenced coastal salt marshes, bays, lagoons, mud and grass flats, and creeks, ranging from Cape Cod, Massachusetts, to Corpus Christi, Texas (Ernst and Lovich 2009). *Malaclemys* is a monotypic genus, and morphological and molecular data provide evidence that *Malaclemys* and *Graptemys*, a genus that includes 12 species of pre-

dominantly riverine turtles, are sister taxa (Lamb and Osentoski 1997; Stephens and Wiens 2003). The two genera, thought to have diverged from a common freshwater ancestor 7 to 11 million years ago, in the late Miocene, share notable life history characteristics such as molluscivory and pronounced sexual dimorphism (Lamb and Osentoski 1997; Ernst and Lovich 2009). *Graptemys*, however, is strictly limited to freshwater environments, whereas *Malaclemys* can exploit waters with a broad range of salinities. For more on the evolutionary history and taxonomy of the diamond-backed terrapin, see Chapters 3 and 4.

Diamond-backed terrapins are able to inhabit the euryhaline salt marsh ecosystem by using a suite of osmoregulatory strategies to control salt and water exchange with the environment and actively regulate body fluid composition. These strategies include behavioral modifications, physiological mechanisms, and morphological adaptations, all of which help to maintain plasma osmotic pressure of one-third to one-half that of environmental water (Gilles-Baillien 1970). The associated energetic cost of these different osmoregulatory strategies ranges from relatively cheap (e.g., avoidance of seawater ingestion; mud burial) to relatively expensive (Na^+/K^+ ATPase pump function; salt gland excretion).

Strategies for Osmoregulation

Integument

Integument (skin and shell) permeability is generally low in reptiles, and this morphological feature minimizes the passive exchange of water and salts between reptiles and their environment (Lillywhite 2006). Turtle integument (epidermis and scales) is made up of multiple layers of α-keratin and β-keratin proteins, the latter of which form thick, mechanically resistant corneous layers that may be mineralized with calcium (Lillywhite 2006). Although integument of many freshwater turtles has relatively low permeability to Na^+, highly aquatic freshwater species can have higher, more variable rates of integumentary water exchange (Dunson 1986; Dunson and Mazzotti 1989). Rates of water loss across the integument of terrapins are much lower than rates documented for freshwater turtles. Fasted terrapins have rates of weight loss, due primarily to water loss (see Bentley et al. 1967), of about 0.3% to 0.6% of mass per day when exposed to seawater, whereas the freshwater turtles *Chrysemys picta* and *Apalone spinifer* experience weight loss of 2.7% to 4.8% of mass per day during seawater exposure (Bentley and Schmidt-Nielsen 1970; Robinson and Dunson 1976; Davenport and Magill 1996). Robinson and Dunson (1976) and Dunson and Mazzotti (1989) described the skin of terrapins as being virtually impermeable to Na^+ and demonstrated that the major route of Na^+ influx was through the soft tissues of the oral cavity. Although rates of water efflux and salt influx across the integument are relatively low for terrapins, prolonged exposure to seawater results in a gradual but significant increase in plasma Na^+ and Cl^- concentrations and weight loss indicative of dehydration (Robinson and Dunson 1976). Periodic access to freshwater is seen as an integral component of osmoregulation in this species (Bentley et al. 1967; Dunson 1970; Dunson and Mazzotti 1989; Davenport and Macedo 1990).

Body Water Balance

Turtles that have a distribution limited to freshwater show poor discrimination in drinking-water preference, whereas species that use brackish water environments show a preference for drinking water with low salinity (Bentley et al. 1967). Laboratory studies and observations of captive terrapins provide strong evidence that they do not drink full-strength seawater but will readily ingest low-salinity (≤20) water or freshwater if given the opportunity to do so (Bentley et al. 1967; Dunson 1970; Cowan 1981; Davenport and Macedo 1990). Terrapins acclimated to seawater under laboratory conditions respond rapidly to vibrations caused by simulated rainfall and are capable of drinking from thin films of freshwater that collect on exposed substrata or on the surface of seawater (as little as 1-6 mm thick; Fig. 9.1) (Davenport and Macedo 1990; Bels et al. 1995). Several lines of evidence support the proposition that terrapins are capable of storing large amounts of ingested water

in their extracellular fluid so that they may draw on this water source during periods of low environmental water availability. Colloquial and scientific sources have noted that terrapins with regular access to freshwater sources exhibit swelling around the base of the limbs (Dunson 1970; Robinson and Dunson 1976), an outwardly visible sign of water storage. Studies of water balance have illustrated that intracellular fluid volume decreases to a lesser extent than extracellular fluid volume during periods of prolonged seawater exposure (73-124 days); furthermore, terrapins injected with an NaCl solution but allowed access to freshwater exhibit a substantial increase in extracellular fluid volume but no concurrent change in intracellular fluid volume compared with freshwater controls (Robinson and Dunson 1976). Taken together, these observations provide evidence that terrapins take advantage of temporally available freshwater sources to replenish extracellular water stores and subsequently use these water stores to maintain constant intracellular fluid volumes during periods of dehydration. Regulation of cell volume in this manner would help ensure that intracellular functions are preserved over a broad range of environmental salinities and precipitation levels.

Diamond-backed terrapins primarily eat mollusks and crustaceans that are isosmotic with the surrounding water (Tucker et al. 1995) and therefore incur considerable salt loads during feeding. During periods of low rainfall when access to freshwater is limited, ingestion of prey may present terrapins with a large osmotic challenge. As illustrated by Davenport and Ward (1993), the appetite of terrapins declined dramatically during exposure to full-strength seawater (34 salinity) with no access to freshwater. After 18 days of seawater exposure, terrapins consumed only 22% to 54% of rations they had fully consumed when freshwater was available. Suppression of appetite at higher salinity levels has also been reported by Dunson (1985) and Holliday et al. (2009). Furthermore, these investigators documented maximum gain in mass at 9 to 10 salinity and a decreasing rate of mass gain as salinity increased above this level. The reduction in feeding of terrapins at higher salinities

Fig. 9.1. Drinking posture employed by diamond-backed terrapins when drinking from thin water films. Note subsidiary meniscus over the snout and small size of the oral aperture. Adapted from Davenport and Macedo (1990), with permission

may reduce the metabolic costs associated with the removal of salt acquired during feeding. Thus, hypophagy appears to be an important contributor to water and salt balance during periods of low freshwater availability. The metabolic consequences of sacrificing energy intake to maintain osmotic balance are worth further investigation.

When preformed drinking water in the environment becomes scarce, terrapins may employ antidiuresis to retain water. Bentley et al. (1967) observed that terrapins produce dilute urine during periods of freshwater exposure, while seawater exposure induces a 5-fold decrease in urine production with a proportionate increase in urine osmotic concentration. Subsequent studies have also noted a much higher mean urine osmotic pressure in seawater-acclimated animals (372–482 mOsm) compared with freshwater-acclimated animals (107–136 mOsm) (Gilles-Baillien 1970; Robinson and Dunson 1976). Urine K^+ concentrations are significantly higher in seawater-acclimated than in freshwater-acclimated terrapins, but there is no difference between groups in urine Na^+ concentrations, which are at very low trace levels (Robinson and Dunson 1976). Furthermore, a positive relationship exists between plasma Na^+ concentration and urine K^+ concentration (Robinson and Dunson 1976). Taken together, these observations indicate that terrapins are capable of a post-renal asymmetric active exchange of Na^+ (reabsorbed) and K^+ (secreted) to promote water retention. Schmidt-Nielsen et al. (1963) suggested that animals with salt glands may reabsorb Na^+ in the cloaca as a mechanism of reclaiming water before excretion and subsequently secrete the Na^+ via the salt gland to maintain osmotic balance. Pidcock et al. (1997) provided support for the existence of such a mechanism in the estuarine crocodile *C. porosus*. The role of antidiuretic hormones such as arginine vasotocin in promoting water retention has not been investigated in terrapins, but research on this topic could provide us with valuable information on endocrine control of osmotic balance in this species.

Solute Transport

Overall volume and composition of body fluids is regulated in large part by controlling the movement and concentration of osmolytes. Mechanisms for solute transport to maintain a gradient between intracellular and extracellular fluid compartments or between body fluids and the external environment require hydrolysis of ATP, thus coming at an energetic cost to the animal. Accumulation of osmolytes in the body fluid of vertebrates is a commonly used mechanism to permit water retention under desiccating environmental conditions. Organic osmolytes are well suited for accumulation at high concentration and are less likely to interfere with cellular machinery and molecular function than are inorganic ions (Hochachka and Somero 2002). Indeed, elevation of intracellular ion concentrations above typical levels can have inhibitory effects on metabolic enzyme activity and protein synthesis (Yancey et al. 1982; Yancey 1994; Hochachka and Somero 2002). In contrast, many types of polyols (glycerol, sorbitol), amino acids and derivatives (serine, glycine, proline, alanine, taurine), and methylamines (betaine) do not have any appreciable effect on cellular function, even when accumulated in high concentrations (Clark and Zounes 1977; Bowlus and Somero 1979). Organic osmolytes that are generally nonperturbing to protein structure and function are referred to as compatible osmolytes (Brown 1978). Other organic osmolytes, such as TMAO, have a stabilizing effect on protein structure that can enhance function (Yancey and Somero 1979; Yancey 1994).

Unlike other organic osmolytes accumulated in response to osmotic stress, the nitrogenous waste product urea has a disruptive effect on protein structure and inhibits enzyme function (Hand and Somero 1982; Yancey 1994; Fuery et al. 1997; Cowan and Storey 2002). Urea, a highly soluble nitrogenous waste product of protein catabolism (Dessauer 1970), is used as the primary blood and tissue osmolyte to maintain osmotic balance in a variety of vertebrate species adapted to high-salinity environments (Griffith 1991). In this case, the accumulation of urea is accompanied by accumulation of compatible or stabilizing organic osmolytes that offset the inhibitory effects of urea on cellular function (Yancey and Somero 1979; Yancey 1994; Withers and Guppy 1996; Treberg et al. 2006). During low water availability and/or high salinity exposure in reptiles, urea may accumulate in blood and tissue due to a combination of increased urea synthesis, decreased renal excretion, and facilitated uptake of urea via urea transporters (Gilles-Baillien 1973; Shoemaker and Nagy 1977; Balinsky 1981).

The few studies that have investigated blood and tissue osmolyte levels in response to salinity in estuarine turtles have provided evidence that urea increases during exposure to progressively higher levels of salinity. Blood and tissue urea content in Chinese softshell turtles (*Pelodiscus sinensis*) exposed to water of 15 salinity for 6 days increased 4- to 7-fold over the control freshwater urea levels (Lee et al. 2006). The increase in tissue urea content

was accompanied by a moderate, 1.5- to 2-fold increase in tissue levels of total free amino acids. Based on blood and tissue urea levels and urea excretion rates for *P. sinensis*, Lee et al. (2006) concluded that the increase in blood and tissue urea during exposure to 15 salinity was partially due to increased urea synthesis, not simply a consequence of urea retention and reabsorption in the bladder.

Laboratory studies show that blood urea levels in terrapins vary depending on whether the turtles are held in freshwater (21 mmol/L) or seawater (115 mmol/L) and that increases in blood osmolality in response to increasing levels of salinity are due primarily to the accumulation of urea (Gilles-Baillien 1970). Urea also accumulates in muscle tissue of terrapins during seawater exposure, along with a concurrent, significant increase in the compatible osmolyte taurine (Gilles-Baillien 1973). Terrapins do not alter the concentration of inorganic ions and amino acids in muscle tissue during the transition from freshwater to seawater (Gilles-Baillien 1973). Plasma urea levels in terrapins acclimated to 17 salinity (30 mmol/L) were considerably higher than levels found in *P. sinensis* at 15 salinity (12 mmol/L) (Lee et al. 2006), and plasma urea levels of terrapins acclimated to full-strength seawater under laboratory conditions are commonly >100 mmol/L (Gilles-Baillien 1970). The relative importance of organic osmolytes in the context of other osmotic strategies for terrapins is not known, but evidence from laboratory studies suggests that regulation of blood and tissue urea levels may play a role in water balance, under conditions of both changing salinity and prolonged mud burial during winter dormancy.

A laboratory study conducted by Gilles-Baillien (1973) on the osmotic status of dormant terrapins found that, in animals held in seawater tanks, blood osmotic pressure gradually increased over the course of winter dormancy and peaked just before the onset of activity in the spring. Urea accumulation in terrapins combined with stable or decreasing inorganic ion concentrations during the winter study suggested that the increase in blood osmolality resulted from the use of urea as an osmoeffector (Table 9.1) (Gilles-Baillien 1973). How-

Table 9.1. Mean monthly mass and plasma concentrations of blood variables in 10 female terrapins, November 2011 to April 2012 (gray rows), and blood variables measured in controlled laboratory experiments on hibernation and osmoregulation in terrapins (white rows)

Month	N	Monthly mass (g)	Plasma osmolality (mOsm)	Plasma level (mmol/L)			
				Urea	Na$^+$	K$^+$	Cl$^-$
Nov 2011	8	529.3±34.4	331.0±8.0	34.2±2.9	145.0±1.4	2.7±0.1	105.1±2.2
Nov	4–6	—	—	85.8±24.4	179.3±19.9	2.9±0.6	134.0±27.3
Dec 2011	9	543.7±27.6	335.4±8.5	37.5±2.3	145.1±1.5	2.6±0.1	103.7±1.8
Dec	3–4	—	—	83.6±29.7	162.7±17.2	2.4±0.3	149.1±16.6
Jan 2012	10	538.0±22.9	345.7±9.1	44.91±3.25	152.8±2.8	2.7±0.1	108.8±3.5
Feb 2012	10	563.0±25.1	332.3±8.9	40.64±3.7	146.6±2.4	2.6±0.1	106.2±2.7
Feb	3	—	—	80.7±11.2	152.4±13.5	4.2±1.1	138.5±3.5
Mar 2012	10	537.9±24.7	318.1±6.1	37.0±3.4	146.0±1.3	2.7±0.1	109.1±2.3
Apr 2012	8	478.1±20.8	329.1±8.3	39.2±3.7	142.9±1.2	3.5±0.1	108.4±1.1
Apr	4–5	—	—	95.5±23.1	199.6±14.3	3.0±0.6	150.5±13.6

Source: Data in gray rows from Harden et al (2015); data in white rows from Gilles-Baillien (1973).
Note: All values are ±SE.

ever, Gilles-Baillien's (1973) exploration of winter osmotic regulation provided terrapins only with aquatic overwintering habitats, whereas free-ranging animals are typically observed buried in intertidal or subtidal mud during dormancy (Fig. 9.2) (Yearicks et al. 1981; Haramis et al. 2011; Harden and Williard 2012; Harden et al. 2015).

In addition to Gilles-Baillien's 1973 study, many of the earlier laboratory studies were conducted on captive animals kept in tanks with constant and/or regulated salinity conditions of freshwater, 50% seawater (~17 salinity), or full-strength seawater (35 salinity) (Bentley et al. 1967; Dunson 1970; Gilles-Baillien 1970, 1973; Robinson and Dunson 1976; Cowan 1981). These stable conditions resulted in progressive (i.e., linear) changes in osmolytes throughout the laboratory-defined winter. Without the option of muddy substrate in which to overwinter, behavioral control over osmotic balance is reduced. More recent experimental field studies conducted by Harden et al. (2015) in North Carolina, in which female adult terrapins were placed in an open-air salt marsh enclosure from October through April, showed little change in overall blood osmolality and both inorganic and organic osmolyte levels (Table 9.1). Furthermore, Harden

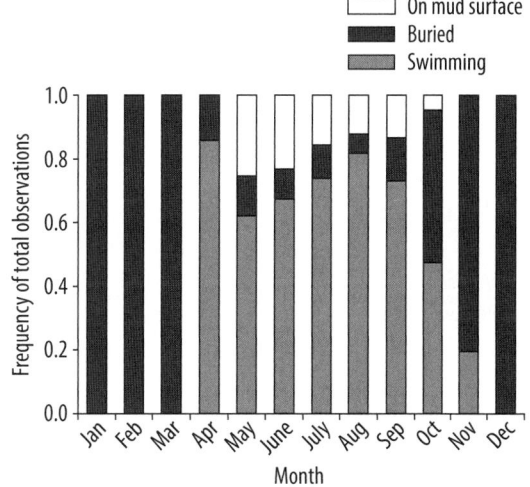

Fig. 9.2. Monthly habitat use and behavior of terrapins at Figure 8 Island and Masonboro Island, North Carolina, documented as frequency of total observations via radio-telemetry. The terrapin's active season of April 1 to September 30 was determined from these data and dormancy entrance/emergence dates. Data from Harden (2013)

et al. (2015) documented plasma urea concentrations lower than those of controlled laboratory studies (Gilles-Baillien 1973; Table 9.1). Differences in osmolyte concentrations and the pattern of the overwintering response between laboratory and field studies may reflect the degree to which terrapins are able to employ behavioral means to maintain osmotic balance. Results obtained from animals in their natural environment suggest that they may be relying heavily on behavioral mechanisms (e.g., habitat use, hypophagy) to control rates of water and salt exchange and regulate body fluid composition and volume while overwintering.

Behavior and Habitat Use

Diamond-backed terrapins inhabit a complex environment of highly variable temperature, salinity, and precipitation. As challenging as the abiotic conditions in the estuarine salt marsh may be, this environment also presents terrapins with multiple options for behavioral means of osmoregulation. Tidal and seasonally influenced shifts between aquatic and terrestrial habitats play an important role in water and salt balance for this species (Coker 1906; Yearicks et al. 1981; Spivey 1998; Butler 2002; Harden et al. 2007; Haramis et al. 2011; Southwood Williard and Harden 2011; Harden and Williard 2012). During the active season, terrapins typically forage on invertebrate prey in shallow, flooded marshes at high tide (Tucker et al. 1995; Whitelaw and Zajac 2002) and either retreat to deeper waters or bury in the mud of the intertidal zone during low tide (Spivey 1998; Harden et al. 2007). Temperature data logger studies show that terrapins buried in exposed mud of the intertidal zone experience very broad fluctuations in temperature, whereas terrapins that remain in water experience a more stable thermal environment (Harden et al. 2007). Given the very broad range of temperatures (<0°C to +40°C) that terrapins are exposed to during mud burial (Southwood Williard and Harden 2011), it is difficult to explain terrestrial habitat use from a solely thermoregulatory perspective. Captive animals maintained in seawater spend a higher proportion of time on basking platforms when deprived of freshwater, even when air temperature (14°C–19°C) is considerably lower than seawater temperature (25°C) and there is no direct source of light above the platform (Davenport and Magill 1996). Rates of water loss in air are greater than rates of water loss in seawater if the air has a relative humidity of <85%; however, it is possible that reduced salt intake and continued excretion of salt via the cloaca and salt glands while on land could provide a more effective means of regulating osmotic pressure of body fluids than might be possible in seawater (Davenport and Magill 1996). Moreover, terrapins often choose muddy salt marsh substratum habitat and thus may experience high relative humidity levels (98%–100%) that could decrease rates of water exchange and further augment their ability to conserve water stores during periods of osmotic stress (Davenport and Magill 1996). Harden et al. (2007) and Harden and Williard (2012) found that terrapins in the Carolinas often bury themselves in the viscous muddy or even sandy substrate of the intertidal zone, which may allow them to create burrows with a small, humid airspace that would reduce water loss.

A shift to upland and intertidal mud habitats may be particularly advantageous during colder winter months, when physiological processes are slowed due to Q_{10} effects and metabolic downregulation (Southwood Williard and Harden 2011), making water and salt balance more difficult to maintain actively (Gilles-Baillien 1973). Radio-telemetry studies in southeastern North Carolina revealed that from October through March, terrapins overwinter in shallow intertidal mud, making only occasional short-distance movements (Fig. 9.2) (Harden and Williard 2012; Harden 2013; Harden et al. 2015). These studies indicated that during the winter, terrapin inactivity (i.e., burial in intertidal mud habitat) and thermal stability were important predictors of blood osmolality and concentration of osmolytes such as Na^+, Cl^-, urea, and uric acid, while abiotic factors such as rainfall and salinity were not good predictors of blood chemical variables. In other words, behavioral modifications appear to be an important component of the osmoregulatory strategy of overwintering terrapins. Burrowing behavior and cessation of feeding could reduce rates of water and salt exchange with the environment and maintain osmotic balance during the winter months. Indeed, stable isotope studies with terrapins show that rates of daily body water flux (as % of total body water [TBW] per day) are considerably lower during winter mud burial (10.52% ± 2.92% TBW per day) than in active periods (21.84% ± 7.30% TBW per day; Harden et al. 2014). Further studies are needed to clarify whether terrapins function as a "closed system" during dormancy (i.e., no salt intake via ingestion of food or drink, and no salt excretion via kidney or salt gland) or are able to maintain the function of ion-regulatory mechanisms (e.g., activation of the kidney and the lachrymal salt gland) during cold exposure; the latter strategies require an energy investment.

The Salt Gland

The reptile kidney is not capable of generating urine that is hyperosmotic to body fluids (Hildebrandt 2001), but many species of estuarine and marine reptiles have cephalic salt glands that facilitate excretion of excess salt and contribute to regulation of fluid osmolality (see the introduction to this chapter). The anatomy of tetrapod salt glands varies among lineages, but the basic functional unit is the secretory tubule. Multiple secretory tubules form a secretory lobule and open into central ductules, which lead to primary ducts and, finally, to main ducts that transport the secreted fluid (Shuttleworth and Hildebrandt 1999; Babonis and Brischoux 2012). The general salt gland ultrastructure of marine birds and reptiles consists of cells with extensive lateral membrane folding, extensive blood capillary networks, abundant mitochondria, and innervation by parasympathetic fibers (Cramp et al. 2007). From here, the ultrastructure becomes more taxon specific.

The structure and mechanism of salt-secreting glands have been best studied in the avian nasal salt gland, the elasmobranch rectal gland, and the teleost chloride cell (Babonis et al. 2009), but there is evidence to indicate that similar mechanisms are at work in the sea snake's posterior sublingual gland and the turtle's lachrymal gland (Shuttleworth and Hildebrandt 1999). Salt glands have the following three key elements that distinguish them from other glands: (1) a compound tubular shape, which probably increases secretory surface area; (2) the presence of ion-transporting proteins within secretory epithelium, specifically the coexpression of a basolateral Na^+/K^+ ATPase (NKA) enzyme pump and a basolateral $Na^+/K^+/Cl^-$ cotransport (NKCC) with an apical Cl^- channel (the cystic fibrosis transmembrane conductance regulator in elasmobranchs and birds); and (3) cell-type homogeneity within the secretory epithelium (Shuttleworth and Hildebrandt 1999; Babonis and Evans 2011). Activation of salt glands results in production of a hypertonic NaCl solution via the active transport of ions. The NKA pump, on the basolateral membrane, utilizes the energy released from hydrolysis of ATP to pump three Na^+ ions out of the secretory cell and two K^+ ions in. The resulting transmembrane electrochemical gradient for Na^+ drives the transport of Cl^- into the cell via the NKCC protein and thus accumulation of Cl^- in the cell. High intracellular concentrations of Cl^- drive the passive diffusion of this ion into the tubule lumen through Cl^- channels on the apical membrane (Silva et al. 1977). The action of the NKA pump also generates localized extracellular regions of high Na^+ concentration, which drives the passive diffusion of Na^+ into the tubule lumen via cation-specific paracellular channels (Shuttleworth and Hildebrandt 1999).

Of estuarine turtles, only the terrapin seems to possess a lachrymal salt gland (Dunson 1969, 1970; Cowan 1981). A combination of salt-secreting and mucus-secreting cells makes up the secretory tubules (Cowan 1969) in terrapin salt glands, a conformation similar to that of the nasal glands of some desert lizards (van Lennep and Komnick 1970). Cowan (1971) tested the hypothesis that the terrapin lachrymal gland, when compared with that of other, closely related stenohaline (i.e., freshwater) emydids (*Graptemys*, *Pseudemys*, and *Chrysemys* spp.), would be the only gland associated with salt secretion. He used electron microscopy and histochemical staining to determine that the ultrastructure of the lachrymal gland was similar among all stenohaline emydids and served a mucoserous function. In contrast, the lachrymal gland of the terrapin was much more complex, with increased basal and lateral cell surface area, longer microvilli, underlying connective tissue, higher density of mitochondria, and innervation (Cowan 1971; Belfry and Cowan 1995); all of these features suggest that the lachrymal gland of terrapins could secrete inorganic electrolytes (e.g., Na$^+$ and Cl$^-$).

The neural and hormonal mechanisms underlying salt gland activation have not been thoroughly investigated in terrapins, but presumably there are similarities with the salt glands of birds and other reptiles. In birds, an increase in plasma Na$^+$ concentration triggers cephalic osmoreceptors (Gerstberger et al. 1984), which leads to stimulation of the salt glands via the autonomic nervous system (Fänge et al. 1958). Anatomical and physiological studies of the terrapin salt gland provide evidence to suggest that ion secretion by the salt gland is primarily under adrenergic and peptidergic control (Cowan 1990; Belfry and Cowan 1995). Circulating levels of steroid hormones influence rates of ion secretion by salt glands in other reptiles (Bentley 2002), so studies to clarify the role of hormones in control of salt gland function in terrapins are warranted.

Diamond-backed terrapins are capable of withstanding prolonged periods (many months) of dehydration in full-strength seawater (Gilles-Baillien 1970), and allow plasma Na$^+$ to reach high levels ($\geq$200 mmol/L) before fully activating their salt gland (Dunson and Dunson 1975). Brischoux et al. (2013) observed hypernatremia (elevated blood Na$^+$ levels) in wild-caught sea kraits (laticaudines) and suggested that tolerance of high plasma Na$^+$ levels and delayed activation of salt glands would decrease the energetic costs associated with osmoregulation. A sufficient supply of ATP to power the NKA transport pumps located in the chloride cells of the gland's epithelial tissue must be available for the gland to effectively secrete highly concentrated salt solutions (Borut and Schmidt-Nielsen 1963; Whittam 1963; Bentley et al. 1967; Shuttleworth and Hildebrandt 1999; Reina et al. 2002). Dunson and Dunson (1975) found that the activity of NKA in salt glands collected from terrapins acclimated to seawater for 90 to 101 days was 1.9 to 3.2 μmol · mg wet wt^{-1} · h^{-1}, whereas NKA activity in freshwater-acclimated animals was 1.3 μmol · mg wet wt^{-1} · h^{-1}. Seawater-acclimated terrapins injected with an NaCl solution to raise plasma Na$^+$ levels above 200 mmol/L had NKA activity of 2.4 to 5.8 μmol · mg wet wt^{-1} · h^{-1} in salt gland tissue, a 3-fold increase above levels observed for freshwater-acclimated animals. Unfed terrapins acclimated to various levels of salinity (10-35) show very low levels of salt gland secretions, but an increase in salt gland secretions occurs rapidly in response to salt loading at all salinity levels (Cowan 1981). In terms of concentration and rate of secretions, the terrapin salt gland is intermediate in secretory capacity between marine and terrestrial reptile species (Table 9.2) (Dunson 1969, 1970; Cowan 1981, 1990; Lutz and Musick 1997).

The overall energetic cost associated with activation of the salt gland in terrapins has not been investigated in great detail. Bentley et al. (1967) reported that the oxygen consumption ($\dot{V}O_2$) of terrapins maintained for 10 days in a solution with osmolarity equivalent to that of seawater was twice the $\dot{V}O_2$ of terrapins maintained in tap water,

Table 9.2. Maximum rates of electrolyte secretion by reptilian salt glands

Species	Habitat	Secretion rates (μmol · 100 g^{-1} · h^{-1})			Reference
		Na$^+$	Cl$^-$	K$^+$	
False iguana (*Ctenosaura*)	Desert	1.0	—	9.4	Templeton 1964
Chuckwalla (*Sauromalus*)	Desert	3.3	—	31.0	Templeton 1964
Swollen-nosed side-blotched lizard (*Uta*)	Intertidal	21.7	26.9	10.7	Hazard et al. 1998
Diamond-backed terrapin (*Malaclemys*)	**Estuarine**	**31.0**	**—**	**1.6**	**Cowan 1990**
Dog-faced water snake (*Cerberus*)	Mangrove	27.5	—	—	Dunson & Dunson 1975
Banded sea snake (*Laticauda*)	Marine	72.6	73.7	3.3	Dunson & Taub 1967
Yellow-bellied sea snake (*Pelamis*)	Marine	218.0	169.0	9.2	Dunson 1968
Marine iguana (*Amblyrhynchus*)	Marine	255.0	237.0	50.7	Dunson 1969
Green sea turtle (*Chelonia*)	Marine	415.0	—	—	Reina & Cooper 2000
Leatherback sea turtle (*Dermochelys*)	Marine	484.0	—	—	Reina et al. 2002

but they expressed doubt that this large difference in $\dot{V}O_2$ could be explained solely by an increase in active transport associated with salt excretion. Other factors, including metabolic responses to alterations in osmolyte levels in other tissues and induction of endocrine responses, may also contribute to changes in organismal metabolism during exposure to different environmental salinities (Bentley et al. 1967). Bentley et al. (1967) did not report activity level for terrapins during the metabolic trials, and variation in activity level could explain some portion of the observed effect on $\dot{V}O_2$. Holliday et al. (2009) reported that terrapins acclimated to 0, 10, 20, or 30 salinity over 6 months (with freshwater exposure every 30 days) exhibited no significant changes in metabolic rate that would indicate elevated energetic costs associated with osmoregulation in high-salinity environments. It is important to note that in Holliday et al.'s study, terrapins in higher-salinity treatments exhibited slower growth rates upon initial exposure to high salinity and achieved smaller body size over the course of the experiment, compared with terrapins in lower-salinity treatments. Taken together, these findings suggest that terrapins may be able to sustain stable levels of energy expenditure over a broad range of salinities by making modifications in energy allocation (i.e., increase energy expended on osmoregulation but decrease energy expended on growth). The metabolic cost associated with maintenance and activation of the salt gland in terrapins is a topic of great interest, given the perceived importance of the salt gland for hydro-mineral balance and the potential energetic consequences of employing this mechanism of salt excretion. While the salt gland is clearly an important adaptation that contributes to terrapins' ability to exploit high-salinity environments, morphological features and behavioral strategies appear to be equally important for maintaining osmotic balance (Dunson and Seidel 1986).

Concluding Thoughts

The diamond-backed terrapin is one of a few species of truly estuarine reptiles and, as such, represents an intermediate form bridging the divide between freshwater and seawater. An understanding of the osmoregulatory strategies employed by terrapins provides insight into the relative importance of behavioral and physiological mechanisms for colonization of estuarine ecosystems and the energetic consequences of exploiting these highly dynamic environments. Simple alterations in behavior, such as selectivity with regard to drinking water and shifts between aquatic and terrestrial habitats, as well as energy-consuming processes related to solute transport, contribute to osmoregulation in terrapins. The balance between behavioral and physiological approaches to osmoregulation and the abiotic and biotic factors that might affect this balance are topics worthy of further investigation, as there are important implications for the energetics and ecology of this species. Energy requirements and patterns of allocation have consequences for individual organisms as well as for population characteristics (growth rates, body size, reproduction; Congdon et al. 1982; Dunham et al. 1989; Chapter 6). Additional research into the energetic cost of osmoregulation in terrapins could help clarify the trade-offs involved in exploitation of brackish and marine habitats and explain geographic variation in terrapin life history patterns and population structure. Furthermore, behavioral and physiological responses to environmental factors such as temperature and salinity are key components underlying seasonal activity patterns. A knowledge of seasonal alterations in aquatic and terrestrial habitat use may contribute to conservation efforts for this species, particularly for preventing incidental capture of terrapins in fishing gear. Trends in terrapin abundance and demography observed by long-term studies in South Carolina (Dorcas et al. 2007) and Maryland (Roosenburg et al. 1997) suggest that bycatch mortality of terrapins resulting from the use of commercial and recreational crab pots may be contributing to shifts in the structure and status of terrapin populations, ultimately causing population declines (Chapter 16). An understanding of terrapin habitat use and the factors contributing to habitat selection can advance efforts to determine the degree of overlap with crab-fishing operations and options for time- and area-based management measures.

REFERENCES

Babonis, L.S. and F. Brischoux. 2012. Perspectives on the convergent evolution of tetrapod salt glands. Integrative and Comparative Biology 52:245-256.

Babonis, L.S. and D.H. Evans. 2011. Morphological and biochemical evidence for the evolution of salt glands in snakes. Comparative Biochemistry and Physiology A 160:400-411.

Babonis, L.S., K.A. Hyndman, H.B. Lillywhite and D.H. Evans. 2009. Immunolocalization of Na+/K+ ATPase and Na+/K+/2Cl− cotransporter in the tubular epithelia of sea snake salt glands. Comparative Biochemistry and Physiology A 154:535-540.

Balinsky, J.N. 1981. Adaptation of nitrogen metabolism to hyperosmotic environment in Amphibia. Journal of Experimental Biology 215:335-350.

Belfry, C.S. and F.B.M. Cowan. 1995. Peptidergic and adrenergic innervation of the lachrymal gland in the euryhaline turtle, *Malaclemys terrapin*. Journal of Experimental Zoology 273:363-375.

Bels, V.L., J. Davenport and S. Renous. 1995. Drinking and water expulsion in the diamondback turtle *Malaclemys terrapin*. Journal of Zoology 236:483-497.

Bentley, P.J. 2002. Endocrines and Osmoregulation: A Comparative Account in Vertebrates. Springer, Berlin. 292 pages.

Bentley, P.J., W.L. Bretz and K. Schmidt-Nielsen. 1967. Osmoregulation in the diamondback terrapin, *Malaclemys terrapin centrata*. Journal of Experimental Biology 46:161-167.

Bentley, P.J. and K. Schmidt-Nielsen. 1970. Comparison of water exchange in two aquatic turtles, *Trionyx spinifer* and *Pseudemys scripta*. Comparative Biochemistry and Physiology 32:363-365.

Borut, A. and K. Schmidt-Nielsen. 1963. Respiration of avian salt-secreting gland in tissue slice experiments. American Journal of Physiology 204:573-581.

Bowlus, R.D. and G.N. Somero. 1979. Solute compatibility with enzyme function and structure: rationales for the selection of osmotic agents and end-products of anaerobic metabolism in marine invertebrates. Journal of Experimental Biology 208:137-151.

Brischoux, F., M.J. Briand, G. Billy and X. Bonnet. 2013. Variations of natremia in sea kraits (*Laticauda* spp.) kept in seawater and fresh water. Comparative Biochemistry and Physiology A 166:333-337.

Brischoux, F., R. Tingley, R. Shine and H.B. Lillywhite. 2012. Salinity influences the distribution of marine snakes: implications for evolutionary transitions to marine life. Ecography 35:994-1003.

Brown, A.D. 1978. Compatible solutes and extreme water stress in eukaryotic micro-organisms. Advances in Microbial Physiology 17:181-242.

Butler, J.A. 2002. Population ecology, home range, and seasonal movements of the Carolina diamondback terrapin, *Malaclemys terrapin centrata*, in northeastern Florida. Florida Fish and Wildlife Conservation Commission Final Report. Tallahassee, FL. 65 pages.

Chan, Y.L., C.E. Hill, J.E. Maldonado and R.C. Fleischer. 2006. Evolution and conservation of tidal-marsh vertebrates: molecular approaches. Studies in Avian Biology 32:54-75.

Clark, M.E. and M. Zounes. 1977. The effects of selected cell osmolytes on the activity of lactate dehydrogenase from the euryhaline polychaete, *Nereis succinea*. Biological Bulletin 153:468-484.

Coker, R.E. 1906. The natural history and cultivation of the diamond-back terrapin with notes on other forms of turtles. North Carolina Geological Survey Bulletin 14:1-67.

Congdon, J.D., A.E. Dunham and D.W. Tinkle. 1982. Energy budgets and life histories of reptiles. Pages 233-271 in C. Gans and F.H. Pough (eds.), Biology of the Reptilia. Volume 13. Physiology D. Physiological Ecology. Academic Press, New York, NY.

Cowan, F.B.M. 1969. Gross and microscopic anatomy of orbital glands of *Malaclemys* and other emydine turtles. Canadian Journal of Zoology 47:723-729.

Cowan, F.B.M. 1971. The ultrastructure of the lachrymal "salt" gland and the Harderian gland in the euryhaline *Malaclemys* and some closely related stenohaline emydines. Canadian Journal of Zoology 49:691-697.

Cowan, F.B.M. 1981. Effects of salt loading on the salt gland function in the euryhaline turtle, *Malaclemys terrapin*. Journal of Comparative Physiology 145:101-108.

Cowan, F.B.M. 1990. Does the lachrymal salt gland of *Malaclemys terrapin* have a significant role in osmoregulation? Canadian Journal of Zoology 68:1520-1524.

Cowan K.J. and K.B. Storey. 2002. Urea and KCl have different effects on enzyme activities in liver and muscle of estivating versus nonestivating species. Biochemical Cell Biology 80:745-755.

Cramp, R.L., N.J. Hudson, A. Holmberg, S. Holmgren and C.E. Franklin. 2007. The effects of saltwater acclimation on neurotransmitters in the lingual salt glands of the estuarine crocodile, *Crocodylus porosus*. Regulatory Peptides 140:55-64.

Davenport, J. and E.A. Macedo. 1990. Behavioural osmotic control in the euryhaline diamondback terrapin *Malaclemys terrapin*: responses to low salinity and rainfall. Journal of Zoology 220:487-496.

Davenport, J. and S.H. Magill. 1996. Thermoregulation or osmotic control? Some preliminary observations on the function of emersion in the diamondback terrapin *Malaclemys terrapin* (Latreille). Herpetological Journal 6:26-29.

Davenport, J. and J.F. Ward. 1993. The effects of salinity and temperature on appetite in the diamondback terrapin, *Malaclemys terrapin* (Latreille). Herpetological Journal 3:95-98.

Davenport, J. and T.M. Wong. 1986. Observations on the water economy of the estuarine turtles *Batagur baska* (Gray) and *Callagur borneoensis* (Schlegel and Muller). Comparative Biochemistry and Physiology 84A:703-707.

Dessauer, H.C. 1970. Blood chemistry of reptiles: physiological and evolutionary aspects. Pages 1-72 in C. Gans and T.S. Parsons (eds.), Biology of the Reptilia. Volume 3. Morphology C. Academic Press, New York, NY.

Dorcas, M.E., J.D. Willson and J.W. Gibbons. 2007. Crab trapping causes population decline and demographic changes in diamondback terrapins over two decades. Biological Conservation 137:334-340.

Dunham, A.E., B.W. Grant and K.L. Overall. 1989. Interfaces between biophysical and physiological ecology and the population ecology of terrestrial vertebrate ectotherms. Physiological Zoology 62:335-355.

Dunson, M.K. and W.A. Dunson. 1975. The relationship between plasma Na$^+$ concentration and salt gland Na-K ATPase content in the diamondback terrapin and the yellow-bellied sea snake. Journal of Comparative Physiology 101:89-97.

Dunson, W.A. 1968. Salt gland secretion in the pelagic sea snake *Pelamis*. American Journal of Physiology 215:1512-1517.

Dunson, W.A. 1969. Electrolyte excretion by the salt gland of the Galapagos marine iguana. American Journal of Physiology 216:995-1002.

Dunson, W.A. 1970. Some aspects of electrolyte and water balance in three estuarine reptiles, the diamondback terrapin, American and "salt water" crocodiles. Comparative Biochemistry and Physiology 32:161-174.

Dunson, W.A. 1979. Control mechanisms in reptiles. Pages 273-322 in R. Gilles (ed.), Mechanisms of Osmoregulation in Animals. Wiley Interscience, New York, NY.

Dunson, W.A. 1985. Effect of water salinity and food salt content on growth and sodium efflux of hatchling diamondback terrapins (*Malaclemys*). Physiological Zoology 58:736-747.

Dunson, W.A. 1986. Estuarine populations of the snapping turtle (*Chelydra*) as a model for the evolution of marine adaptations in reptiles. Copeia 1986:741-756.

Dunson, W.A. and F.J Mazzotti. 1989. Salinity as a limiting factor in the distribution of reptiles in Florida Bay: a theory for the estuarine origin of marine snakes and turtles. Bulletin of Marine Science 44:229-244.

Dunson, W.A. and E. Moll. 1980. Osmoregulation in sea water of hatchling emydid turtles, *Callagur borneoensis*, from a Malaysian sea beach. Journal of Herpetology 14:31-36.

Dunson, W.A. and M.E. Seidel. 1986. Salinity tolerance of estuarine and insular emydid turtles (*Pseudemys nelson* and *Trachemys decussata*). Journal of Herpetology 20:237-245.

Dunson, W.A. and A.M. Taub. 1967. Extrarenal salt excretion in sea snakes (*Laticauda*). American Journal of Physiology 213:975-982.

Dunson, W.A. and J. Travis. 1994. Patterns in the evolution of physiological specialization in salt marsh animals. Estuaries 17:102-110.

Ernst, C.H. and J.E. Lovich. 2009. Turtles of the United States and Canada. 2nd edition. Johns Hopkins University Press, Baltimore, MD. 827 pages.

Fänge, R., K. Schmidt-Nielsen and H. Osaki. 1958. The salt gland of the herring gull. Biological Bulletin 115:162-171.

Fuery, C.J., P.V. Attwood, P.C. Withers, P.H. Yancey, J. Baldwin and M. Guppy. 1997. Effects of urea on M4-lactate dehydrogenase from elasmobranchs and urea-accumulating Australian desert frogs. Comparative Biochemistry and Physiology B 117:143-150.

Gerstberger, R., S. Oppermann and R. Kaul. 1984. Cephalic osmoreceptor control of salt gland activation and inhibition in the salt adapted duck. Journal of Comparative Physiology B 154:449-456.

Gilles-Baillien, M. 1970. Urea and osmoregulation in the diamondback terrapins *Malaclemys centrata centrata* (Latreille). Journal of Experimental Biology 52:691-697.

Gilles-Baillien, M. 1973. Hibernation and osmoregulation in the diamondback terrapin *Malaclemys centrata centrata* (Latreille). Journal of Experimental Biology 59:45-51.

Griffith, R.W. 1991. Guppies, toadfish, lungfish, coelacanths and frogs: a scenario for the evolution of urea retention in fishes. Environmental Biology of Fishes 32:199-218.

Hand, S.C. and G.N. Somero. 1982. Urea and methylamine effects on rabbit muscle phosphofructokinase. Journal of Biological Chemistry 257:734-741.

Haramis, G.M, P.P.F. Henry and D.D. Day. 2011. Using scrape fishing to document diamondback terrapins in hibernacula in Chesapeake Bay. Herpetological Review 42:170-177.

Harden, L.A. 2013. Seasonal variation in ecology and physiology of diamondback terrapins (*Malaclemys terrapin*) in North Carolina. PhD dissertation, University of North Carolina Wilmington, Wilmington, NC. 151 pages.

Harden, L.A., N.A. DiLuzio, M.E. Dorcas and J.W. Gibbons. 2007. The spatial ecology and thermal ecology of diamondback terrapins (*Malaclemys terrapin*) in a South Carolina salt marsh. Journal of the North Carolina Academy of Sciences 123:154-162.

Harden, L.A., K.A. Duernberger, T.T. Jones and A.S. Williard. 2014. Total body water and water turnover rates in the estuarine diamondback terrapin (*Malaclemys terrapin*) during the transition from dormancy to activity. Journal of Experimental Biology 217:4406-4413.

Harden, L.A., S.R. Midway and A.S. Williard. 2015. The blood biochemistry of overwintering diamondback terrapins (*Malaclemys terrapin*). Journal of Experimental Marine Biology and Ecology 466:34-41.

Harden, L.A. and A.S. Williard. 2012. Using spatial and behavioral data to evaluate the seasonal bycatch risk of diamondback terrapins *Malaclemys terrapin* in crab pots. Marine Ecology Progress Series 467:207-217.

Hart, K.M. and D.S. Lee. 2006. The diamondback terrapin: the biology, ecology, cultural history, and conservation status of an obligate estuarine turtle. Studies in Avian Biology 32:206-213.

Hazard, L.C., V.H. Shoemaker and L.L. Grismer. 1998. Salt gland secretion by an intertidal lizard, *Uta tumidarostra*. Copeia 1998:231-234.

Heatwole, H. 1999. Sea Snakes. Krieger Publishing Company, Malabar, FL. 148 pages.

Hildebrandt, J.P. 2001. Coping with excess salt: adaptive functions of extrarenal osmoregulatory organs in vertebrates. Zoology 104:209-220.

Hirayama, R. 1998. Oldest known sea turtle. Nature 392:705-708.

Hochachka, P.W. and G.N. Somero. 2002. Biochemical Adaptation: Mechanism and Process in Physiological Evolution. Oxford University Press, New York, NY. 480 pages.

Holliday, D., A. Elskus and W.M. Roosenburg. 2009. Impacts of multiple stressors on growth and metabolic rate of *Malaclemys terrapin*. Environmental Toxicology and Chemistry 28:338-345.

Holmes, W.N. and R.L. McBean. 1964. Some aspects of electrolyte excretion in the green turtle, *Chelonia mydas mydas*. Journal of Experimental Biology 41:81-90.

Kinneary, J.J. 1993. Salinity relations of *Chelydra serpentina* in a Long Island estuary. Journal of Herpetology 27:441-446.

Kirschner, L.B. 1970. The study of NaCl transport in aquatic animals. American Zoologist 10:365-376.

Kuchel, L.J. and C.E. Franklin. 1998. Kidney and cloaca function in the estuarine crocodile (*Crocodylus porosus*) at different salinities: evidence for solute-linked water uptake. Comparative Biochemistry and Physiology A 119:825-831.

Lamb, T. and M.F. Osentoski. 1997. On the paraphyly of *Malaclemys*: a molecular genetic assessment. Journal of Herpetology 31:258-265.

Lee, S.M.L., W.P. Wong, K.C. Hiong, A.M. Loong, S.F. Chew and Y.K. Ip. 2006. Nitrogen metabolism and excretion in the aquatic Chinese soft-shelled turtle, *Pelodiscus sinensis*, exposed to a progressive increase in ambient salinity. Journal of Experimental Zoology 305A:995-1009.

Lillywhite, H.B. 2006. Water relations of tetrapod skin. Journal of Experimental Biology 209:202-226.

Lillywhite, H.B., L.S. Babonis, C.M. Sheeny III and M.C. Tu. 2008. Sea snakes (*Laticauda* spp.) require fresh drinking water: implication for the distribution and persistence of populations. Physiological and Biochemical Zoology 81:785-796.

Lillywhite, H.B. and T.M. Ellis. 1994. Ecophysiological aspects of the coastal-estuarine distribution of acrochordid snakes. Estuaries 17:53-61.

Lutz, P.L. and J.A. Musick. 1997. The Biology of Sea Turtles. CRC Press, Boca Raton, FL. 432 pages.

Marshall, A.T. and P.D. Cooper. 1988. Secretory capacity of the lachrymal salt gland of hatchling sea turtles, *Chelonia mydas*. Journal of Comparative Physiology 157:821-827.

Pidcock, S., L.E. Taplin and G.C. Grigg. 1997. Differences in renal-cloacal function between *Crocodylus porosus* and *Alligator mississippiensis* have implications for crocodilian evolution. Journal of Comparative Physiology B 167:153-158.

Reina, R.D. and P.D. Cooper. 2000. Control of salt gland activity in the hatchling green sea turtle, *Chelonia mydas*. Journal of Comparative Physiology B 170:27-35.

Reina, R.D., T.T. Jones and J.R. Spotila. 2002. Salt and water regulation by the leatherback sea turtle *Dermochelys coriacea*. Journal of Experimental Biology 205:1853-1860.

Robinson, G.D. and W.A. Dunson. 1976. Water and sodium balance in the estuarine diamondback terrapin (*Malaclemys*). Journal of Comparative Physiology 105:129-152.

Roosenburg, W.M., W. Cresko, M. Modesitte and M.B. Robbins. 1997. Diamondback terrapin (*Malaclemys terrapin*) mortality in crab pots. Conservation Biology 11:1166-1172.

Schmidt-Nielsen, K., A. Borut, P. Lee and E. Crawford, Jr. 1963. Nasal salt excretion and the possible function of the cloaca in water conservation. Science 142:1300-1301.

Schmidt-Nielsen, K. and R. Fänge. 1958. Salt glands in marine reptiles. Nature 182:783-785.

Shoemaker, V.H. and K.A. Nagy. 1977. Osmoregulation in amphibians and reptiles. Annual Review of Physiology 39:449-471.

Shoemaker, V.H. and K.A. Nagy. 1984. Osmoregulation in the Galapagos marine iguana, *Amblyrhynchus cristatus*. Physiological Zoology 57:291-300.

Shuttleworth, T.J. and J.P. Hildebrandt. 1999. Vertebrate salt glands: short- and long-term regulation of function. Journal of Experimental Zoology 283:689-701.

Silva, P., R. Solomon, K. Spokes and F.H. Epstein. 1977. Ouabain inhibition of gill Na-K-ATPase: relationship to active chloride transport. Journal of Experimental Zoology 199:419-426.

Southwood Williard, A. and L.A. Harden. 2011. Seasonal changes in thermal environment and metabolic enzyme activity in the diamondback terrapin (*Malaclemys terrapin*). Comparative Biochemistry and Physiology A 158:477-484.

Spivey, P.B. 1998. Home range, habitat selection, and diet of the diamondback terrapin (*Malaclemys terrapin*) in a North Carolina estuary. MS thesis, University of Georgia, Athens, GA. 80 pages.

Stephens, P.R. and J.J. Wiens. 2003. Explaining species richness from continents to communities: the time for speciation effect in emydid turtles. American Naturalist 161:112-128.

Taplin, L.E. and G.C. Grigg. 1981. Salt glands in the tongue of the estuarine crocodile, *Crocodylus porosus*. Science 212:1045-1047.

Taplin, L.E., G.C. Grigg, L.A. Beard and T. Pulsford. 1999. Osmoregulatory mechanisms of the Australian freshwater crocodile, *Crocodylus johnstoni*, in freshwater and estuarine habitats. Journal of Comparative Physiology B 169:215-233.

Templeton, J.R. 1964. Nasal salt excretion in terrestrial lizards. Comparative Biochemistry and Physiology 11:223-229.

Treberg, J. R., B. Speers-Roesch, P. Piermarini, Y.K. Ip, J.S. Ballantyne and W.R. Driedzic. 2006. The accumulation of methylamine counteracting solutes in elasmobranchs with differing levels of urea: comparison of marine and freshwater species. Journal of Experimental Biology 209:860-870.

Tucker, A.D., N. Fitzsimmons and J.W. Gibbons. 1995. Resource partitioning by the estuarine turtle, *Malaclemys terrapin*: trophic, spatial, and temporal foraging constraints. Herpetologica 51:167-181.

van Lennep, E.W. and H. Komnick. 1970. Fine structure of the nasal gland in the desert lizard *Uromastyx acanthinurus*. Cytobiologie 2:47-67.

Whitelaw, D.M. and R.N. Zajac. 2002. Assessment of prey availability for diamondback terrapins in a Connecticut salt marsh. Northeastern Naturalist 9:407-418.

Whittam, R. 1963. The interdependence of metabolism and active transport. Pages 139-154 in J.F. Hoffman (ed.), The Cellular Function of Membrane Transport. Prentice Hall, Upper Saddle River, NJ.

Withers, P.C. and M. Guppy. 1996. Do Australian desert frogs co-accumulate counteracting solutes with urea during aestivation? Journal of Experimental Biology 199:1809-1816.

Yancey, P.H. 1994. Compatible and counteracting solutes. Pages 81-110 in K. Strange (ed.), Cellular and Molecular Physiology of Cell Volume Regulation. CRC Press, Boca Raton, FL.

Yancey, P.H., M.E. Clark, S.C. Hand, R.D. Bowlus and G.N. Somero. 1982. Living with water stress: evolution of osmolyte systems. Science 217:1214-1222.

Yancey, P.H. and G.N. Somero. 1979. Counteraction of urea destabilization of protein structure by methylamine osmoregulatory compounds of elasmobranch fishes. Biochemistry Journal 182:317-323.

Yearicks, E.F., R.C. Wood and W.S. Johnson. 1981. Hibernation of the northern diamondback terrapin, *Malaclemys terrapin terrapin*. Estuaries 4:78-81.

10

Temperature-Dependent Sex Determination

THANE WIBBELS

TAYLOR ROBERGE

ALLEN R. PLACE

The diamond-backed terrapin (*Malaclemys terrapin*) possesses temperature-dependent sex determination (Jeyasuria et al. 1994; Roosenburg and Kelley 1996; Jeyasuria and Place 1997; Burke and Calichio 2014). Temperature-dependent sex determination (TSD) can produce a wide variety of sex ratios, ranging from equality to highly biased (Mrosovsky et al. 1984; Mrosovsky 1994; Wibbels 2003), and therefore has significant implications for a species' ecology, evolution, and conservation (Mrosovsky and Yntema 1980; Roosenburg 1996; Shine 1999; Wibbels 2003). For example, from ecological and evolutionary viewpoints, sex ratios can affect reproductive success (Fisher 1930; Janzen 1995; Kvarnemo and Ahnesjo 1996); from a conservation perspective, sex ratios can affect the recovery rate of endangered species (Coyne and Landry 2007). Further, TSD is of particular interest in considering the effects of global climate change, which could alter sex ratios and the reproductive ecology of species with TSD (Janzen 1994; Hawkes et al. 2007; Schwanz and Janzen 2008; Mitchell and Janzen 2010). In this chapter we provide an overview of TSD relative to the biology and conservation of the diamond-backed terrapin.

Sex Determination in Amniotic Vertebrates

A variety of sex determination systems occurs in amniotic vertebrates (Bull 1981), including male heterogamety in mammals (e.g., XX/XY), female heterogamety in birds (e.g., ZZ/ZY), and TSD (Marshall Graves and Shetty 2001; Marshall Graves and Peichel 2010). All three types of sex determination systems are represented among reptiles (Janzen and Krenz 2004). Of evolutionary interest, many ancestral groups of extant reptiles (crocodilians, most turtles, and tuataras) possess TSD, suggesting that it represents a basal form of sex determination in amniotic vertebrates. Consequently, TSD probably gave rise to genotypic sex determination (GSD) in other amniotic vertebrate groups, including birds and mammals (Janzen and Paukstis 1991; Marshall Graves and Shetty 2001; Janzen and Krenz 2004). However, the phylogenetic distribution of TSD and GSD in reptiles reveals that closely related species can differ in their sex determination mechanism (Janzen and Krenz 2004). For example, the lizard *Gekko japonicus* has TSD, whereas *G. gecko* has GSD (Solleder and Schmid 1984; Tokunaga 1985; Janzen and Krenz 2004; Pokorna and Kratochvíl 2009). Thus, TSD and GSD in reptiles may be similar, and Sarre et al. (2004) proposed that they should be viewed not as distinctly different systems but as a continuum. Several studies suggest that differences between TSD and GSD relate to the mechanism(s) controlling a dosage-dependent effect from sex gene(s) (Quinn et al. 2007, 2011; Holleley et al. 2015) and that changes in dosage control mechanisms for those gene(s) can occur over a relatively short evolutionary time (Ezaz et al. 2009; Holleley et al. 2015). Genetic and phylogenetic studies of TSD and GSD in reptiles suggest that transitions from TSD to GSD, or vice versa, have occurred numerous times in reptiles (Janzen and Krenz 2004; Quinn et al. 2007, 2011; Ezaz et al. 2009; Holleley et al. 2015).

Phylogenetic analysis of TSD among reptiles indicates that the majority of turtle species have TSD, but some have GSD (Bull et al. 1985; Ewert and Nelson 1991; Janzen and Paukstis 1991; Ewert et al. 1994; Janzen and Krenz 2004). In the family Emydidae, which includes the diamondbacked terrapin, almost all species examined have TSD and show a similar thermosensitive pattern of sex determination (Ewert and Nelson 1991; Janzen and Paukstis 1991; Ewert et al. 1994; Janzen and Krenz 2004). One exception, *Clemmys insculpta*, has GSD and with its congener, *C. guttata*, which has TSD, exemplifies the relatedness and transitions between GSD and TSD (Bull et al. 1985; Ewert and Nelson 1991).

The TSD Reaction Norm in Terrapins

The terrapin has a male:female, or type Ia (Ewert and Nelson 1991), TSD reaction norm: cool temperatures produce males, and warm temperatures produce females (Jeyasuria et al. 1994; Roosenburg and Kelley 1996; Burke and Calichio 2014). Eggs collected from terrapin populations in Maryland (Jeyasuria et al. 1994) and New York (Burke and Calichio 2014) were incubated in constant-temperature laboratory incubators. The study by Jeyasuria et al. (1994) found that constant incubation temperatures of 26°C and 27°C produced all males, 29.5°C produced a mixed sex ratio (13% male), and 31°C and 32°C produced all females. Burke and Calichio (2014) used 10 different temperatures ranging from 24°C to 34°C and analyzed the results using TSD curve-fitting software (Girondot 1999; Godfrey et al. 2003). They estimated the "pivotal temperature" (temperature producing a 1:1 sex ratio) to be 28.16°C and the "transitional range of temperatures" (TRT; temperature range from maximal temperature producing 100% males to minimal temperature producing 100% females) to be 25.88°C to 30.45°C. They also used the TSD curve-fitting software to analyze the data from Jeyasuria et al. (1994), which used eggs from Maryland, and estimated a pivotal temperature of 29.25°C and TRT of 28.85°C to 29.64°C. Collectively, these data indicate that the estimated pivotal temperatures for the terrapin (28.2°C for New York study site, 29.3°C for Maryland) are within the ranges of those reported for other emydid turtles. However, 28.2°C is lower than most other reported pivotal temperatures (Bull et al. 1982b; Etchberger et al. 1991; Ewert and Nelson 1991; Wibbels 2003; Burke and Calichio 2014), and Burke and Calichio's results suggest a substantial difference in pivotal temperatures between New York and Maryland.

Geographic variation in pivotal temperatures has been suggested in other emydid turtles (Wibbels et al. 1991c) and requires further investigation; it may be associated with the genetic structure observed along the terrapin's broad range (Hauswalt and Glen 2005; Hart et al. 2014; Converse et al. 2015). However, because the studies by Jeyasuria et al.

(1994) and Burke and Calichio (2014) were conducted at different laboratories at different times, differences in experimental protocol (e.g., incubators, humidity) could have contributed to the reported differences in pivotal temperatures. Further, clutch effects on pivotal temperature have been reported for emydid turtles (Bull et al. 1982a; Ewert et al. 1994; Rhen and Lang 1998; Dodd et al. 2006).

Collectively, these studies indicate that terrapins show a M:F pattern of TSD, with pivotal temperatures that fall within the range reported for other emydid turtles. The wide latitudinal distribution of the terrapin makes it an ideal species for evaluating intraspecific variation in pivotal temperature and TRT.

Embryonic Staging

Earlier studies determining the thermosensitive period (TSP) for turtles used the staging characters described by Yntema (1968). Jeyasuria (1997) compared the morphological developmental stages of terrapins with those of the snapping turtle (*Chelydra serpentina*) at various days of incubation and found minor staging differences; he then reevaluated the morphology of embryonic development in the terrapin. He monitored development by candling eggs and dissecting three eggs approximately twice a week to verify specific developmental stages, based on criteria described for the snapping turtle (Yntema 1968). Shown in Fig. 10.1 are growth curves at 26.5°C and 30.5°C, based on embryo wet mass during a two year study. As expected, terrapin embryos grow faster at female-producing temperatures than male-producing temperatures, and there is a high concordance between years for each temperature. A plot of stage vs. days of incubation is shown in Fig. 10.2, with the corresponding regression curve for staging based on days of incubation. The TSP is shaded, on the basis of data obtained for the red-eared slider (*Trachemys scripta*; Wibbels et al. 1991a). Greenbaum's (2002) more recent comparison of the embryonic stages of *T. scripta* with those of *C. serpentina* (Yntema 1968) highlights the importance of fore-

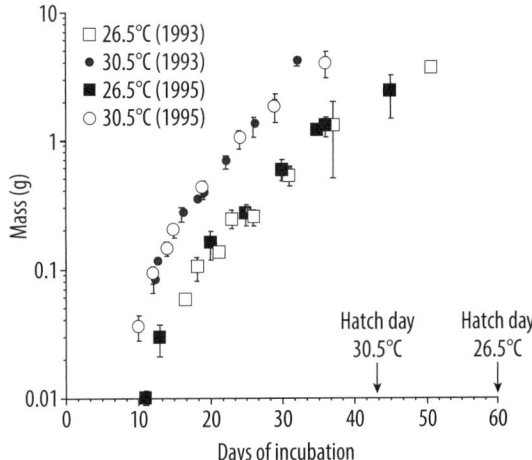

Fig. 10.1. Terrapin embryonic growth shown as embryo wet mass vs. days of incubation. The study used two incubation temperatures, 26.5°C (a male-producing temperature) and 30.5°C (a female-producing temperature). Eggs from two different nesting seasons (1993 and 1995) were used in the study.

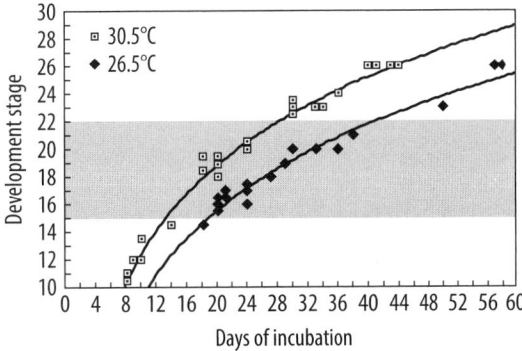

Fig. 10.2. Terrapin embryonic growth shown as embryonic stage vs. days of incubation. The study used two incubations temperature, 26.5°C (a male-producing temperature) and 30.5°C (a female-producing temperature). Embryonic staging was based on Yntema (1968).

limb and claw morphology in staging. A key observation by Jeyasuria (1997) for the terrapin was that the forelimb is longer and wider at stage 13 than stage 12, with the beginning of the paddle-shaped forelimb. This is a stage earlier than described for *T. scripta* (Greenbaum 2002). The remaining characters that are descriptive for later stages are consistent between the terrapin and *T. scripta*.

Thermosensitive Period of Sex Determination

The TSP of sex determination has been examined in a number of turtles, typically using constant-temperature incubators. In those experiments, eggs were shifted from male-producing to female-producing incubation temperatures (or vice versa) at various stages of embryonic development. Sex determination is sensitive to temperature during the approximate middle third of embryonic development (Yntema 1979; Bull and Vogt 1981; Pieau and Dorrizi 1981; Yntema and Mrosovsky 1982; Bull 1987; Webb et al. 1987; Wibbels et al. 1991a; Merchant-Larios et al. 1997; Rhen et al. 2015).

In terrapins, Burke and Calichio (2014) determined the TSP using two sets of shift experiments. In the first set of experiments, they incubated eggs at a baseline male-producing temperature (25°C), then shifted the eggs to a female-producing temperature (31°C) for 7-day periods at various days of incubation. There was temperature sensitivity between days 28 and 35, representing an incubation time ranging from 40.3% to 50.4% of total incubation time. Embryos at the beginning of the shift (day 28) were at stage 16 (Greenbaum 2002), based on data for the emydid turtle *T. scripta*. This stage is typically reported to be near the early portion of the TSP in other emydid turtles. Burke and Calichio (2014) conducted a second set of experiments in which they used a female-producing temperature baseline (31°C) and shifted to a male-producing temperature (25°C) for 7-day periods. They found temperature sensitivity between days 30 and 35, which represented 69.4% to 80.0% of total incubation time. Evaluation of embryos indicated that they were at approximately stage 23 when shifted on day 30, a stage typically later than normal TSPs reported for emydid turtles. In both temperature shift experiments (25°C and 31°C baselines), only one 7-day period in each of the entire series of shifts produced a significant change in sex ratios. A 7-day period represents significantly less than one-third of total embryonic development at both 25°C and 31°C incubation temperatures, and therefore, depending on when the shift occurred, the 7-day period could have been affecting the sex

determination mechanism but may not have been sufficient to cause sex reversal. This could account for the contrast between these results and the longer TSPs reported in studies of other turtles, as noted above.

The findings by Burke and Calichio (2014) provide important insight into the temporal aspects of temperature sensitivity, but future studies may be necessary to confirm or refine data on the TSP in terrapins. Such studies should try a variety of temperature-shift protocols that include a range of shift durations and shift temperatures, because both factors can affect sex reversal (Wibbels et al. 1991a).

Natural Incubation Temperature and Modeling of Nest Sex Ratios

Initial studies of TSD in turtles used constant incubation temperatures to characterize parameters such as pivotal temperature, TRT, and TSP (Mrosovsky and Pieau 1991). However, under natural conditions, turtle eggs are exposed to fluctuating temperatures during incubation, including daily cycles and seasonal weather-related changes. Therefore, many researchers have attempted to evaluate and model the effects of fluctuating temperatures on TSD (Valenzuela 2001; Georges et al. 2004; Delmas et al. 2008; Bowden et al. 2014). Although the effects of fluctuating incubation temperature have not been studied in the terrapin, information from other reptiles can provide a general model of such effects on TSD. We review studies that provide a framework for designing and evaluating future studies of TSD in the terrapin under natural fluctuating temperature regimes. An understanding of the effects of fluctuating incubation temperatures on TSD is a prerequisite for accurately estimating hatchling sex ratios based on natural nest temperatures.

Fluctuating incubation temperature has a range of effects on the sex of turtles with TSD, depending on the mean temperature, magnitude of temperature fluctuation, and whether the range of temperatures exceeds those at which embryonic development can occur. Painted turtle (*Chrysemys picta*) and red-eared slider eggs incubated under temperatures fluctuating around a mean of 28.5°C ± 3.0°C produced significantly more females than did eggs incubated at a constant 28.5°C (Les et al. 2007). In general, as the magnitude of fluctuation increases around male-producing or pivotal temperatures, a greater proportion of females are produced in species with pattern Ia TSD (Bull 1985; Georges et al. 1994; Paitz et al. 2010; Neuwald and Valenzuela 2011). Increasing incubation temperature fluctuations around a female-producing mean (31.0°C ± 5.0°C) resulted in the production of mostly male hatchlings (Neuwald and Valenzuela 2011). It is hypothesized that the embryos were experiencing a decreased developmental rate at extreme temperatures, which enhanced male sex determination (Georges et al. 2005; Les et al. 2009). Thus, in nests with relatively large temperature fluctuations, it may be difficult to predict sex ratios based on information extrapolated from constant-temperature incubations. Developing a model that can accurately predict the sex ratio could describe the complex effects of fluctuating incubation temperature on the underlying mechanisms of TSD. Additionally, such a model could be used as a nonlethal method for predicting hatchling sex ratios in natural nests of species of conservation concern.

Historically, a simple method was used to predict and describe the outcome of temperature on sex. The mean temperature during the TSP was compared with TSD reaction norms, informed by data on constant incubation temperature (Marcovaldi et al. 1997; Hanson et al. 1998; Godfrey et al. 1999). However, nests experiencing relatively large fluctuations in temperature did not always produce sex ratios as predicted from mean incubation temperature during the TSP (Pieau 1982). Several authors suggested that the effect of temperature on developmental rate may explain the difference between predicted and observed sex ratios in naturally incubated nests (Pieau 1982; Bull 1985; Georges 1989). In general, embryonic developmental rate is positively associated with temperature, within limits (Yntema 1968; Wibbels et al. 1991a; Les et al. 2009). As such, an embryo spending an equal proportion of time above and below pivotal temperature would undergo proportionally more embryonic development at temperatures above

pivotal. Thus, it was hypothesized that the best predictor of sex is the proportion of embryonic development accumulated above vs. below the pivotal temperature (Georges 1989).

Georges (1989) developed the constant temperature equivalent (CTE) value, based on a degree-hour model, which represents the constant incubation temperature that would produce the same observed sex ratios as in nests undergoing sinusoidal temperature fluctuations around a stationary mean. Support for the degree-hour model was found through reanalysis of data from field-incubated *Graptemys* nests (Bull 1985) that experienced temperature variance and sex ratios that were not explained by mean temperature alone (Georges 1989). Georges et al. (1994) tested this model in *Caretta caretta* and found that the model was more accurate than mean temperature in predicting nest sex ratios. However, the degree-hour approach was not without its limitations. Some assumptions of the model may be violated by irregular fluctuations in diel temperature over the incubation period, and the model may perform poorly when nest temperatures are extreme (De Souza and Vogt 1994; Valenzuela et al. 1997; for a discussion, see Georges et al. 2004).

Curvilinear models that more accurately describe the relationship between temperature and embryonic development have increased the accuracy of the degree-hour model (Georges et al. 2005); however, the parameters may be difficult to estimate, and a more simplistic "critical thermal maximum" may be used (Telemeco et al. 2013). Delmas et al. (2008) developed a mechanistic model of TSD for *Emys orbicularis* that was based on underlying physiological processes of TSD and did not rely on sinusoidal variation around a stationary mean. Limitations of this model remain, however, due to lack of information on the effect of extreme temperatures on developmental rates and the physiology of TSD outside the range of nonlethal constant incubation temperatures required for successful embryogenesis.

Although these models were relatively accurate at predicting sex ratios, all are still based on data from constant incubation temperatures for the species of interest. These data are not always readily available, can be time consuming to acquire, and require the sampling (i.e., dissection) of many individuals at multiple temperatures. As an alternative, Girondot and Kaska (2014) developed and tested a model to predict the thermal reaction norm of embryonic growth in turtles incubated under natural conditions. Their model uses time-series temperature data recorded from naturally incubated nests, between the date of oviposition and date of hatching, to model embryonic growth for the species of interest. The resulting thermal reaction norm of embryonic growth can more accurately determine the middle third of development (i.e., TSP) in natural nests. The "embryogrowth" model is available as an R package (http://cran.r-project.org/web/packages/embryogrowth). Future studies should validate this model for terrapin nest temperatures by determining hatchling sex ratios from a subset of nests to compare with the model's predictions. The accuracy of all these models is dependent on understanding the effect of incubation temperature on embryonic development and sex determination. Future studies that provide a better understanding of the physiological processes involved in TSD could enhance the accuracy of the models.

Physiological and Genetic Basis

Place and Lance (2004) summarized the known genes of the mammalian sex determination/differentiation network and their orthologs in reptiles (see also Rhen and Schroeder 2010). Ten years later, genome sequencing of several turtles (Shaffer et al. 2013; Wang et al. 2013) identified new potential genes involved in the TSD cascade. Previous studies addressed the physiological basis of TSD in turtles and provided insights into genes implicated in the physiological cascades leading to either ovarian or testicular differentiation. However, a comprehensive understanding of the molecular mechanisms underlying TSD remains elusive. Mammals have a sex-determining gene (*SRY*), but there is no homolog to the *SRY* gene in reptiles or birds (Marshall Graves and Shetty 2001; Marshall Graves and Peichel 2010). Due to the limited number of TSD studies conducted on the terrapin, the following is

a brief review of genes and hormones implicated as putative components of the physiological cascades underlying TSD and sex differentiation in turtles. The review also provides a conservative framework for planning future studies to investigate the molecular basis of TSD and sex differentiation in the terrapin.

Many of the initial studies of the physiology underlying TSD focused on the effects and roles of steroid hormones. Subsequently, a wide variety of studies addressed candidate sex determination genes involved in the molecular basis of TSD. Researchers have long hypothesized that female-producing temperature initiates the production of endogenous estrogen in the developing gonad, and the estrogen stimulates ovarian differentiation (Bull et al. 1988; reviewed by Pieau and Dorizzi 2004; Yao and Capel 2005; Shoemaker and Crews 2009). This hypothesis was initially developed due to a series of studies in which turtle eggs treated with exogenous estrogen developed as females even though incubated at male-producing temperatures (Raynaud and Pieau 1985; Crews et al. 1989; Dorizzi et al. 1991, Wibbels et al. 1991b; Jeyasuria et al. 1994). Those studies were followed by experiments in which turtle eggs, including those of the terrapin, were treated with drugs that blocked aromatase (the enzyme that catalyzes production of estradiol), and these produced male embryos (Crews and Bergeron 1994; Dorizzi et al. 1994; Rhen and Lang 1994; Wibbels and Crews 1994; Richard-Mercier et al. 1995; Place et al. 2001). Several studies also recorded higher aromatase activity and/or higher estrogen levels in gonads or adrenal-kidney-gonad complexes from embryonic turtles incubated at female-producing temperatures (Dorizzi et al. 1991; Desvages and Pieau 1992; Desvages et al. 1993; Place et al. 2001). Additionally, temperature-specific expression of aromatase mRNA has been reported in a number of turtles, including the terrapin (Jeyasuria and Place 1997, 1998; Murdock and Wibbels 2003). Collectively, these studies suggest aromatase activity and endogenous estrogen production as putative components in the female sex determination cascade.

Not all turtle TSD studies consistently support the estrogen hypothesis (reviewed by Lance 2009).

For example, in several studies, no sex-specific differences in aromatase activity and/or aromatase gene expression occurred, or they occurred after the TSP (Willingham et al. 2000; Murdock and Wibbels 2003; Valenzuela and Shikano 2007). Sex-specific differences in aromatase activity or gene expression could have been masked by the use of tissue from adrenal-kidney-gonad complexes rather than only gonads (Ramsey and Crews 2007; Lance 2009). Although the results from all studies do not consistently support the estrogen hypothesis (Lance 2009), the production of endogenous estrogen at female-producing temperatures remains a viable candidate as a significant component of the female sex determination cascade in turtles with TSD.

While many other putative sex determination genes in terrapins have not been evaluated, the expression of a wide variety of potential sex determination/sex differentiation genes has been examined in other turtles with TSD. An exhaustive review of these studies is beyond the scope of this chapter, but several genes consistently emerge as leading candidates for playing major roles in sex determination because (1) they show sexually dimorphic expression during sex determination and/or gonadal differentiation, and (2) many appear to be conserved components in the sex determination of amniotic vertebrates. In an effort to highlight genes that may be of prime interest in future studies targeting the physiological basis of TSD in the terrapin, we review studies of several genes that have consistently shown sex-specific expression.

Two genes seem to be consistently expressed relatively early in the TSP in female embryos and seem to be conserved elements in the sex determination of amniotic vertebrates: *RSPO1* and *FOXL2*. At female-producing temperature in the red-eared slider, *RSPO1* is expressed in the gonad during the TSP (Smith et al. 2008). The relatively early onset of its sexually dimorphic expression qualifies this gene as a leading candidate as a female sex-determining gene (Smith et al. 2008; Shoemaker and Crews 2009). In mammals, *RSPO1* activates a canonical beta-catenin signaling system via the *Wnt4* gene, and activation of this system is needed

for ovarian differentiation (Tomizuka et al. 2008). Another gene, *FOXL2*, has been the focus of several studies of turtles with TSD because it is expressed at higher levels at female-producing incubation temperatures (Rhen et al. 2007; Shoemaker et al. 2007a) and potentially upregulates the aromatase gene (Smith and Sinclair 2004; Pannetier et al. 2006; Rhen et al. 2007; Wang et al. 2007). Thus *FOXL2* could represent a key regulatory element in the production of estrogen during female sex determination and/or ovarian differentiation. Finally, Rhen and Schroeder (2010) identified a gene, *CIRBP*, that is expressed in the gonad at significantly higher levels at female-producing temperature in the snapping turtle. A more recent study identifies a potential genetic basis for the temperature-dependent expression of the *CIRBP* gene (Schroeder et al. 2016).

Several genes stand out as potential sex determination or sex differentiation genes in male turtle embryos. In particular, three genes—*Dmrt1*, *SOX9*, and the gene for antimullerian hormone, *AMH*—have sexually dimorphic expression patterns, and all three appear to be conserved components in the sex determination systems of amniotic vertebrates. *Dmrt1* is of particular interest because it is produced relatively early during the TSP at male-producing temperatures, and its expression remains relatively high during testicular differentiation (Kettlewell et al. 2000; Maldonado et al. 2002; Murdock and Wibbels 2003; Rhen et al. 2007; Shoemaker et al. 2007b; Bieser and Wibbels 2014). The sex-specific expression of this gene during testis differentiation also seems to be conserved in amniotic vertebrates and has been reported in mammals (Raymond et al. 2000) and birds (Smith et al. 2009). Further, in the chicken (*Gallus gallus domesticus*) it is linked to the Z sex chromosome, and its expression may be a primary regulator of male sex determination (Smith et al. 2009). Another gene expressed in the developing testes of turtles with TSD is *SOX9* (Spotila et al. 1994a, 1998; Shoemaker-Daly et al. 2010; Bieser and Wibbels 2014). Some researchers have hypothesized that *Dmrt1* upregulates *SOX9* in the developing gonad (Rhen et al. 2007; Bagheri-Fam et al. 2010; Bieser and Wibbels 2014). As with *Dmrt1*, *SOX9* appears to be a conserved component in testicular differentiation in other amniotic vertebrates, including mammals (da Silva et al. 1996) and birds (Sinclair and Smith 2009). In mammals, *SOX9* is a regulator of Sertoli cell differentiation and function in the differentiating testis, including upregulation of the *AMH* gene (De Santa Barbara et al. 1998; Wilhelm et al. 2005; Sekido and Lovell-Badge 2009). Expression of *AMH*, a third gene in the sex differentiation cascade, increases during the latter portion of the TSP in turtles and remains relatively high during a period when the Müllerian ducts are regressing in the male embryo (Wibbels et al. 1999; Shoemaker et al. 2007b; Bieser and Wibbels 2014). The *AMH* gene also seems to be a conserved element in the male sex differentiation cascade in amniotic vertebrates, in both mammals and birds, and stimulates Müllerian duct regression in mammals (Lee and Donahoe 1993; Smith et al. 1999; Allard et al. 2000).

Evolutionary Implications

The retention of TSD in many species of reptiles while GSD systems evolved in other reptiles, birds, and mammals suggests that, in some circumstances, TSD may have an evolutionary advantage (Charnov and Bull 1977). A variety of studies suggest the possible adaptive significance of TSD (Charnov and Bull 1977; Bull 1980; Ewert and Nelson 1991; Janzen and Paukstis 1991; Burke 1993; Reinhold 1998; Girondot and Pieau 1999; Shine 1999; Janzen and Phillips 2006; Warner and Shine 2008). Possible adaptive models include: (1) differential fitness, (2) sib avoidance, and (3) group-structured adaptation. Ewert and Nelson (1991) and Shine (1999) also suggested that TSD may have survived due to phylogenetic inertia. Shine (1999) provided a general overview of the differential fitness hypothesis and included six models: (1) matching sex to egg size, (2) matching sex to time of hatching, (3) nest site philopatry, (4) matching sex to phenotype, (5) sex and temperature interaction for egg or hatchling survival, and (6) sex and temperature interaction for phenotype. As an example, Ewert and Nelson (1991) suggested that TSD could match sex with specific growth rates that might be optimal for each sex. Despite the variety of hypotheses, only a few studies of

terrapins have addressed the adaptive significance of TSD.

Roosenburg and Kelley (1996) found that terrapin eggs incubated at female-producing temperatures produced hatchlings that grew significantly faster than hatchlings incubated at male-producing temperatures. Additionally, egg size correlated with nest location, with smaller eggs laid at cooler sites. They suggested that this could be an adaptive advantage because adult females are larger and take longer to mature than males (Roosenburg 1996). Females did grow faster, but the results were not able to distinguish whether the increased growth rate was linked to sex or to incubation temperature. Rhen and Lang (1994) conducted a similar study in the snapping turtle and found that growth rate was linked to incubation temperature rather than sex. Steyermark and Spotila (2001a, b) found that incubation temperature affected hatching success and hatchling size but did not affect hatchling growth rate. There are examples in other species of reptiles with TSD in which faster growth rates are associated with the larger sex, even if the larger sex is the male; these include the snapping turtle (Brooks et al. 1991; McKnight and Gutzke 1993; Bobyn and Brooks 1994), the desert tortoise (Spotila et al. 1994b), and the American alligator (*Alligator mississippiensis*; Webb et al. 1987).

Another hypothesis is that TSD and/or incubation temperature could affect fitness by controlling hatchling size. Roosenburg and Kelley (1996) did not find a difference between the mass of terrapin hatchlings incubated at male-producing vs. female-producing temperatures when egg size was used as a covariate, but Petrov (2014) found that terrapin eggs incubated at a male-producing temperature (26°C) produced larger hatchlings than did eggs incubated at a female-producing temperature (31°C). This finding was consistent with those from previous studies of the painted turtle (Gutzke et al. 1987) and the snapping turtle (Packard et al. 1987), indicating that cooler, male-producing temperatures produced larger hatchlings with less residual yolk than hatchlings from warmer, female-producing temperatures. Again, these studies were unable to distinguish between the effects of temperature and of sex on hatchling size. Hatchling size may be of ecological and evolutionary significance because several studies with turtles have suggested that larger hatchlings may have better survival (Barrows and Swartz 1895; Bustard 1979; Swingland and Coe 1979; Janzen 1993; Janzen et al. 2000a, b). For example, larger size could allow greater locomotor performance and predator avoidance (Janzen et al. 2000a, b), and incubation temperature affects locomotor performance in a variety of turtles, lizards, and snakes (Booth 2006).

Population Sex Ratios: Ecological and Evolutionary Implications

Before the discovery of TSD in turtles, Gibbons (1970) noted that many studies reported female-biased sex ratios in turtle populations; Carr (1952) also indicated that, in many species, females outnumber males. The ability of TSD to affect sex ratios makes it a subject of ecological and evolutionary interest (Janzen and Paukstis 1991; Shine 1999; Mrosovsky 1994; Wibbels 2003). Evolutionary theory suggests that sex ratios within a population should approximate a 1:1 ratio if parental input is equal for both sexes (Fisher 1930). However, population sex ratios in reptiles with TSD are known to range from significant male biases to significant female biases (Seigel 1984; Bull and Charnov 1988, 1989; Lovich and Gibbons 1990; Ewert and Nelson 1991; Lang and Andrews 1994; Mrosovsky 1994; Chaloupka and Limpus 2001; Wibbels 2003). Such reports have prompted a variety of hypotheses that could account for differential parental input to male vs. female offspring, as well as other hypotheses explaining biased sex ratios (Charnov and Bull 1977; Ewert and Nelson 1991; Janzen and Paukstis 1991; Reinhold 1998; Shine 1999; Valenzuela 2004; Janzen and Phillips 2006). The most frequently suggested hypothesis is the "differential fitness" model, in which the adaptive significance of TSD allows the embryo to develop into the sex that is best suited to the specific incubation conditions of the egg. Shine (1999) gave a variety of scenarios by which the differential fitness model would provide an adaptive advantage and how it could lead to biased sex ratios in lizards (Shine et al. 1995; Warner and Shine 2005, 2008).

Results of studies on hatchling sex ratios in terrapins in relation to TSD vary widely, depending on the specific location and thermal qualities of the nesting area. In contrast, when examining population sex ratios, the adult sex ratio represents a condensation of many years of hatchling production and also may reflect factors such as differential mortality (i.e., sex-specific mortality). The adult sex ratios in terrapin populations range from significant male biases to significant female biases, as well as near 1:1 ratios (Table 10.1). The causal basis for this variability is unknown, but it could be due to a variety of natural factors as well as factors related to sampling bias. For example, Roosenburg et al. (1997) indicated that sampling bias could confound sex ratio studies and reported a male bias (3M:2F) in terrapins captured in crab pots when sampling a population with a predicted female bias (1M:2F). They attributed the biased sampling to the inability of large females to enter the crab pots. Bishop (1983) also noted that male-biased sex ratios recorded in two South Carolina estuaries may have been affected by sampling bias due to the restriction of large females from pots. Additionally, males may be more aggressive than females when trying to enter pots, thus enhancing their capture rate over that of females (Hildebrand 1928; Bishop 1983). Natural factors related to the ecology of the terrapin could also contribute to the variability. The specific location of sampling could affect sex ratio (Bishop 1983) because females may be more abundant in one sampling location than another. Sex ratios also can vary during the year in specific locations, with an increasing percentage of males in the mating season (Seigel 1984). Hurd et al. (1979) also noted an increasing male bias over the summer sampling period. Thus, sex-specific migration patterns could produce biased sampling, depending on its timing and location, as in other turtles (Parker 1984; Gibbons 1970; Wibbels 2003). With regard to life history factors that could affect sex ratio, Lovich and Gibbons (1990) suggested that earlier age to maturity for males may have contributed to the male-biased sex ratio they recorded in Georgia. Differential age to maturity and the sexual size dimorphism of adults could also confound sex ratio estimates, as suggested by Gibbons (1970),

Table 10.1. Examples of sex ratios reported in diamond-backed terrapin populations

Reference	Location	N	Sex ratio (F:M)
Roberge et al., unpub. data	Alabama (Cedar Point Marsh)	78	1.1:1.0
Roosenburg et al. 1997	Maryland (Patuxent River)	5,002	2.0:1.0
Lovich & Gibbon 1990	South Carolina (Kiawah Island)	414	0.6:1.0
Seigel 1984	Florida (Banana River)	47	8.4:1.0
Seigel 1984	Florida (Indian River)	125	8.6:1.0
Bishop 1983	South Carolina (Ashley River Estuary)	65	1.9:1.0
Bishop 1983	South Carolina (Wando River Estuary)	216	0.2:1.0
Hurd et al. 1979	Delaware (Canary Creek Marsh)	2,033	1.4:1.0
Cagle 1952	Louisiana (Dulac)	70	0.2:1.0

since relatively large immature males could be mistaken for mature females.

Despite the range of factors that can potentially confound estimations of sex ratios in terrapin populations, some of the more comprehensive studies may have avoided these problems. For example, Roosenburg et al. (1997) used a variety of sampling techniques in the Patuxent River, Maryland, and were able to sample a relatively large number of individuals ($N = 5,002$). They reported a 2F:1M sex ratio, which did include data from crab pots and the potential sampling bias. Seigel (1984) sampled throughout most of the year in several locations in the Indian and Banana rivers, Florida, using seines and hand captures of turtles basking or swimming near shore; he found an overall sex ratio of 8.5 F:1M ($N = 172$). Seigel indicated that the female-biased sex ratios recorded in that study should have avoided most sampling biases. However, Lovich and Gibbons (1990) suggested that the strong female bias reported by Seigel (1984) could have been influenced by the sampling location. Lovich and Gibbons used a variety of capture methods (that did not include crab pots) over multiple years at Kiawah Island, Georgia, and found a sex ratio of 0.6 F:1.0 M ($N = 414$). Collectively, these findings suggest that sex ratios may differ significantly among populations of terrapins and that a wide range of sex ratios may occur. The findings also highlight the need for continued and enhanced studies to determine the range of sex ratios in terrapin populations and the causal basis for sex ratios that are not consistent across different populations.

Assuming such sex ratio estimates are relatively free of sampling bias, the variation in sex ratios reported throughout the terrapin's range could reflect natural or anthropogenic-induced differences in the environment and ecology of terrapin populations. Evolutionary theory predicts that sex ratios are adaptive, and similar ecology and environments across different populations should select for similar sex ratios (Fisher 1930; Shine 1999). The wide geographic distribution and the habitat differences among terrapin populations provide environmental variation that may affect their ecology, TSD, and sex ratios. Furthermore, anthropogenic influences such as coastal development and fisheries could have altered the sex ratios in some terrapin populations over the past century. For example, crab pots significantly increase mortality within terrapin populations (Bishop 1983; Burger 1989; Roosenburg et al. 1997; Wood 1997; Hoyle and Gibbons 2000; Roosenburg and Green 2000; Roosenburg 2004; Dorcas et al. 2007; Grosse et al. 2009), and, as noted above, crab pots may selectively capture more adult males than females (Roosenburg et al. 1997). However, the effect of crab pots on sex ratio may also be population specific because adult female size can vary from one population to another, and females in some populations may be more vulnerable than those in others (Coleman et al. 2014). Loss of salt marsh habitat for terrapins (Teal and Teal 1969; Wood and Herlands 1997; Roosenburg et al. 1999; Gibbons et al. 2001) due to coastal development, hurricanes, pollution, etc., reduces not only foraging habitat but also nesting habitat, and this could potentially limit or restrict the range of incubation temperatures available to a population.

Nest site selection significantly affects nest temperature and sex ratios in turtles with TSD (Janzen 1994; Roosenburg 1996; Valenzuela and Janzen 2001;

Kolbe and Janzen 2002). Nest site selection may also have additional effects on adult sex ratio because it can affect hatchling fitness, growth rates, and survival (as described above). Such factors could contribute to the variation in sex ratio among populations. Therefore, when examining sex ratios in a population, it is advisable to also evaluate major ecological aspects of the habitat that could be affecting the population sex ratio (e.g., nesting locations, nest temperatures, potential historical changes in habitat and nesting areas, fisheries in the area).

Another topic that should be considered when addressing the ecology and evolution of sex ratios is the "operational sex ratio," typically defined as the ratio of males to females that are ready to mate in a population at a given time (Emlen and Oring 1977; Kvarnemo and Ahnesjo 1996). In some species of turtles, this ratio can differ significantly from the overall adult sex ratio. For example, in sea turtle species, females typically have a multiyear remigration interval (i.e., breeding and nesting does not occur every year), whereas males often breed yearly, suggesting that an overall female bias in a population could result in a near 1:1 operational sex ratio (Hays et al. 2010, 2014). Both male and female terrapins are generally believed to have an annual breeding cycle, as was initially documented in captive terrapins (Hildebrand 1928). Roosenburg and Dunham (1997) showed that females nest annually with up to three clutches per season and with approximately 15 days between nestings. Males are generally assumed to be reproductively active annually (Hildebrand 1928), but this is not well documented. Assuming this is the case, the adult sex ratio in a terrapin population should approximate the operational sex ratio. Additionally, female terrapins are known to store sperm for up to 4 years (Hildebrand 1928), which could potentially affect operational sex ratios. This would depend on sperm storage affecting the mating behavior of the female, which is currently unknown.

Global Climate Change

The International Panel on Climate Change currently predicts an increase in environmental tem-

perature of approximately 1.8°C to 3.4°C by the year 2100, depending on the projection model (Solomon 2007). Given that the TRT of terrapins is relatively narrow and that temperatures of approximately 31°C or above produce 100% females (Jeyasuria et al. 1994; Burke and Calichio 2014), a global climate change of that magnitude could easily result in the production of extreme female biases. While the effect of sex ratio on reproductive output in a population is not well understood, extreme female biases could potentially decrease populations' reproductive output if males become limiting. The pivotal temperature and TRT as part of the TSD system might not evolve quickly enough to compensate for temperature increases associated with global warming. Terrapins could alter their reproductive timing and nesting site choice in response to increasing environmental temperature, compensating for long-term warming. For example, they could nest earlier in the nesting season and/or nest in cooler locations, if available. However, we wonder whether these adaptive changes would be enough to offset the projected increase in environmental temperature. Regardless, this projected increase highlights the immediate need for long-term studies that document the timing of nesting, nesting location, thermal profiles of nesting areas, nest temperatures, and hatchling sex ratios. A variety of studies have begun to evaluate the effect of climate change on freshwater turtles (Janzen, 1994; Schwanz and Janzen 2008; Telemeco et al. 2013) and marine turtles (Hawkes et al. 2007, 2009; Hamann et al. 2013).

Summary and Future Research

Temperature-dependent sex determination provides the diamond-backed terrapin with a flexibility that generates a wide variety of scientific questions regarding physiology, ecology, evolution, and conservation. Our understanding of TSD has advanced significantly in the past few decades, and we have summarized here the current state of knowledge regarding TSD in terrapins and highlighted some knowledge gaps and questions. There is a pressing need for range-wide ecological studies evaluating the thermal characteristics of nesting areas, nest temperatures, and hatchling sex ratios. This baseline information is a prerequisite for evaluating the ecological and evolutionary implications of TSD. Such information is also paramount for evaluating the effect of global climate change on TSD and whether and how the terrapin will adapt.

ACKNOWLEDGMENTS

We acknowledge the Alabama Department of Conservation and Natural Resources, the Birmingham Audubon Society, the Alabama Academy of Sciences, and the University of Alabama at Birmingham Department of Biology for their support of research on the diamond-backed terrapin in Alabama. This chapter is contribution UMCES 5175 from the University of Maryland Center for Environmental Science and IMET 16-174 from the Institute of Marine and Environmental Technology.

REFERENCES

Allard, S., P. Adin, L. Gouedard, N. Di Clemente, N. Josso, M.C. Orgebin-Crist, J.Y. Picard and F. Xavier. 2000. Molecular mechanisms of hormone-mediated Mullerian duct regression: involvement of beta-catenin. Development 127:3349-3360.

Bagheri-Fam, S., A.H. Sinclair, P. Koopman and V.R. Harley. 2010. Conserved regulatory modules in the *Sox9* testis-specific enhancer predict roles for SOX, TCF/LEF, Forkhead, DMRT, and GATA proteins in vertebrate sex determination. International Journal of Biochemistry & Cell Biology 42:472-477.

Barrows, W.B. and E.A. Schwarz. 1895. The Common Crow of the United States. US Department of Agriculture, Division of Ornithology and Mammalogy, Washington, DC. 98 pages.

Bieser, K.L. and T. Wibbels. 2014. Chronology, magnitude and duration of expression of putative sex-determining/differentiation genes in a turtle with temperature-dependent sex determination. Sexual Development 8:364-375.

Bishop, J.M. 1983. Incidental capture of diamondback terrapin by crab pots. Estuaries 6:426-430.

Bobyn, M.L. and R.J. Brooks. 1994. Incubation conditions as potential factors limiting the northern distribution of snapping turtles, *Chelydra serpentina*. Canadian Journal of Zoology 72:28-37.

Booth, D.T. 2006. Influence of incubation temperature on hatchling phenotype in reptiles. Physiological and Biochemical Zoology 79:274-281.

Bowden, R.M., A.W. Carter and R.T. Paitz. 2014. Constancy in an inconstant world: moving beyond constant temperatures in the study of reptilian incubation. Integrative and Comparative Biology 54:830-840.

Brooks, R.J., M.L. Bobyn, D.A. Galbraith, J.A. Layfield and E.G. Nancekivell. 1991. Maternal and environmental influences on growth and survival of embryonic and hatchling snapping turtles (*Chelydra serpentina*). Canadian Journal of Zoology 69:2667-2676.

Bull, J.J. 1980. Sex determination in reptiles. Quarterly Review of Biology 55:3-21.

Bull, J.J. 1981. Evolution of environmental sex determination from genotypic sex determination. Heredity 47:173-184.

Bull, J.J. 1985. Sex ratio and nest temperature in turtles: comparing field and laboratory data. Ecology 66:1115-1122.

Bull, J.J. 1987. Temperature-sensitive periods of sex determination in a lizard: comparison with turtles and crocodiles. Journal of Experimental Zoology 241:143-148.

Bull, J.J. and E.L. Charnov. 1988. How fundamental are Fisherian sex ratios? Oxford Surveys in Evolutionary Biology 5:96-135.

Bull, J.J. and E.L. Charnov. 1989. Enigmatic reptilian sex ratios. Evolution 43:1561-1566.

Bull, J.J., W.H. Gutzke and D. Crews. 1988. Sex reversal by estradiol in three reptilian orders. General and Comparative Endocrinology 70:425-428.

Bull, J.J., J.M. Legler and R.C. Vogt. 1985. Non-temperature dependent sex determination in two suborders of turtles. Copeia 1985:784-786.

Bull, J.J. and R.C. Vogt. 1981. Temperature-sensitive periods of sex determination in emydid turtles. Journal of Experimental Zoology 218:435-440.

Bull, J.J., R.C. Vogt and M.G. Bulmer. 1982a. Heritability of sex ratio in turtles with environmental sex determination. Evolution 36:333-341.

Bull, J.J., R.C. Vogt and C.J. McCoy. 1982b. Sex determining temperatures in turtles: a geographic comparison. Evolution 36:326-332.

Burger, J. 1989. Diamondback terrapin protection. Plastron Papers 19:35-40.

Burke, R.L. 1993. Adaptive value of sex determination mode and hatching sex ratio bias in reptiles. Copeia 1999:854-859.

Burke, R.L. and A.M. Calichio. 2014. Temperature-dependent sex determination in the diamond-backed terrapin (*Malaclemys terrapin*). Journal of Herpetology 48:466-470.

Bustard, H.R. 1979. Population dynamics of sea turtles. Pages 523-540 in M. Harless and H. Morlock (eds.),

Turtles: Perspectives and Research. Wiley and Sons, New York, NY.

Cagle, F.R. 1952. A Louisiana terrapin population (*Malaclemys*). Copeia 1952:74-76.

Carr, A.F. 1952. Handbook of Turtles: The Turtles of the United States, Canada, and Baja California. Cornell University Press, Ithaca, NY. 542 pages.

Chaloupka, M. and C. Limpus. 2001. Trends in the abundance of sea turtles resident in southern Great Barrier Reef waters. Biological Conservation 102:235-249.

Charnov, E.L. and J.J. Bull. 1977. When is sex environmentally determined? Nature 266:828-830.

Coleman, A.T., T. Roberge, T. Wibbels, K. Marion, D. Nelson and J. Dindo. 2014. Size-based mortality of adult female diamond-backed terrapins (*Malaclemys terrapin*) in blue crab traps in a Gulf of Mexico population. Chelonian Conservation and Biology 13:140-145.

Converse, P.E., S.R. Kuchta, W.M. Roosenburg, P.F. Henry, G.M. Haramis and T.L. King. 2015. Spatio-temporal analysis of gene flow in Chesapeake Bay diamondback terrapins (*Malaclemys terrapin*). Molecular Ecology 24:5864-5876.

Coyne, M. and A.M. Landry Jr. 2007. Population sex ratio and its impact on population models. Pages 191-211 in P.T. Plotkin (ed.), Biology and Conservation of Ridley Sea Turtles. Johns Hopkins University Press, Baltimore, MD.

Crews, D. and J.M. Bergeron. 1994. Role of reductase and aromatase in sex determination in the red-eared slider (*Trachemys scripta*), a turtle with temperature-dependent sex determination. Journal of Endocrinology 143:279-289.

Crews, D., T. Wibbels and W.H.N. Gutzke. 1989. Action of sex steroid hormones on temperature-induced sex determination in the snapping turtle (*Chelydra serpentina*). General and Comparative Endocrinology 76:159-166.

da Silva, S.M., A. Hacker, V.R. Harley, P. Goodfellow, A. Swain and R. Lovell-Badge. 1996. SOX9 expression during gonadal development implies a conserved role for the gene in testis differentiation in mammals and birds. Nature Genetics 14:62-68.

Delmas, V., A.C. Prevot-Julliard, C. Pieau and M. Girondot. 2008. A mechanistic model of temperature-dependent sex determination in a chelonian: the European pond turtle. Functional Ecology 22:84-93.

De Santa Barbara, P., N. Bonneaud, B. Boizet, M. Desclozeaux, B. Moniot, P. Sudbeck, G. Scherer, F. Poulat and P. Berta. 1998. Direct interaction of *SRY*-related protein SOX9 and steroidogenic factor 1 regulates transcription of the human anti-Müllerian hormone gene. Molecular and Cellular Biology 18:6653-6665.

De Souza, R.R. and R.C. Vogt. 1994. Incubation temperature influences sex and hatchling size in the neotropical turtle *Podocnemis unifilis*. Journal of Herpetology 28:453-464.

Desvages, G., M. Girondot and C. Pieau. 1993. Sensitive stages for the effects of temperature on gonadal aromatase activity in embryos of the marine turtle *Dermochelys coriacea*. General and Comparative Endocrinology 92:54-61.

Desvages, G. and C. Pieau. 1992. Aromatase activity in gonads of turtle embryos as a function of the incubation temperature of eggs. Journal of Steroid Biochemistry and Molecular Biology 41:851-853.

Dodd, K.L., C. Murdock and T. Wibbels. 2006. Interclutch variation in sex ratios produced at pivotal temperature in the red-eared slider, a turtle with temperature-dependent sex determination. Journal of Herpetology 40:544-549.

Dorcas, M.E., J.D. Willson and J.W. Gibbons. 2007. Crab trapping causes population decline and demographic changes in diamondback terrapins over two decades. Biological Conservation 137:334-340.

Dorizzi, M., T.-M. Mignot, A. Guichard, G. Desvages and C. Pieau. 1991. Involvement of estrogens in sexual differentiation of gonads as a function of temperature in turtles. Differentiation 47:9-17.

Dorizzi, M., N. Richard-Mercier, G. Desvages, M. Girondot and C. Pieau. 1994. Masculinization of gonads by aromatase inhibitors in a turtle with temperature-dependent sex determination. Differentiation 58:1-8.

Emlen, S.T. and L.W. Oring. 1977. Ecology, sexual selection, and the evolution of mating systems. Science 197:215-223.

Etchberger, C.R., J.B. Phillips, M.A. Ewert, C.E. Nelson and H.D. Prange. 1991. Effects of oxygen concentration and clutch on sex determination and physiology in red-eared slider turtles (*Trachemys scripta*). Journal of Experimental Zoology 258:394-403.

Ewert, M.A., D.R. Jackson and C.E. Nelson. 1994. Patterns of temperature-dependent sex determination in turtles. Journal of Experimental Zoology 270:3-15.

Ewert, M.A. and C.E. Nelson. 1991. Sex determination in turtles: diverse patterns and some possible adaptive values. Copeia 1991:50-69.

Ezaz, T., A.E. Quinn, S.D. Sarre, D. O'Meally, A. Georges and J.A. Marshall Graves. 2009. Molecular marker suggests rapid changes of sex-determining mechanisms in Australian dragon lizards. Chromosome Research 17:91-98.

Fisher, R.A. 1930. The Genetical Theory of Natural Selection. Clarendon Press, Oxford. 286 pages.

Georges, A. 1989. Female turtles from hot nests: is it duration of incubation or proportion of development at high temperatures that matters? Oecologia 81:323-328.

Georges, A., K. Beggs, J.E. Young and J.S. Doody. 2005. Modelling development of reptile embryos under fluctuating temperature regimes. Physiological and Biochemical Zoology 78:18-30.

Georges, A., S. Doody, K. Beggs and J. Young. 2004. Thermal models of TSD under laboratory and field conditions. Pages 79-89 in N. Valenzuela and V.A. Lance (eds.), Temperature-Dependent Sex Determination in Vertebrates. Smithsonian Institution Scholarly Press, Washington, DC.

Georges, A., C. Limpus and R. Stoutjesdijk. 1994. Hatchling sex in the marine turtle *Caretta caretta* is determined by proportion of development at a temperature, not daily duration of exposure. Journal of Experimental Zoology 270:432-444.

Gibbons, J.W. 1970. Sex ratio in turtles. Researches on Population Ecology 12:252-254.

Gibbons, J.W., J.E. Lovich, A.D. Tucker, N.N. Fitzsimmons and J.L. Greene. 2001. Demographic and ecological factors affecting conservation and management of the diamondback terrapin (*Malaclemys terrapin*). Chelonian Conservation and Biology 4:66-74.

Girondot, M. 1999. Statistical description of temperature-dependent sex determination using maximum likelihood. Evolutionary Ecology Research 1:479-486.

Girondot, M. and Y. Kaska. 2014. A model to predict the thermal reaction norm for the embryo growth rate from field data. Journal of Thermal Biology 45:96-102.

Girondot, M. and C. Pieau. 1999. A fifth hypothesis for the evolution of TSD in reptiles. Trends in Ecology and Evolution 14:359-360.

Godfrey, M.H., A.F. D'Amato, M.A. Marcovaldi and N. Mrosovsky. 1999. Pivotal temperature and predicted sex ratios for hatchling hawksbill turtles from Brazil. Canadian Journal of Zoology 77:1465-1473.

Godfrey, M.H., V. Delmas and M. Girondot. 2003. Assessment of patterns of temperature-dependent sex determination using maximum likelihood model selection. Ecoscience 10:265-272.

Greenbaum, E. 2002. A standardized series of embryonic stages for the emydid turtle *Trachemys scripta*. Canadian Journal of Zoology 80:1350-1370.

Grosse, A.M., J.D. Dijk, K.L. Holcomb and J.C. Maerz. 2009. Diamondback terrapin mortality in crab pots in a Georgia tidal marsh. Chelonian Conservation and Biology 8:98-100.

Gutzke, W.H., G.C. Packard, M. Packard and T.J. Boardman. 1987. Influence of the hydric and thermal environments on eggs and hatchlings of painted turtles (*Chrysemys picta*). Herpetologica 43:393-404.

Hamann, M., M.M.P.B. Fuentes, N.C. Ban and V.J.L. Mocellin. 2013. Climate change and marine turtles. Pages 353-373 in J. Wyneken, K.J. Lohmann and J.A. Musick (eds.), The Biology of Sea Turtles. Volume III. CRC Press, Boca Raton, FL.

Hanson, J., T. Wibbels and R.E. Martin. 1998. Predicted female bias in sex ratios of hatchling loggerhead sea turtles from a Florida nesting beach. Canadian Journal of Zoology 76:1850-1861.

Hart, K.M., M.E. Hunter and T.L. King. 2014. Regional differentiation among populations of the diamondback terrapin (*Malaclemys terrapin*). Conservation Genetics 15:593-603.

Hauswalt, J.S. and T.C. Glen. 2005. Population genetics of the diamond-backed terrapin (*Malaclemys terrapin*). Molecular Ecology 14:723-732.

Hawkes, L.A., A.C. Broderick, M.H. Godfrey and B.J. Godley. 2007. Investigating the potential impacts of climate change on a marine turtle population. Global Change Biology 13:923-932.

Hawkes, L.A., A.C. Broderick, M.H. Godfrey and B.J. Godley. 2009. Climate change and marine turtles. Endangered Species Research 7:137-154.

Hays, G.C., S. Fossette, K.A. Katselidis, G. Schofield and M.B. Gravenor. 2010. Breeding periodicity for male sea turtles, operational sex ratios, and implications in the face of climate change. Conservation Biology 24:1636-1643.

Hays, G.C., J.A. Mortimer, D. Ierodiaconou and N. Esteban. 2014. Use of long-distance migration patterns of an endangered species to inform conservation planning for the world's largest marine protected area. Conservation Biology 28:1636-1644.

Hildebrand, S.F. 1928. Review of experiments on artificial culture of diamondback terrapin. Bulletin of the US Bureau of Fisheries 45:25-70.

Holleley, C.E., D. O'Meally, S.D. Sarre, J.A. Marshall Graves, T. Ezaz, K. Matsubara, B. Azad, X. Zhang and A. Georges. 2015. Sex reversal triggers the rapid transition from genetic to temperature-dependent sex. Nature 523:79-82.

Hoyle, M.E. and J.W. Gibbons. 2000. Use of a marked population of diamondback terrapins (*Malaclemys terrapin*) to determine impacts of recreational crab pots. Chelonian Conservation and Biology 3:735-737.

Hurd, L., G. Smedes and T. Dean. 1979. An ecological study of a natural population of diamondback terrapins (*Malaclemys t. terrapin*) in a Delaware salt marsh. Estuaries 2:28-33.

Janzen, F.J. 1993. An experimental analysis of natural selection on body size of hatchling turtles. Ecology 74:332-341.

Janzen, F.J. 1994. Climate change and temperature-dependent sex determination in reptiles. Proceedings of the National Academy of Sciences 91:7487-7490.

Janzen, F.J. 1995. Experimental evidence for the evolutionary significance of temperature-dependent sex determination. Evolution 49:864-873.

Janzen, F.J. and J.G. Krenz. 2004. Which was first, TSD or GSD? Pages 121-130 in N. Valenzuela and V.A. Lance (eds.), Temperature-Dependent Sex Determination in Vertebrates. Smithsonian Institution Scholarly Press, Washington, DC.

Janzen, F.J. and G.L. Paukstis. 1991. Environmental sex determination in reptiles: ecology, evolution, and experimental design. Quarterly Review of Biology 66:149–179.

Janzen, F.J. and P.C. Phillips. 2006. Exploring the evolution of environmental sex determination, especially in reptiles. Journal of Evolutionary Biology 19:1775–1784.

Janzen, F.J., J.K. Tucker and G.K. Paukstis. 2000a. Experimental analysis of an early life-history stage: avian predation selects for larger body size of hatchling turtles. Journal of Evolutionary Biology 13:947–954.

Janzen, F.J., J.K. Tucker and G.L. Paukstis. 2000b. Experimental analysis of an early life-history stage: selection on size of hatchling turtles. Ecology 81:2290–2304.

Jeyasuria, P. 1997. The embryonic brain-gonadal axis in the temperature-dependent sex determination of *Malaclemys terrapin*: a role for P450 aromatase (*CYP19*). PhD dissertation, University of Maryland, College Park, MD. 524 pages.

Jeyasuria, P. and A.R. Place. 1997. Temperature-dependent aromatase expression in developing diamondback terrapin (*Malaclemys terrapin*) embryos. Journal of Steroid Biochemistry and Molecular Biology 61:415–425.

Jeyasuria, P. and A.R. Place. 1998. The embryonic brain-gonadal axis in temperature-dependent sex determination of reptiles: a role for P450 aromatase (*CYP19*). Journal of Experimental Zoology 281:455–462.

Jeyasuria, P., W.M. Roosenburg and A.R. Place. 1994. Role of P-450 aromatase in sex determination of the diamondback terrapin, *Malaclemys terrapin*. Journal of Experimental Zoology 270:95–111.

Kettlewell, J., C. Raymond and D. Zarkower. 2000. Temperature-dependent expression of turtle Dmrt1 prior to sexual differentiation. Genesis 26:174–178.

Kolbe, J.J. and F.J. Janzen. 2002. Impact of nest-site selection on nest success and nest temperature in natural and disturbed habitats. Ecology 83:269–281.

Kvarnemo, C. and I. Ahnesjo. 1996. The dynamics of operational sex ratios and competition for mates. Trends in Ecology and Evolution 11:404–408.

Lance, V.A. 2009. Is regulation of aromatase expression in reptiles the key to understanding temperature-dependent sex determination? Journal of Experimental Zoology Part A: Ecological Genetics and Physiology 311:314–322.

Lang, J.W. and H.V. Andrews. 1994. Temperature-dependent sex determination in crocodilians. Journal of Experimental Zoology 270:28–44.

Lee, M.M. and P.K. Donahoe. 1993. Mullerian inhibiting substance: a gonadal hormone with multiple functions. Endocrine Reviews 14:152–64.

Les, H.L., R.T. Paitz and R.M. Bowden. 2007. Experimental test of the effects of fluctuating incubation temperatures on hatchling phenotype. Journal of Experimental Zoology Part A: Ecological Genetics and Physiology 307:274–280.

Les, H.L., R.T. Paitz and R.M. Bowden. 2009. Living at extremes: development at the edges of viable temperature under constant and fluctuating conditions. Physiological and Biochemical Zoology 82:105–112.

Lovich, J.E. and J.W. Gibbons. 1990. Age at maturity influences adult sex ratio in the turtle *Malaclemys terrapin*. Oikos 59:126–134.

Maldonado, L.T., A.L. Piedra, N.M. Mendoza, A.M. Valencia, A.M. Martinez and H.M. Larios. 2002. Expression profiles of *Dax1*, *Dmrt1*, and *Sox9* during temperature sex determination in gonads of the sea turtle *Lepidochelys olivacea*. General and Comparative Endocrinology 129:20–26.

Marcovaldi, M.A., M.H. Godfrey and N. Mrosovsky. 1997. Estimating sex ratios of loggerhead turtles in Brazil from pivotal incubation durations. Canadian Journal of Zoology 75:755–770.

Marshall Graves, J.A. and C.L. Peichel. 2010. Are homologies in vertebrate sex determination due to shared ancestry or to limited options? Genome Biology 11:1–12.

Marshall Graves, J.A. and S. Shetty. 2001. Sex from W to Z: evolution of vertebrate sex chromosomes and sex determining genes. Journal of Experimental Zoology 290:449–462.

McKnight, C.M. and W.H. Gutzke. 1993. Effects of the embryonic environment and of hatchling housing conditions on growth of young snapping turtles (*Chelydra serpentina*). Copeia 1993:475–482.

Merchant-Larios, H., S. Ruiz-Ramirez, N. Moreno-Mendoza and A. Marmolejo-Valencia. 1997. Correlation among thermosensitive period, estradiol response, and gonad differentiation in the sea turtle *Lepidochelys olivacea*. General and Comparative Endocrinology 107:373–385.

Mitchell, N. and F.J. Janzen. 2010. Temperature-dependent sex determination and contemporary climate change. Sexual Development 4:129–140.

Mrosovsky, N. 1994. Sex ratios of sea turtles. Journal of Experimental Zoology 270:16–27.

Mrosovsky, N., S.R. Hopkins-Murphy and J.I. Richardson. 1984. Sex ratio of sea turtles: seasonal changes. Science 225:739–741.

Mrosovsky, N. and C. Pieau. 1991. Transitional range of temperature, pivotal temperatures and thermosensitive stages for sex determination in reptiles. Amphibia-Reptilia 12:169–179.

Mrosovsky, N. and C.L. Yntema. 1980. Temperature dependence of sexual differentiation in sea turtles: implications for conservation practices. Biological Conservation 18:271-280.

Murdock, C. and T. Wibbels. 2003. Cloning and expression of aromatase in a turtle with temperature-dependent sex determination. General and Comparative Endocrinology 130:109-119.

Neuwald, J.L. and N. Valenzuela. 2011. The lesser known challenge of climate change: thermal variance and sex-reversal in vertebrates with temperature-dependent sex determination. PloS One 6:e18117. DOI 10.1371/journal.pone.0018117.

Packard, G.C., M.J. Packard, K. Miller and T.J. Boardman. 1987. Influence of moisture, temperature, and substrate on snapping turtle eggs and embryos. Ecology 68:983-993.

Paitz, R.T., S.G. Clairardin, A.M. Griffin, M.C. Holgersson and R.M. Bowden. 2010. Temperature fluctuations affect offspring sex but not morphological, behavioral, or immunological traits in the Northern painted turtle (*Chrysemys picta*). Canadian Journal of Zoology 88:479-486.

Pannetier, M., S. Fabre, F. Batista, A. Kocer, L. Renault, G. Jolivet, B. Mandon-Pépin, C. Cotinot, R. Veitia and E. Pailhoux. 2006. *FOXL2* activates P450 aromatase gene transcription: towards a better characterization of the early steps of mammalian ovarian development. Journal of Molecular Endocrinology 36:399-413.

Parker, W.S. 1984. Immigration and dispersal of slider turtles *Pseudemys scripta* in Mississippi farm ponds. American Midland Naturalist 112:280-293.

Petrov, T.L.D. 2014. Biology, ecology, and conservation of hatchling and post-hatchling diamondback terrapin (*Malaclemys terrapin pileata*). MS thesis, University of Alabama at Birmingham, Birmingham, AL. 57 pages.

Pieau, C. 1982. Modalities of the action of temperature on sexual differentiation in field-developing embryos of the European pond turtle, *Emys orbicularis*. Journal of Experimental Zoology Part A: Ecological Genetics and Physiology 220:353-360.

Pieau, C. and M. Dorizzi. 1981. Determination of temperature sensitive stages for sexual differentiation of the gonads in the embryos of the turtle, *Emys orbicularis*. Journal of Morphology 170:373-382.

Pieau, C. and M. Dorizzi. 2004. Oestrogens and temperature-dependent sex determination in reptiles: all is in the gonads. Journal of Endocrinology 181:367-377.

Place, A.R. and V.A. Lance. 2004. The temperature-dependent sex determination drama—same cast, different stars. Pages 99-110 in N. Valenzuela and V.A. Lance (eds.), Temperature-Dependent Sex Determination. Smithsonian Institution Scholarly Press, Washington, DC.

Place, A.R., J. Lang, S. Gavasso and P. Jeyasuria. 2001. Expression of P450$_{arom}$ in *Malaclemys terrapin* and *Chelydra serpentina*: a tale of two sites. Journal of Experimental Zoology 290:673-690.

Pokorna, M. and L. Kratochvíl. 2009. Phylogeny of sex-determining mechanisms in squamate reptiles: are sex chromosomes an evolutionary trap? Zoological Journal of the Linnean Society 156:168-183.

Quinn, A.E., A. Georges, S.D. Sarre, F. Guarino, T. Ezaz and J.A. Marshall Graves. 2007. Temperature sex reversal implies sex gene dosage in a reptile. Science 316:411.

Quinn, A.E., S.D. Sarre, T. Ezaz, J.A. Marshall Graves and A. Georges. 2011. Evolutionary transitions between mechanisms of sex determination in vertebrates. Biology Letters. DOI 10.1098/rsbl.2010.1126.

Ramsey, M. and D. Crews. 2007. Adrenal-kidney-gonad complex measurements may not predict gonad-specific changes in gene expression patterns during temperature-dependent sex determination in the red-eared slider turtle (*Trachemys scripta elegans*). Journal of Experimental Zoology Part A: Ecological Genetics and Physiology 307:463-470.

Raymond, C.S., M.W. Murphy, M.G. O'Sullivan, V.J. Bardwell and D. Zarkower. 2000. *DMRT1*, a gene related to worm and fly sexual regulators, is required for mammalian testis differentiation. Genes and Development 14:2587-2595.

Raynaud, A. and C. Pieau. 1985. Embryonic development of the genital system. Pages 149-300 in C. Gans and F. Billett (eds.), Biology of the Reptilia. Volume 15. Development B. John Wiley and Sons, New York, NY.

Reinhold, K. 1998. Nest-site philopatry and selection for environmental sex determination. Evolutionary Ecology 12:245-250.

Rhen, T., R. Fagerlie, A. Schroeder, D.A. Crossley II and J.W. Lang. 2015. Molecular and morphological differentiation of testes and ovaries in relation to the thermosensitive period of gonad development in the snapping turtle, *Chelydra serpentina*. Differentiation 89:31-41.

Rhen, T. and J.W. Lang. 1994. Temperature-dependent sex determination in the snapping turtle: manipulation of the embryonic sex steroid environment. General and Comparative Endocrinology 96:243-254.

Rhen, T. and J.W. Lang. 1998. Among-family variation for environmental sex determination in reptiles. Evolution 52:1514-1520.

Rhen, T., K. Metzger, A. Schroeder and R. Woodward. 2007. Expression of putative sex-determining genes during the thermosensitive period of gonad development in the snapping turtle, *Chelydra serpentina*. Sexual Development 1:255-270.

Rhen, T. and A. Schroeder. 2010. Molecular mechanisms of sex determination in reptiles. Sexual Development 4:16-28.

Richard-Mercier, N., M. Dorizzi, G. Desvages, M. Girondot and C. Pieau. 1995. Endocrine sex reversal of gonads by the aromatase inhibitor Letrozole (CGS 20267) in *Emys orbicularis*, a turtle with temperature-dependent sex determination. General and Comparative Endocrinology 100:314-326.

Roosenburg, W.M. 1996. Maternal condition and nest site choice: an alternative for the maintenance of environmental sex determination? American Zoologist 36:157-168.

Roosenburg, W.M. 2004. The impact of crab pot fisheries on terrapin (*Malaclemys terrapin*) populations: where are we and where do we need to go? Pages 23-30 in C. Swarth, W.M. Roosenburg and E. Kiviat (eds.), Conservation and Ecology of Turtles of the Mid-Atlantic Region: A Symposium. Bibliomania, Salt Lake City, UT.

Roosenburg, W.M., W. Cresko, M. Modesitte and M.B. Robbins. 1997. Diamondback terrapin (*Malaclemys terrapin*) mortality in crab pots. Conservation Biology 11:1166-1172.

Roosenburg, W.M. and A.E. Dunham. 1997. Allocation of reproductive output: egg and clutch-size variation in the diamondback terrapin. Copeia 1997:290-297.

Roosenburg, W.M. and J.P. Green. 2000. Impact of a bycatch reduction device on diamondback terrapin and blue crab capture in crab pots. Ecological Applications 10:882-889.

Roosenburg, W.M., K.L. Haley and S. McGuire. 1999. Habitat selection and movements of diamondback terrapins, *Malaclemys terrapin*, in a Maryland estuary. Chelonian Conservation and Biology 3:425-429.

Roosenburg, W.M. and K.C. Kelley. 1996. The effect of egg size and incubation temperature on growth in the turtle, *Malaclemys terrapin*. Journal of Herpetology 30:198-204.

Sarre, S.D., A. Georges and A. Quinn. 2004. The ends of a continuum: genetic and temperature-dependent sex determination in reptiles. BioEssays 26:639-645.

Schroeder, A.L., K.J. Metzger, A. Miller and T. Rhen. 2016. A novel candidate gene for temperature-dependent sex determination in the common snapping turtle. Genetics 203:557-571.

Schwanz, L.E. and F.J. Janzen. 2008. Climate change and temperature-dependent sex determination: can individual plasticity in nesting phenology prevent extreme sex ratios? Physiological and Biochemical Zoology: Ecological and Evolutionary Approaches 81:826-834.

Seigel, R.A. 1984. Parameters of two populations of diamondback terrapins (*Malaclemys terrapin*) on the Atlantic coast of Florida. Pages 77-87 in R.A. Seigel,

L.E. Hunt, J.L. Knight, L. Malaret and N.L. Zuschlag (eds.), Vertebrate Ecology and Systematics: A Tribute to Henry S. Fitch. Museum of Natural History, University of Kansas, Lawrence, KS.

Sekido, R. and R. Lovell-Badge. 2009. Sex determination and *SRY*: down to a wink and a nudge? Trends in Genetics 25:19-29.

Shaffer, H.B. and 58 coauthors. 2013. The western painted turtle genome, a model for the evolution of extreme physiological adaptations in a slowly evolving lineage. Genome Biology 14:R28. DOI 10.1186/gb-2013-14-3-r28

Shine, R. 1999. Why is sex determined by nest temperature in many reptiles? Trends in Ecology and Evolution 14:186-189.

Shine, R., M.J. Elphick and P.S. Harlow. 1995. Sisters like it hot. Nature 378:451-452.

Shoemaker, C.M. and D. Crews. 2009. Analyzing the coordinated gene network underlying temperature-dependent sex determination in reptiles. Seminars in Cell and Developmental Biology 20:293-303.

Shoemaker, C.M., J. Queen and D. Crews. 2007a. Response of candidate sex-determining genes to changes in temperature reveals their involvement in the molecular network underlying temperature-dependent sex determination. Molecular Endocrinology 21:2750-2763.

Shoemaker, C., M. Ramsey, J. Queen and D. Crews. 2007b. Expression of *Sox9*, *Mis*, and *Dmrt1* in the gonad of a species with temperature-dependent sex determination. Developmental Dynamics 236:1055-1063.

Shoemaker-Daly, C.M., K. Jackson, R. Yatsu, Y. Matsumoto and D. Crews. 2010. Genetic network underlying temperature-dependent sex determination is endogenously regulated by temperature in isolated cultured *Trachemys scripta* gonads. Developmental Dynamics 239:1061-1075.

Sinclair, A. and C. Smith. 2009. Females battle to suppress their inner male. Cell 139:1051-1053.

Smith, C.A., K.N. Roeszler, T. Ohnesorg, D.M. Cummins, P.G. Farlie, T.J. Doran and A.H. Sinclair. 2009. The avian Z-linked gene *DMRT1* is required for male sex determination in the chicken. Nature 461:267-271.

Smith, C.A., C.M. Shoemaker, K.N. Roeszler, J. Queen, D. Crews and A.H. Sinclair. 2008. Cloning and expression of R-Spondin1 in different vertebrates suggests a conserved role in ovarian development. BMC Developmental Biology. DOI 10.1186/1471-213X-8-72.

Smith, C.A. and A.H. Sinclair. 2004. Sex determination: insights from the chicken. BioEssays 26:120-132.

Smith, C.A., M.J. Smith and A.H. Sinclair. 1999. Gene expression during gonadogenesis in the chicken embryo. Gene 234:395-402.

Solleder, E. and M. Schmid. 1984. XX/XY-sex chromosomes in *Gekko gecko* (Sauria, Reptilia). Amphibia-Reptilia 5:339-345.

Solomon, S., D. Qin, M. Manning, Z. Chen, M. Marquis, K. Averyt, M. Tignor and H. Miller (eds.). 2007. IPCC Climate Change 2007: The Physical Science Basis. Contribution of Working Group I to the Fourth Assessment Report of the Intergovernmental Panel on Climate Change. Cambridge University Press, Cambridge and New York, NY. 996 pages.

Spotila, L.D., N.F. Kaufer, E. Theriot, K.M. Ryan, D.N. Penick and J.R. Spotila. 1994a. Sequence analysis of the *ZFY* and *SOX* genes in the turtle, *Chelydra serpentina*. Molecular Phylogenetics and Evolution 3:1-9.

Spotila, L.D., J.R. Spotila and S.E. Hall. 1998. Sequence and expression analysis of *WT1* and *Sox9* in the red-eared slider turtle, *Trachemys scripta*. Journal of Experimental Zoology 281:417-427.

Spotila, J.R., L.C. Zimmerman, C.A. Binckley, J.S. Grumbles, D.C. Rostal, A. List Jr., E.C. Beyer, K.M. Phillips and S.J. Kemp. 1994b. Effects of incubation conditions on sex determination, hatching success, and growth of hatchling desert tortoises, *Gopherus agassizii*. Herpetological Monographs 8:103-116.

Steyermark, A.C. and J.R. Spotila. 2001a. Effects of maternal identity and incubation temperature on hatching and hatchling morphology in snapping turtles, *Chelydra serpentina*. Copeia 2001:129-135.

Steyermark, A.C. and J.R. Spotila. 2001b. Effects of maternal identity and incubation temperature on snapping turtle (*Chelydra serpentina*) growth. Functional Ecology 15:624-632.

Swingland, I.R. and M.J. Coe. 1979. The natural regulation of giant tortoise populations on Aldabra Atoll: recruitment. Philosophical Transactions of the Royal Society of London B: Biological Sciences 286:177-188.

Teal, J. and M. Teal. 1969. Life and Death of the Salt Marsh. Ballantine Books, New York, NY. 274 pages.

Telemeco, R.S., K.C. Abbott and F.J. Janzen. 2013. Modeling the effects of climate change–induced shifts in reproductive phenology on temperature-dependent traits. American Naturalist 181:637-648.

Tokunaga, S. 1985. Temperature-dependent sex determination in *Gekko japonicus* (Gekkonidae, Reptilia). Development, Growth, and Differentiation 27:117-120.

Tomizuka, K., K. Horikoshi, R. Kitada, Y. Sugawara, Y. Iba, A. Kojima, A. Yoshitome, K. Yamawaki, M. Amagai, A. Inoue, T. Oshima and M. Kakitani. 2008. R-spondin1 plays an essential role in ovarian development through positively regulating *Wnt-4* signaling. Human Molecular Genetics 17:1278-1291.

Valenzuela, N. 2001. Constant, shift, and natural temperature effects on sex determination in *Podocnemis expansa* turtles. Ecology 82:3010-3024.

Valenzuela, N. 2004. Evolution and maintenance of temperature-dependent sex determination. Pages 131-147 in N. Valenzuela and V. Lance (eds.), Temperature-Dependent Sex Determination in Vertebrates. Smithsonian Institution Scholarly Press, Washington, DC.

Valenzuela, N., R. Botero and E. Martínez. 1997. Field study of sex determination in *Podocnemis expansa* from Colombian Amazonia. Herpetologica 53:390-398.

Valenzuela, N. and F.J. Janzen. 2001. Nest-site philopatry and the evolution of temperature-dependent sex determination. Evolutionary Ecology Research 3:779-794.

Valenzuela, N. and T. Shikano. 2007. Embryological ontogeny of aromatase gene expression in *Chrysemys picta* and *Apalone mutica* turtles: comparative patterns within and across temperature-dependent and genotypic sex-determining mechanisms. Development Genes and Evolution 217:55-62.

Wang, D.S., T. Kobayashi, L.Y. Zhou, B. Paul-Prasanth, S. Ijiri, F. Sakai, K. Okubo, K.I. Morohashi and Y. Nagahama. 2007. *Foxl2* up-regulates aromatase gene transcription in a female-specific manner by binding to the promoter as well as interacting with *Ad4* binding protein/steroidogenic factor 1. Molecular Endocrinology 21:712-725.

Wang, Z. and 33 coauthors. 2013. The draft genomes of soft-shell turtle and green sea turtle yield insights into the development and evolution of the turtle-specific body plan. Nature Genetics 45:701-706.

Warner, D.A. and R. Shine. 2005. The adaptive significance of temperature-dependent sex determination: experimental tests with a short-lived lizard. Evolution 59:2209-2221.

Warner, D. and R. Shine. 2008. The adaptive significance of temperature-dependent sex determination in a reptile. Nature 451:566-568.

Webb, G.J.W., S.C. Manolis and P.J. Whitehead. 1987. Wildlife Management: Crocodiles and Alligators. Surrey Beatty and Sons, Chipping Norton, NSW, Australia. 552 pages.

Wibbels, T. 2003. Critical approaches to sex determination in sea turtles. Pages 103-134 in P.L. Lutz, J.A. Musick and J. Wyneken (eds.), The Biology of Sea Turtles. CRC Press, Boca Raton, FL.

Wibbels, T., J.J. Bull and D. Crews. 1991a. Chronology and morphology of temperature-dependent sex determination. Journal of Experimental Zoology 260:371-381.

Wibbels, T., J.J. Bull and D. Crews. 1991b. Synergism between temperature and estradiol: a common pathway in turtle sex determination? Journal of Experimental Zoology 260:130-134.

Wibbels, T. and D. Crews. 1994. Putative aromatase inhibitor induces male sex determination in a female unisexual lizard and in a turtle with temperature-dependent sex determination. Journal of Endocrinology 141:295-299.

Wibbels, T., F.C. Killebrew and D. Crews. 1991c. Sex determination in Cagle's map turtle: implications for evolution, development, and conservation. Canadian Journal of Zoology 69:2693-2696.

Wibbels, T., C. Wilson and D. Crews. 1999. Müllerian duct development and regression in a turtle with temperature-dependent sex determination. Journal of Herpetology 33:149-152.

Wilhelm, D., F. Martinson, S. Bradford, M.J. Wilson, A.N. Combes, A. Beverdam, J. Bowles, H. Mizusaki and P. Koopman. 2005. Sertoli cell differentiation is induced both cell-autonomously and through prostaglandin signaling during mammalian sex determination. Developmental Biology 287:111-124.

Willingham, E., R. Baldwin, J. Skipper and D. Crews. 2000. Aromatase activity during embryogenesis in the brain and adrenal-kidney-gonad of the red-eared slider turtle, a species with temperature-dependent sex determination. General and Comparative Endocrinology 119:202-207.

Wood, R.C. 1997. The impact of commercial crab traps on northern diamondback terrapins, *Malaclemys terrapin terrapin*. Pages 21-27 in J. Van Abbema (ed.), Proceedings: Conservation, Restoration, and Management of Tortoises and Turtles—An International Conference. New York Turtle and Tortoise Society, State University of New York, Purchase, NY.

Wood, R.C. and R. Herlands. 1997. Turtles and tires: the impact of roadkills on northern diamondback terrapin, *Malaclemys terrapin terrapin*, populations on the Cape May Peninsula, southern New Jersey, USA. Pages 46-53 in J. Van Abbema (ed.), Proceedings: Conservation, Restoration, and Management of Tortoises and Turtles—An International Conference. New York Turtle and Tortoise Society, State University of New York, Purchase, NY.

Yao, H.H. and B. Capel. 2005. Temperature, genes, and sex: a comparative view of sex determination in *Trachemys scripta* and *Mus musculus*. Journal of Biochemistry 138:5-12.

Yntema, C.L. 1968. A series of stages in the embryonic development of *Chelydra serpentina*. Journal of Morphology 125:219-251.

Yntema, C.L. 1979. Temperature levels and periods of sex determination during incubation of eggs of *Chelydra serpentina*. Journal of Morphology 159:17-27.

Yntema, C.L. and N. Mrosovsky. 1982. Critical periods and pivotal temperatures for sexual differentiation in loggerhead sea turtles. Canadian Journal of Zoology 60:1012-1016.

11

Foraging Ecology and Habitat Choice

ANTON D. TUCKER
RUSSELL L. BURKE
DIANE C. TULIPANI

The diamond-backed terrapin (*Malaclemys terrapin*) is a sister taxon to the map and sawback turtles (*Graptemys*) and shares the trait of molluscivory. Adults of these genera have strong sexual size dimorphism, with females growing larger bodies and wider heads with crushing plates. Enlarged head size (macrocephaly) in species of *Malaclemys* and *Graptemys* determines the maximum size and hardness of any prey they can consume (Lindeman 2013); dietary studies reveal that increasing head width increases crushing ability. The divergence in head size between the sexes can promote habitat partitioning based on differences in the size or type of prey eaten.

As predators, terrapins are top-level consumers in tidal estuaries of the western Atlantic Ocean and Gulf of Mexico. They forage in salt marshes, adjacent to barrier islands, seagrass meadows, mangroves, muddy creeks, rivers, bays, and harbors, and near sandy beaches. Such a variety of foraging habits yields a diverse prey base. Terrapins occupy a wide latitudinal range, yet researchers typically characterize foraging at a spatial scale visible from the deck of a small boat. Consequently, descriptions of predator-prey systems are site specific and would be expected to vary across the terrapin's broad latitudinal distribution. Prey types can vary in robustness, whether from shell architecture, organic composition, or carapace brittleness that yields to the terrapin's developed mandibular morphology and crushing jaw force.

Early dietary reports cataloged firsthand observations/anecdotes (e.g., Coker 1906) or restated general knowledge (Carr 1952; Pope 1971; Ernst and Lovich 2009), as developed through the commercial farming of terrapins. Husbandry documented relative growth rates in captivity when terrapins were fed a variety of seafood scraps or supplemental foods (Coker 1906; Hildebrand 1928). In the 1990s, quantitative field studies of terrapins produced an ecological understanding of the species' role in salt marsh and mangrove ecosystems as a transboundary predator in terrestrial, intertidal, and subtidal zones, depending on ontogenetic stage and season.

Comparative Review

We reviewed quantitative studies of terrapin diets chronologically (Table 11.1) to provide a comparative synthesis of earlier studies and generate insights for future work. Early work had revealed that for wild or captive terrapins, bivalves (B), crustaceans (C), and gastropods (G) constituted their primary diet, with a variety of other (O) prey species eaten in smaller quantities. We employed a convenient shorthand, BCGO, to build a ranked order that summarizes the results from each study—e.g., B if a study found only bivalves; CGB to indicate crustaceans as a main prey item, followed in decreasing order by gastropods and bivalves (Table 11.1). Some prey species (e.g., different plant types) may not have been reported in some studies if they were considered incidental or accidental prey.

The published and gray literature through mid-2016 included 27 field projects, 7 laboratory or captive projects, and 4 projects combining field and laboratory work. We categorized the literature by the regional management units developed by Hart et al. (2014). The list tallied 31 sources from the Coastal Mid-Atlantic, 6 from Florida, and 2 from the Gulf of Mexico (Table 11.1). A lack of work across the entire terrapin range suggests that we know less about terrapin diets than is generally presumed. This summary distills a baseline for new investigations, such as dietary changes associated with urbanization, climate change, invasive species, and marsh restoration; annual variation in prey availability; and learning and habituation in prey choice.

Spectrum of Prey, Foraging Modes, and Habitats

Diamond-backed terrapins feed on a variety of marsh crabs, snails, mussels, clams, barnacles, and plant species. Although direct field observations of intertidal and subtidal feeding remain sparsely documented, the dietary compositions (listed in Table 11.1) imply that terrapins use multiple foraging modes: sit and wait, cropping, direct consumption, substrate excavation, scavenging, opportunistic encounter, and solitary foraging. These modes are likely to be influenced by prey density and availability.

Terrapin dietary studies have encompassed terrestrial, intertidal, and subtidal zones, but often for only a single life cycle stage at a study site. The foraging habits of terrestrial phases of hatchling and juvenile terrapins have been studied in a single population (King 2007). Intertidal foraging has mostly been characterized in salt marshes, with only a recent interest in mangrove habitats (Denton et al. 2016). Subtidal foraging habitats may range from mud or sand flats (Butler et al. 2012) to submerged aquatic vegetation (SAV) and seagrass beds (Spivey 1998; Tulipani 2013). Terrapins in salt marsh and seagrass habitats in the lower Chesapeake Bay preferred foraging in seagrasses when available, contributing to prey differences between small and large terrapins (Tulipani 2013).

Although terrapin diets are assumed to be broadly similar within a local study area, prey species can show gradients reflecting geographic distribution. Interpopulation differences in primary prey consumed are probably a function of prey availability (Kaplan 1988; Gosner 1999) along latitudinal and salinity gradients. Marsh periwinkles (*Littoraria irrorata*) (Fig. 11.1) range across the terrapin's distribution, but another gastropod, the mud snail (*Ilyanassa obsoleta*), occurs from Canada to northern Florida, yet some terrapin populations do not eat them. Blue crabs (*Callinectes sapidus*) occur throughout the terrapin's range, but their local density can be influenced by commercial and recreational crab harvest (Bishop 1983) rather than by terrapin predation.

In Long Island Sound, New York, the mud snail was the most common terrapin prey (Petrochic 2009), whereas mature females from nearby Jamaica Bay ate primarily bivalves, even though mud snails were abundant (Erazmus 2012). In Virginia, small terrapins mostly ate barnacles—

Table 11.1. Chronology of studies on foraging ecology of the diamond-backed terrapin through mid-2016

Reference	State	Management unit*	Diet[†]	Setting	Relevant findings
Coker 1906	NC	CMA	GCO	Captive, field	Qualitative list of prey
Hildebrand 1928	NC, MD	CMA	GCO	Captive, field	Qualitative list of prey
Cagle 1952	LA	GULF	GB	Field	Qualitative list of prey; repeats Coker 1906
Carr 1952	NC	CMA, FL	GCO	Captive, field	Qualitative list of prey. FL location is *M. t. centrata* of undisclosed location
Allen & Littleford 1955	NC	CMA	CBGO	Field	Qualitative list of prey
Pope 1971	NC	CMA	GCO	Captive	Qualitative list of prey
Hurd et al. 1979	DE	CMA	B	Field	Qualitative list of prey
Middaugh 1981	SC	CMA	O	Field	Terrapins listed as predators of silverside minnows (*Menidia menidia*) in spring spawning run
Lovich et al 1991	SC	CMA	—	Field	Hatchling preference for terrestrial margin of salt marsh
Davenport et al. 1992; Davenport & Ward 1993	—	CMA	C	Lab	Foraging mode, lab trials, cropping or injuring strategies by terrapin size
Roosenburg 1994	MD	CMA	G or B	Field	Qualitative statement that 4 terrapin populations have different habitats and prey availabilities, characterized as G in marshes or B in open water
Tucker et al. 1995	SC	CMA	GCOB	Field	Sex differences in gape, tidal differences, spatial differences in prey, energy content of prey, telemetry to follow movements, resource partitioning, habitat partitioning
Tucker et al. 1997	—	CMA	G	Lab	Tests of prey accessibility
Spivey 1998	NC	CMA	GCB	Field	Small: GCB (4 species), large: C (11 species); spatial differences in prey availability
Bels et al. 1998	—	CMA	CB	Lab	Kinematics of feeding
Roosenburg et al. 1999	MD	CMA	B	Field	Qualitative statements about bivalves in open sandy bottoms
Levesque 2000	SC	CMA	G	Field	Field exclosure test of salt marsh predator influences on snail grazer densities
Silliman & Bertness 2002	GA	CMA	GC	Field	Trophic cascades theory that marsh predators control grazers, influencing salt marsh community structure
Whitelaw & Zajac 2002	CT	CMA	CGB	Field	Prey availability study: all varied spatially, but creek with terrapins had greatest dietary diversity of prey species
Ehret & Werner 2004	NJ	CMA	O	Field	Dermestid beetles, fish bones as suggestion of scavenging
Butler et al. 2006	FL	FL	B	Field	No prey partitioning by turtle size or over time
King 2007	NY	CMA	CGO	Field	Insects, hatchling study, seasonal variation, size variation
Harden et al. 2007	NC	CMA	—	Field	Defines a thermal window of activity for feeding/foraging
Kinneary 2008	—	CMA	—	Lab	Captive growth of hatchlings, mixed diet
Petrochic 2009	NY	CMA	GOBC	Field, lab	12 prey species, seasonal differences in prey, kinematics of feeding, lab tests of bite force for prey species
Bachman & Kays 2010	NJ	CMA	CO	Field	Contrasts of females from 2 areas with different diets, water temperature, pollution, and prey availabilities; 15 prey species, wide crossover, but unique prey assemblages at each site

(continued)

Table 11.1. (continued)

Reference	State	Management unit*	Diet[†]	Setting	Relevant findings
Craven & Calvo 2010	—	CMA	C	Lab	Cafeteria feeding trials for captive reared terrapins on pellet food; fiddler crabs (*Uca*) preferred over *Littorina*, *Ilyanassa*, and *Palaemonetes*
Boykin 2011	FL	FL	G	Field	Fecal observations
Basile et al. 2011		CMA	B	Field	Mussels (*Mytilus edulis*) identified as prey in pollutant study
Williard & Harden 2011	NC	CMA	—	Field	Defined annual thermal window of activity for feeding/foraging
Butler et al. 2012	FL	FL	BGC	Field	Male-female differences in FL, differences between nesters and non-nesters, local differences in prey availability
Erazmus 2012	NY	CMA	BOCG	Field	Multiyear study, diets vary with time but always strongly plant and bivalve dominated, consumed non-native crab species; sea lettuce (*Ulva lactuca*) incidentally consumed with bivalves may suggest nitrogen loading in system or decreasing habitats
Denton et al. 2013	FL	FL	CGO	Field	Preliminary diet results from different locations in southwest FL, stable isotope analysis of blood and scute; carbon isotope differences occurred among sites, and nitrogen isotope differences more pronounced in males than females
Orridge et al. 2013	NY	CMA	BCO	Field	Multiyear study with switching from BC to vegetation; marked differences between 2 adjacent populations
Tulipani 2013	VA	CMA	OCBG	Field	Habitat differences between seagrass and salt marsh
Underwood et al. 2013	—	CMA	C	Lab	Prey handling times, bite force and diet preferences, sexual dimorphism
Alleman 2015	TX	GULF	GCO	Field	Diets differ by sex, location, and season
Denton et al. 2016	FL	FL	GCBO	Field	Multivariate analysis revealed differences in food sources based on habitat more than terrapin size class

*Management units follow Hart et al. (2014). CMA = Coastal Mid-Atlantic; FL = Florida; Gulf = Gulf of Mexico.
[†]Predominant dietary items, given in rank order. B = bivalves; C = crustaceans; G = gastropods; O = other.

bay barnacle (*Balanus improvisus*) and ivory barnacle (*B. eburneus*)—whereas medium and large terrapins ate bivalves (Tulipani 2013). In North Carolina, small terrapins ate salt marsh snails (*Melampus bidentatus*), whereas medium and large terrapins ate blue crabs (Spivey 1998). South Carolina terrapins ate marsh periwinkles (Tucker et al. 1995) and ignored mud snails, presumably because they are more difficult to process due to their harder shell (Tucker et al. 1997). There are no terrapin dietary studies for populations north of New York or on the southeastern or Gulf coasts of Florida or the Gulf of Mexico.

Some dietary variation may be attributable to assortment of prey size and prey assemblages across saline gradients (Roosenburg et al. 1999; Butler et al. 2012; Tulipani 2013). Microhabitat differences in prey availability are expected when prey are zoned specifically to terrestrial, upper intertidal, or subtidal foraging habitats, characteristic of estuarine habitats under tidal influence. Bachman and Kays (2010) noted prey differences between eutrophic and unpolluted regions of salt marsh that differed in salinity and water temperature but separated by only a few kilometers. Prey availability varied greatly between salt marshes with and without terrapins

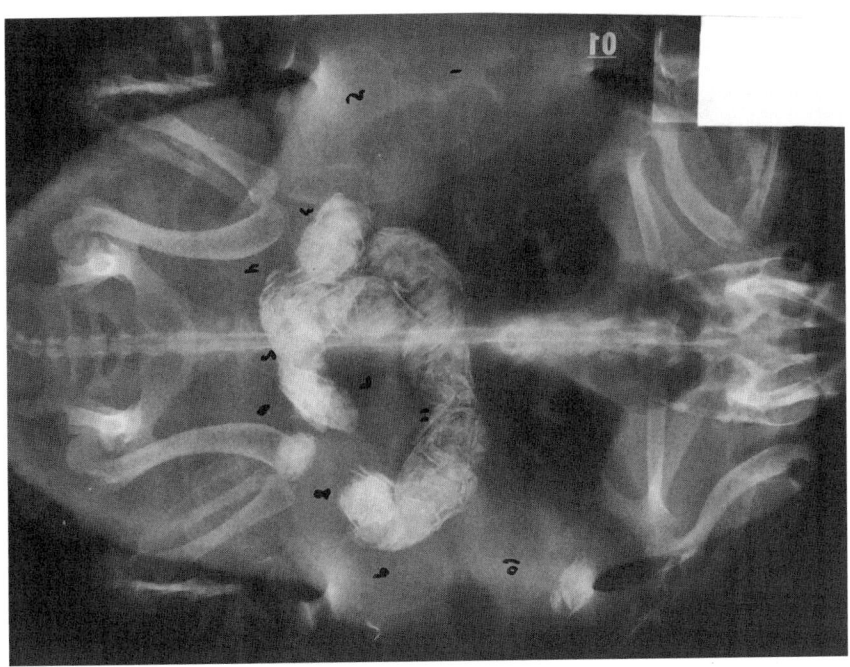

Fig. 11.1. Radiograph of a gravid *Malaclemys terrapin terrapin* from Patuxent River, Maryland, showing evidence of crushed mollusk shells throughout the intestinal tract and oval outlines of shelled terrapin eggs (numbered). Image courtesy of Willem Roosenburg

(Whitelaw and Zajac 2002). In Florida, prey species of mature female terrapins in tidal creeks differed from those in areas adjacent to nesting beaches (Butler et al. 2012). In lower Chesapeake Bay, terrapins preferred readily available epifauna living on eelgrass over the plentiful marsh periwinkles and mud snails found in nearby salt marshes (Tulipani 2013). Subtidal eelgrass epifauna were always available to terrapins, whereas the availability of snails in upper reaches of salt marshes was restricted by the tidal cycle (Tulipani 2013). Within Chesapeake Bay, piers and other structures generate epifaunal communities that attract terrapins. Hooked mussels (*Ischadium recurvum*) and barnacles are common colonizing species on dock structures and are consumed by terrapins from pilings, rocks, and bulkheads (W. Roosenburg, Ohio University, pers. comm.). Spivey (1998) found higher blue and fiddler crab abundance in low than in high marsh and a patchy distribution of *Melampus* snails. However, it was unclear whether patterns of prey distribution or lack of terrapin predation pressure were associated with the size of crabs consumed. In general, terrapins select the same prey groups throughout their range, though sometimes novel organisms are ingested (Table 11.1) (Tulipani 2013).

For Texas terrapins (Alleman 2015), the most common prey species were horn snails (*Cerithidea pliculosa*) and fiddler crabs (*Uca*), though male and female terrapins had different diets. Females consumed more periwinkle snails (*L. irrorata*) and total gastropods than did males, while males consumed more blue crabs and total decapods than did females. Dietary diversity was higher in males than in females. Seasonal dietary differences were noted with total gastropods, *C. sapidus*, and *Uca*.

The variation in prey consumed by terrapins is probably driven mostly by habitat-dependent prey availability, but longitudinal studies that track terrapin foraging movements with changes in prey availability are needed to explore this question (Tulipani 2013). Sometimes multiple habitats are in close proximity, affording an expanded choice of hard-shelled invertebrate species (e.g., eelgrass epibiota vs. salt marsh snails). Conversely, widely distributed habitats with different resources could promote increased terrapin movement rates and home range sizes, although there are as yet no such studies.

Chesapeake Bay has extensive seagrass beds of eelgrass (*Zostera marina*) and widgeon grass (*Ruppia maritima*) adjacent to *Spartina alterniflora*-dominated salt marshes (Reay and Moore 2009; Orth et al. 2010, 2011). These grasses occur on many North Atlantic coastlines and provide critical habitat that structures invertebrate and vertebrate biodiversity and anchors sediment. Terrapins that feed on barnacles and other eelgrass epibiota will ingest and disperse eelgrass seeds in their fecal material during May and early June, when the seeds are present (Tulipani and Lipcius 2014). Erazmus (2012) found seeds in terrapin fecal material, but whether these were from subtidal or salt marsh plants is unknown.

An approximation of potential terrapin foraging ground and habitat can be estimated from the coastal lengths of states within the terrapin's range. More than 69,040 linear km of coastline occurs in the 16 states that host terrapins (NOAA 2011). Although linear coastline is not a metric of occupied habitat, it defines a starting point for more localized spatial analyses to evaluate potential habitats across large spatial scales.

Bilkovic et al. (2012) used a quantitative GIS (geographic information systems) model to define habitat metrics that characterize terrapin habitat suitability within Chesapeake Bay. First, they quantified the distribution of shallow coastal waters. Second, they quantified the occurrence and distribution of essential shoreline characteristics to determine preferred terrapin habitats, such as mesohaline to polyhaline tidal marshes and potential nesting beaches. Third, they quantified riparian land use by access to high marsh and high-quality, adjacent upland habitat; forested lands within 10 m of the shore were considered to preclude access to high marsh and upland nesting habitats and diminish land-water connectivity. Lastly, SAV was included as a measure of supplementary feeding habitats that may enhance terrapin growth and survival. These preliminary steps may need refinement. For example, forested areas near beaches may not preclude nesting, or terrapins may travel farther from water or traverse thick upland vegetation to reach open nesting areas. However, the steps provide a preliminary tool to evaluate habitat and resource availability, as later extended by Isdell et al. (2015). Combining this GIS approach of occupancy modeling with ground

truth studies for presence/absence of terrapins or their prey would be a preliminary step in analyzing spatial proximity to threats, including habitat fragmentation, habitat suitability, and anthropogenic habitat alteration.

Biotelemetry to Define Activity Spaces

Recent biotelemetry studies clarified that the foraging behavior of terrapins is reduced when water temperatures are below 16°C (Harden et al. 2007; Williard and Harden 2011). Fitted with acoustic transmitters, terrapins did not aggregate or remain surface active once temperature declined below 20°C in North Carolina. Radio-telemetry (limited to above-waterline transmissions) indicated where terrapins foraged at high and low tides (Tucker et al. 1995; Harden et al. 2007). Estep (2005), combining radio-telemetry and acoustic tracking to determine activity periods and relationships to water temperature, documented that habitat use and site fidelity were influenced by season, time of day, tide, and reproductive status. With acoustic telemetry, Tulipani (2013) confirmed that small terrapins (males and juvenile females) had smaller home ranges than adult females and stayed mostly in shallow, nearshore water in seagrass beds, whereas large females entered these areas only on flood and high tides. Small terrapins were less likely than large females to move out of an area, but males made infrequent long-distance trips (unidirectional distance >4 km). Both sexes and size classes frequently moved between nonconnected *Z. marina* beds.

Telemetry studies can clarify tidal influences on movements among feeding areas (Spivey 1998; Tulipani 2013). Males and juvenile females feed within the nearshore intertidal and subtidal zones (Roosenburg et al. 1999; Tulipani 2013). Mature females were observed most frequently in deeper water farther from shore (Roosenburg et al. 1999). *Zostera marina* is found in the subtidal zone, typically within 300 m of shore, where water depth <2 m (Reay and Moore 2009), and its most abundant epifauna are barnacles (Tulipani 2013). In eelgrass habitats, barnacles were a preferred prey of small terrapins. Most ingested barnacles were attached to pieces of eelgrass leaves and spathes (seed pods). Terrapins ingested viable eelgrass seeds incidental to consuming *Z. marina* epifauna, and small terrapins were more likely than large terrapins to consume seeds because of habitat preference and prey choice (Tulipani 2013).

Possible Effects of Range Expansions and Invasive Species

Novel dietary prey may appear with biological invasions or with changing thermal or current conditions. The close morphological and caloric resemblance of non-native prey items to those in the terrapin's native diet suggests that these non-native prey might be taken. For example, the primary prey species in the diet of map turtles (*Graptemys*) in Texas (Lindeman 2000a), in Pennsylvania (J. Patterson and Lindeman 2009), and throughout the Great Lakes were found to be invasive bivalves, zebra mussel (*Dreissena polymorpha*) and quagga mussel (*D. rostriformis*).

It remains an open question whether non-native gastropods, bivalves, and crustaceans that now occupy terrapins' range will appear in their diets. The green crab (*C. maenas*) is a non-native crustacean with an established range from New England to mid-Virginia (Klassen and Locke 2007) and is similar in size to native blue crabs. Although green crabs were consumed by terrapins in captive feeding trials (Davenport et al. 1992), we know of no studies that indicate that terrapins feed on them in the wild. The salinity preferences of green crabs are stenohaline (>35 salinity), so their intolerance of lower salinities may limit overlap with terrapins, which occupy 0 to 34 salinity (Robinson and Dunson 1976). The Chinese mitten crab (*Eriocheir sinensis*) has been spreading through estuaries in New York and Chesapeake Bay since its discovery in 2005. The Asian shore crab (*Hemigrapsus sanguineus*) spread from New Jersey in 1988, northward to Maine and southward to North Carolina. It is approximately the size of the native crabs, *Uca* and *Sesarma*, ingested in many parts of the terrapin distribution. The gastropod *Littorina littorea* was introduced to North America in the 1880s and today is a predominant mollusk from New Jersey

northward. It is unclear whether *L. littorea* has been incorporated into terrapin diets, because the two relevant studies (Petrochic 2009; Erazmus 2012) were unable to distinguish the species of *Littorina* consumed. The nonindigenous brown mussel (*Perna perna*) is now widespread in the Gulf of Mexico, as is the Asian green mussel (*P. viridis*). Both mussels have broad thermal and salinity tolerances and may become established across the terrapin's distribution (Segnini de Bravo et al. 1998). The northern extent of *Rangia cuneata* into Chesapeake Bay (Hopkins and Andrews 1970) may now overlap with the clams *Macoma balthica*, *M. mitchelli*, and *M. tenta* in the Bay, as well as with the invasive clam *Corbicula fluminea*. Clearly, the potential of invasive alternative prey or competition with native prey is not hypothetical.

Temporal Changes in Prey Availability or Foraging Activity

Seasonal changes in water temperature affect terrapins' foraging dynamics (Harden et al. 2007; Williard and Harden 2011) as ectothermic predators. Monthly tidal fluctuations from neap to spring tides influence dietary changes (Tucker et al. 1995) and the tidal synchronicity of feeding behavior (Muehlbauer 1987). Patch dynamics and prey density gradients are associated with tidal amplitude when spring high tides flood marshes or mangroves and provide access to otherwise inaccessible prey (Tucker et al. 1995). The higher tides facilitate terrapins' access to snails seeking refuge at the upper levels of *Spartina* stems. Dietary shifts also may occur across the months before and after nesting seasons (Petrochic 2009; Butler et al. 2012). During the synchronous premolt stage for blue crabs (i.e., peeler runs), there appears to be a dietary shift as terrapins increase consumption of crabs (W. Roosenburg, pers. comm.), perhaps suggesting selective feeding on prey items that are more vulnerable and have higher fat content. Diet composition changes from year to year, presumably concomitant with changes in prey abundance or availability (Erazmus 2012; Orridge et al. 2013).

Sex and Size Differences

Sexual size dimorphism and its associated gape limitation cause sex-based prey differences in terrapins (Tucker et al. 1995; Petrochic 2009; Butler et al. 2012; Tulipani 2013; Underwood et al. 2013; Denton et al. 2016). Sex differences in habitat use and foraging are well known in the related *Graptemys* complex (Lindeman 2000a, b, 2013).

Tulipani (2013) summarized general patterns of dietary overlap based on terrapin size. There was substantial overlap in prey choice among habitats, though further analyses determined selectivity for barnacles by small terrapins in seagrass beds. In marsh-mudflat habitats, small terrapins consumed more periwinkles than did large terrapins or all sizes of terrapins from marsh-SAV habitats. The size of ingested periwinkles was related to terrapin size, with snails ingested by mature females presenting a bimodal distribution.

Multiple studies have posed basic questions about dietary or habitat partitioning by size-dimorphic terrapin populations (as listed in Table 11.1). We expect novel insights to emerge as terrapin dietary studies add additional levels of habitat complexity, document whether prey switching occurs for novel prey types, and visit unstudied range distributions with as yet limited data.

Moving Beyond a "Classic" Dietary Study

"Classic" dietary studies can be combined with contemporary analyses or methods. Classic dietary analyses inspect terrapin fecal material voided over a short period of captivity. Also, stomach lavages might reveal softer-bodied prey consumed by smaller terrapins residing above the tide line. However, lavages are highly invasive, and large terrapins crush hard-shelled prey into small sharp remains that render a lavage questionable both ethically and logistically. The standard enumeration of prey remains and abundance data can benefit from multivariate analysis beyond the standard biodiversity indices (e.g., Denton et al. 2016). We recommend use of null model approaches for diversity and overlap or the use of random effects

models to evaluate prey biodiversity composition in salt marsh food webs (see Erazmus 2012; Tulipani 2013).

Conduct Stable Isotope Analyses of Dietary Composition

The use of stable isotope analysis (SIA) to analyze the composition of the terrapin diet is still rare (e.g., Denton et al. 2013; Burke, unpub. data), although there are abundant parallels in SIA studies of sea turtles (Jones and Seminoff 2013; see Frye 2006 for an extensive overview of ecological applications with SIA). Ironically, it may be simpler to characterize niches for terrapins than for sea turtles, given the year-round access to terrapins and their site fidelity relative to wide-ranging sea turtles. Stable isotope techniques offer the potential to investigate the "lost years" of difficult-to-locate 0- to 3-year-old terrapins in terrestrial habitats (Muldoon and Burke 2012; Duncan 2014), with working hypotheses that $\delta^{13}C$ isotopes may distinguish terrestrial from estuarine prey types and that ontogenetic dietary shifts should manifest a difference in $\delta^{15}N$ isotopes. Stable isotopes can also be used to explore the importance of plant material in the terrapin's diet, and other isotopes may trace pollution from anthropogenic sources that can modify a local baseline for nitrogen isotopes. Isotope researchers should anticipate how the variable salinities of estuarine systems will affect the interpretation of nitrogen values (Olin et al. 2013). Therefore, future isotope studies that contrast different terrapin habitats should also evaluate baseline nitrogen signals or use compound-specific stable isotope analyses.

Develop Experiments on Dietary Choice and Grazing Pressure

Field experiments should elucidate the terrapin's roles as a coastal consumer and in structuring salt marsh ecosystems. Levesque (2000) conducted field manipulations with excluders to measure trophic-level effects of terrapin foraging. Unfortunately, her experiment was compromised by a hurricane, but she found that terrapins grazed down gastropod densities in microcosms. Such experiments should be replicated at other sites. Additional field studies should be placed in a context of contemporary terrapin population densities that are depleted from historical densities.

In controlled-diet studies, Erazmus (2012) revealed definite preferences for prey species that were not always the most common species in natural diets. Tulipani (2013) developed prey choice experiments in mesocosms that showed that terrapins had a preference for juvenile blue crabs over periwinkles and mottled dog whelks (*Nassarius vibex*). Tulipani (2013) also found that terrapins were less successful in finding and consuming blue crabs as the density of SAV increased.

Terrapins are believed to be attracted by the baits in crab pots (Bishop 1983), but they may actually be lured by opportunities to feed on entrapped crabs (Tulipani 2013). Researchers have recorded that unbaited traps will sometimes catch terrapins (Bishop 1983; Bilkovic et al. 2012; Tulipani, pers. obs.; Burke, pers. obs.). Can captive studies of terrapin feeding behavior (Davenport et al. 1992) be extrapolated to terrapin and prey inside crab pots? If terrapins are curious enough to enter unbaited pots, this raises two related questions: why do terrapins enter crab pots, and how do they enter? The former problem relates to ghost pots, and the latter has an apparent solution with terrapin excluder devices (Chapters 2 and 16).

Document Habitat Changes during Climate Change, Sea Level Rise, and Coastal Development

Coastal habitats occupied by terrapins are modified by biological and anthropogenic (climate change and sea level rise) influences. Black mangroves (*Avicennia*) are projected to expand their latitudinal range northward along the Gulf and Atlantic coasts in response to global climate change (Comeaux et al. 2012; Cavanaugh et al. 2013) and might encroach on or replace current salt marshes (C. Patterson et al. 1993; Duke et al. 1998). Coastlines are complicated by habitat loss resulting from armored shores, marsh drainage, and developed coasts (Winters et al. 2016). Coastal sea level rises involve a complicated interplay among changing sediment accretion rates, marsh subsidence rates, and the projected rate of increase and height of water levels (Reed 1990, 1995). The site fidelity of terrapins and their reliance on coastal habitats make them vulnerable to sea level rise (Woodland et al. 2017).

Human-driven degradation of terrapin foraging habitats by drainage and conversion to uplands for urbanization and agriculture has been ceaseless since the mid-1800s. Four (Florida, Louisiana, New Jersey, Texas) of five states with the highest levels of estuarine wetland losses (Tiner 1984) collectively include 67% of the terrapin's total range. Urban and agricultural runoff and salt marsh draining are linked to nitrogen enrichment and altered sediment flows, which contribute to declining salt marsh health. For example, urban nitrogen pollution in Jamaica Bay is associated with salt marsh loss and degradation of foods (Rafferty et al. 2011; Wigand et al. 2014). Coastal eutrophication led to a collapse of creek banks due to reduced *Spartina* root biomass and an increased microbial decomposition of organic matter (Deegan et al. 2012).

Terrapins also consume prey that accumulates PAHs (polycyclic aromatic hydrocarbons), but they metabolize PAHs at a lower rate than other vertebrates (Holliday et al. 2008; Chapter 12). Therefore, terrapins offer a unique toxicological model for monitoring oil contamination, given their food consumption in a restricted coastal range and their high site fidelity, long lives, and slow growth. Prime examples of dose-response studies include data obtained on crabs and terrapins before and after an oil spill in the New York–New Jersey harbor estuary (Burger 1994). Anthropogenic contamination also may affect prey availability, such as the losses of periwinkles in Louisiana salt marshes heavily oiled during the Deepwater Horizon oil spill and cleanup efforts (Silliman et al. 2012; Zengel et al. 2016).

What Structures Salt Marsh Food Webs?

Scientists have studied salt marsh dieback for decades, and they still debate its causes (Pennings 2012). High salt marsh dieback includes two abiotic factors, framed as waterlogging (Goodman and Williams 1961) and salinity factors (Brown and Pezeshki 2007). However, a biological controversy remains over the respective trophic roles of gastropods and their predators in structuring salt marsh food webs. Some studies suggested terrapins as the predators with top-down control on snails (Tucker et al. 1997; Levesque 2000). Other studies viewed blue crabs as a keystone predator of plant-grazing snails (Silliman and Bertness 2002). Both views must be calibrated against the shrinking baseline of terrapin populations and depleted habitats relative to historical levels. Despite differing views about a top trophic link, there is a consensus that salt marsh health is declining.

Looking Forward

The foraging ecology of the diamond-backed terrapin has reached a key stage. A refined understanding of dietary choice and foraging ecology should refocus questions on the remaining knowledge gaps. Research themes and technology have developed sufficiently to allow major progress in the coming decades. However, many detailed studies remain as unpublished theses or in the gray literature, such as abstracts from past terrapin workshops. Investigators and students should strive to move these almost-there products into the published literature.

The following topics will reward renewed attention and study:

- Address gaps in the terrapin's range where nothing is known about diet or foraging ecology. A pressing need remains for dietary studies from underrepresented regions.
- Quantify prey availability, because we know little about prey choice. Combine studies of prey choice and stable isotope analyses.
- Compare and contrast broader geospatial areas in a range-wide synthesis. Marsh and mangrove habitats differ dramatically, and greater dietary diversity is expected as more of the terrapin's range is included, particularly in areas with intergrading distributions of mangrove and salt marsh habitats.
- Culminate with a meta-analysis of dietary variability that synthesizes information from all habitats and regional genetic groups.

- Conduct latitudinal contrast studies across the Florida coastlines for the Gulf and Atlantic populations.
- Evaluate shifts in prey distributions at thermal boundaries defined by the Gulf Stream currents near Cape Hatteras.
- Use multifactor discriminant analysis across large tracts of terrapin habitat to tease out cofactors such as dominant SAV species, latitude, tidal range, turbidity, and salinity.
- Investigate temporal shifts in diet both within and across years as a function of changing prey availability. Also, determine any ontogenetic dietary shifts, particularly comparing terrestrial and aquatic life cycle stages.
- Evaluate experimentally the terrapin's potential as a keystone predator in habitat studies that measure effects across multiple trophic levels.
- Investigate terrapins as toxicological models where oil spills are affecting coastal ecosystems.

ACKNOWLEDGMENTS

ADT thanks the University of Georgia Savannah River Ecology Laboratory and Earthwatch Institute volunteers for support of the Kiawah Island terrapin project coordinated by J. W. Gibbons. RLB thanks A. Kanonik, K. Erazmus, L. Peyer, and numerous Hofstra University undergraduates for their help in collecting and sorting through terrapin fecal samples. DCT thanks R. Lipcius and students and staff of Virginia Institute of Marine Science, College of William and Mary, including funding sources NOAA-NERR (NA10NOS4200026, 2010–2012) and NSF GK-12 (DGE-0840804, 2012–2013). This chapter is VIMS contribution no. 3573.

REFERENCES

Alleman, B.J. 2015. Diet analysis and assessment of habitat and prey availability associated with Texas diamond-backed terrapin (*Malaclemys terrapin littoralis*). MS thesis, University of Houston, Clear Lake, TX. 110 pages.

Allen, J.F. and R.A. Littleford. 1955. Observations on the feeding habits and growth of immature diamondback terrapins. Herpetologica 11:77-80.

Bachman, C. and D. Kays. 2010. Comparative analysis of *Malaclemys terrapin terrapin* diet at two different sites in New Jersey. In Abstracts, Fifth Symposium on the Ecology, Status, and Conservation of the Diamondback Terrapin. Diamondback Terrapin Working Group, Jacksonville, FL.

Basile, E.R., H.W. Avery, W.F. Bien and J.M. Keller. 2011. Diamondback terrapins as indicator species of persistent organic pollutants: using Barnegat Bay, New Jersey as a case study. Chemosphere 82:137-144.

Bels, V.L., J. Davenport and S. Renous. 1998. Food ingestion in the estuarine turtle *Malaclemys terrapin*: comparison with the marine leatherback turtle *Dermochelys coriacea*. Journal of the Marine Biological Association of the United Kingdom 78:953-972.

Bilkovic, D.M., R. Chambers, M. Leu, K. Havens and T. Russell. 2012. Diamondback terrapin bycatch reduction strategies for commercial and recreational blue crab fisheries. Final Report to Virginia Sea Grant, National Oceanic and Atmospheric Administration (NOAA) Agency Award no. NA10OAR4170085. Silver Spring, MD. 21 pages.

Bishop, J.M. 1983. Incidental capture of diamondback terrapin by crab pots. Estuaries 6:426-430.

Boykin, C. 2011. The status and demography of the ornate diamondback terrapin (*Malaclemys terrapin macrospilota*) within the Saint Martins Marsh Aquatic Preserve. Report of the Department of Environmental Protection Florida. Tallahassee, FL. 47 pages.

Brown, C.E. and S.R. Pezeshki. 2007. Threshold for recovery in the marsh halophyte *Spartina alterniflora* grown under the combined effects of salinity and soil drying. Journal of Plant Physiology 164:274-282

Burger, J. 1994. Before and After an Oil Spill: The Arthur Kill. Rutgers University Press, New Brunswick, NJ. 305 pages.

Butler, J.A., G.L. Heinrich and M.L. Mitchell. 2012. Diet of the Carolina diamondback terrapin (*Malaclemys terrapin centrata*) in northeastern Florida. Chelonian Conservation and Biology 11:124-128.

Butler, J.A., R.A. Seigel and B.K. Mealey. 2006. *Malaclemys terrapin*—diamondback terrapin. Pages 279-295 in P. Meylan (ed.), Biology and Conservation of Florida Turtles. Chelonian Research Monographs. Volume 3. Chelonian Research Foundation, Lunenburg, MA.

Cagle, F.R. 1952. A Louisiana terrapin population (*Malaclemys*). Copeia 1952:74-76.

Carr, A.F. 1952. Handbook of Turtles: Turtles of the United States, Canada, and Baja California. Cornell University Press, Ithaca, NY. 542 pages.

Cavanaugh, K.C., J.R. Kellner, A.J. Forde, D.S. Gruner, J.D. Parker, W. Rodriguez and I.C. Feller. 2013. Poleward expansion of mangroves is a threshold response to decreased frequency of extreme cold events. Proceedings of the National Academy of Sciences 111:723-727.

Coker, R.E. 1906. The natural history and cultivation of the diamond-back terrapin with notes on other forms of turtles. North Carolina Geological Survey Bulletin 14:1-67.

Comeaux, R.S., M.A. Allison and T.S. Bianchi. 2012. Mangrove expansion in the Gulf of Mexico with climate change: implications for wetland health and resistance to rising sea levels. Estuarine, Coastal, and Shelf Science 96:81-95.

Craven, K.S. and M. Calvo. 2010. Feeding preferences in captive diamondback terrapins, *Malaclemys terrapin centrata*. In Abstracts, Fifth Symposium on the Ecology, Status, and Conservation of the Diamondback Terrapin. Diamondback Terrapin Working Group, Jacksonville, FL.

Davenport, J., M. Spikes, S.M. Thornton and B.O. Kelly. 1992. Crab-eating in the diamondback terrapin *Malaclemys terrapin*: dealing with dangerous prey. Journal of the Marine Biological Association of the United Kingdom 72:835-848.

Davenport, J. and J.F. Ward. 1993. The effects of salinity and temperature on appetite in the diamondback terrapin, *Malaclemys terrapin* (Latreille). Herpetological Journal 3:95-98.

Deegan, L.A., D.S. Johnson, R.S. Warren, B.J. Peterson, J.W. Fleeger, S. Fagherazzi and W.M. Wolheim. 2012. Coastal eutrophication as a driver of salt marsh loss. Nature 490:388-392.

Denton, M., A. Demopoulos, K. Hart, A. Aleinik and J. Baldwin. 2013. Diet and foraging ecology of mangrove diamondback terrapins (*Malaclemys terrapin rhizophorarum*) in protected habitats of South Florida. In Abstracts, Sixth Symposium on the Ecology, Status, and Conservation of the Diamondback Terrapin. Diamondback Terrapin Working Group, Jacksonville, FL.

Denton, M.J., K.M. Hart, A.W.J. Demopoulos, A. Oleinik and J.D. Baldwin. 2016. Diet of diamondback terrapins (*Malaclemys terrapin*) in subtropical mangrove habitats in South Florida. Chelonian Conservation and Biology 15:54-61.

Duke, N.C., M.C. Ball and J.K.C. Ellison. 1998. Factors influencing biodiversity and distributional gradients in mangroves. Global Ecology and Biogeography Letters 7:27-47.

Duncan, N.P. 2014. Dispersal of newly-emerged diamondbacked terrapin (*Malaclemys terrapin*) hatchlings at Jamaica Bay, NY. MS thesis, Hofstra University, Hempstead, NY. 38 pages.

Ehret, D.J. and R.E. Werner. 2004. *Malaclemys terrapin terrapin* (northern diamondback terrapin) diet. Herpetological Review 35:265.

Erazmus, K.R. 2012. Diet and prey choice of female diamond-backed terrapins (*Malaclemys terrapin*) in Jamaica Bay, New York. MS thesis, Hofstra University, Hempstead, NY. 46 pages.

Ernst, C.E. and J.E. Lovich. 2009. Turtles of the United States and Canada. 2nd edition. John Hopkins University Press, Baltimore, MD. 827 pages.

Estep, R.L. 2005. Seasonal movements and habitat pattern use of a diamondback terrapin (*Malaclemys terrapin*) population. MS thesis, College of Charleston, Charleston, SC. 102 pages.

Frye, B. 2006. Stable Isotope Ecology. Springer, New York, NY. 308 pages.

Goodman, P.J. and W.T. Williams. 1961. Investigations into "die-back" in *Spartina townsendii*. Journal of Ecology 49:391-398.

Gosner, K.L. 1999. A Field Guide to the Atlantic Seashore: From the Bay of Fundy to Cape Hatteras. Houghton-Mifflin, Boston, MA. 329 pages.

Harden, L.A. N.A. Diluzio, J.W. Gibbons and M.E. Dorcas. 2007. Spatial and thermal ecology of the diamondback terrapin (*Malaclemys terrapin*) in a South Carolina salt marsh. Journal of the North Carolina Academy of Sciences 123:154-162.

Hart, K.M., M.E. Hunter and T.L. King. 2014. Regional differentiation among populations of diamondback terrapin (*Malaclemys terrapin*). Conservation Genetics 15:593-603.

Hildebrand, S.F. 1928. Review of the experiments on artificial culture of the diamond-back terrapin. Bulletin of the US Bureau of Fisheries 45:25-70.

Holliday, D.K, W.M. Roosenburg and A.A. Elskus. 2008. Spatial variation in polycyclic aromatic hydrocarbon concentrations in eggs of diamondback terrapins, *Malaclemys terrapin*, from the Patuxent River, Maryland. Bulletin of Environmental Contamination and Toxicology 80:119-122.

Hopkins, S.H. and J.D. Andrews. 1970. *Rangia cuneata* on the East Coast: thousand mile range extension, or resurgence? Science 167:868-869.

Hurd, L.E., G.W. Smedes and T.A. Dean. 1979. An ecological study of a natural population of diamond-back terrapins (*Malaclemys t. terrapin*) in a Delaware salt marsh. Estuaries 2:28-33.

Isdell, R.E., R.M. Chambers, D.M. Bilkovic and M. Leu. 2015. Effects of terrestrial-aquatic connectivity on an estuarine turtle. Diversity and Distributions 21:643-653.

Jones, T.T. and J.A. Seminoff. 2013. Feeding biology—advances from field-based observations, physiological studies, and molecular techniques. Pages 211-247 in J. Wyneken, K.J. Lohmann and J.A. Musick (eds.), The Biology of Sea Turtles. Volume 3. CRC Press, Boca Raton, FL.

Kaplan, E.H. 1988. A Field Guide to the Southeastern and Caribbean Sea Shores. Houghton-Mifflin, Boston, MA. 480 pages.

King, T.M. 2007. The diet of northern diamondback terrapins (Order Testudines; *Malaclemys terrapin terrapin*). MS thesis, C.W. Post Long Island University, Brookville, NY. 35 pages.

Kinneary, J.J. 2008. Observation on terrestrial behavior and growth in diamondback terrapin (*Malaclemys*) and snapping turtle (*Chelydra*) hatchlings. Chelonian Conservation and Biology 7:118-119.

Klassen, G. and A. Locke. 2007. A biological synopsis of the European green crab, *Carcinus maenus*. Canadian Manuscript Report of Fisheries and Aquatic Sciences 2818:1-75.

Levesque, E.M. 2000. Distribution and ecology of the diamondback terrapin (*Malaclemys terrapin*) in South Carolina salt marshes. MS thesis, College of Charleston, Charleston, SC. 69 pages.

Lindeman, P.V. 2000a. Evolution of the relative width of the head and alveolar surfaces in map turtles (Testudines: Emydidae: *Graptemys*). Biological Journal of the Linnaean Society 69:549-576.

Lindeman, P.V. 2000b. Resource use of five sympatric turtle species: effects of competition, phylogeny, and morphology. Canadian Journal of Zoology 78:992-1008.

Lindeman, P.V. 2013. The Map Turtle and Sawback Atlas: Ecology, Evolution, Distribution, and Conservation. University of Oklahoma Press, Norman, OK. 460 pages.

Lovich, J.H., A.D. Tucker, D.E. Kling, J.W. Gibbons and T. D. Zimmerman. 1991. Behavior of hatchling diamond-back terrapins (*Malaclemys terrapin*) released in a South Carolina marsh. Herpetological Review 22:81-83.

Middaugh, D.P. 1981. Reproductive ecology and spawning periodicity of the Atlantic silverside, *Menidia menidia* (Pisces: Atherinidae). Copeia 1981:766-769.

Muehlbauer, E. 1987. Field and laboratory studies of tidal activity in the turtle *Malaclemys terrapin terrapin*. PhD dissertation, New York University, New York, NY. 60 pages.

Muldoon, K.A. and R.L. Burke. 2012. Hatchling diamond-backed terrapin (*Malaclemys terrapin*) movements, overwintering, and mortality at Jamaica Bay, New York. Canadian Journal of Zoology 90:651-662.

NOAA Office of Ocean and Coastal Resource Management. 2011. States and territories working with NOAA on ocean and coastal management. https://En.wikipedia.org/wiki/List_of_U.S._states_by_coastline.

Olin, J.A., N.E. Husse, S.A. Rus, G.R. Poulaki, C.A. Simpfendorfer, M.R. Heupel and A.T. Fisk. 2013.

Seasonal variability in stable isotopes of estuarine consumers under different freshwater flow regimes. Marine Ecology Progress Series 487:55-69.

Orridge, J., K.E. Napolitano, A. Kanonik and R. Burke. 2013. Diet differences within a single bay: diamondback terrapin diets in eastern and central Jamaica Bay, New York. In Abstracts, Sixth Symposium on the Ecology, Status, and Conservation of the Diamondback Terrapin. Diamondback Terrapin Working Group, Jacksonville, FL.

Orth, R.J., S.R. Marion, K.A. Moore and D.J. Wilcox. 2010. Eelgrass (*Zostera marina* L.) in the Chesapeake Bay region of mid-Atlantic coast of the USA: challenges in conservation and restoration. Estuaries and Coasts 33:139-150.

Orth, R.J., D.J. Wilcox, J.R. Whiting, L. Nagey, A.L. Owens and A.K. Kenne. 2011. 2010. Distribution of submerged aquatic vegetation in Chesapeake Bay and coastal bays. Virginia Institute of Marine Science Special Report no. 153. Gloucester Point, VA. http://web.vims.edu/bio/sav/sav10/index.html.

Patterson, C.S., I.A. Mendelssohn and E.M. Swenson. 1993. Growth and survival of *Avicennia germinans* seedlings in a mangal/salt marsh community in Louisiana, USA. Journal of Coastal Research 9:801-810.

Patterson, J.C. and P.V. Lindeman. 2009. Effects of zebra and quagga mussel (*Dreissena* spp.) invasion on the feeding habits of *Sternotherus odoratus* (stinkpot) on Presque Isle, northwestern Pennsylvania. Northeastern Naturalist 16:365-374.

Pennings, S.C. 2012. Ecology: the big picture of marsh loss. Nature 490:352-353.

Petrochic, S.L. 2009. Feeding ecology of the northern diamondback terrapin, *Malaclemys terrapin terrapin*. MS thesis, C.W. Post University of Long Island University, Brookville, NY. 66 pages.

Pope, C.H. 1971. Turtles of the United States and Canada. Alfred A. Knopf, New York, NY. 343 pages.

Rafferty, P., J. Castagna and D. Adamo. 2011. Building partnerships to restore an urban marsh ecosystem at Gateway National Recreation Area. Park Science 27:34-41.

Reay, W.G. and K.A. Moore. 2009. Introduction to the Chesapeake Bay National Estuarine Research Reserve in Virginia. Journal of Coastal Research 57:1-9.

Reed, D.J. 1990. The impact of sea-level rise on coastal salt marshes. Progress in Physical Geography 14:465-481.

Reed, D.J. 1995. The response of coastal marshes to sea level rise: survival or submergence? Earth Surface Processes and Landforms 20:39-48.

Robinson, G.D. and W.A. Dunson. 1976. Water and sodium balance in the estuarine diamondback terrapin (*Malaclemys*). Journal of Comparative Physiology 105:129-152.

Roosenburg, W.M. 1994. Nesting habitat requirements of the diamondback terrapin: a geographic comparison. Wetlands Journal 6(2):9-12.

Roosenburg, W.M., K.L. Haley and S. McGuire.1999. Habitat selection and movements of diamondback terrapins, *Malaclemys terrapin*, in a Maryland estuary. Chelonian Conservation and Biology 3:425-429.

Segnini de Bravo, M.I.S., K.S. Chung, and J.E. Perez. 1998. Salinity and temperature tolerances of the green and brown mussels, *Perna viridis* and *Perna perna* (Bivalvia: Mytilidae). Revista de Biologica Tropica 46:121-125.

Silliman, B.R. and M.D. Bertness. 2002. A trophic cascade regulates salt marsh primary productivity. Proceedings of the National Academy of Sciences 99:10,500-10,505.

Silliman, B.R., J. van de Koppel, M.W. McCoy, J. Diller, G.N. Kasozi, K. Earl, P.N. Adams and A.R. Zimmerman. 2012. Degradation and resilience in Louisiana salt marshes after the BP-Deepwater Horizon oil spill. Proceedings of the National Academy of Sciences 109:11,234-11,239.

Spivey, P.B. 1998. Home range, habitat selection, and diet of the diamondback terrapin (*Malaclemys terrapin*) in a North Carolina estuary. MS thesis, University of North Carolina, Wilmington, NC. 80 pages.

Tiner, R.W. Jr. 1984. Wetlands of the United States: current status and recent trends. US Fish and Wildlife Service Report no. 3386. US Government Printing Office, Washington, DC. 59 pages. http://www.arlis.org/docs/vol2/hydropower/APA_DOC_no._2417.pdf.

Tucker, A.D., N.N. Fitzsimmons and J.W. Gibbons. 1995. Resource partitioning by the estuarine turtle *Malaclemys terrapin*: trophic, spatial and temporal foraging constraints. Herpetologica 51:167-181.

Tucker, A.D., R.E. Yeomans and J.W. Gibbons. 1997. Shell strength of mud snails (*Ilyanassa obsoleta*) may deter foraging by diamondback terrapins (*Malaclemys terrapin*). American Midland Naturalist 138:224-229.

Tulipani, D.C. 2013. Foraging ecology and habitat use of the northern diamondback terrapin (*Malaclemys terrapin terrapin*) in southern Chesapeake Bay. PhD dissertation, College of William and Mary, Williamsburg, VA. 224 pages.

Tulipani, D.C. and R.N. Lipcius. 2014. Evidence of eelgrass (*Zostera marina*) seed dispersal by northern diamondback terrapin (*Malaclemys terrapin terrapin*) in lower Chesapeake Bay. PLoS One 9(7):e103346.

Underwood, E.B., S. Bowers, J.C. Guzy, J.E. Lovich, C.A. Taylor, J.W. Gibbons and M.E. Dorcas. 2013. Sexual dimorphism and feeding ecology of diamond-backed terrapins (*Malaclemys terrapin*). Herpetologica 69:397-404.

Whitelaw, D.M. and R.N. Zajac. 2002. Assessment of prey availability for terrapins in a Connecticut salt marsh. Northeastern Naturalist 9:407-418.

Wigand, C., C.T. Roman, E. Davey, M. Stolt, R. Johnson, A. Hanson, E.B. Watson, S.B. Moran, D.R. Cahoon, J.C. Lynch and P. Rafferty. 2014. Below the disappearing marshes of an urban estuary: historic nitrogen trends and soil structure. Ecological Applications 24:633-649.

Williard, A.S. and L.A. Harden. 2011. Seasonal changes in thermal environment and metabolic enzyme activity in the diamondback terrapin (*Malaclemys terrapin*). Comparative Biochemistry and Physiology A 158:477-484.

Winters, J.M., W.C. Carruth, J.R. Spotila, D.C. Rostal and H.W. Avery. 2016. Endocrine indicators of a stress response in nesting diamondback terrapins to shoreline barriers in Barnegat Bay, New Jersey. General and Comparative Endocrinology 235:136-141.

Woodland, R.J., C.L. Rowe and P.F.P. Henry. 2017. Changes in habitat availability for multiple life stages of diamondback terrapins (*Malaclemys terrapin*) in Chesapeake Bay in response to sea level rise. Estuaries and Coasts. DOI 10.1007/s12237-017-0209-2.

Zengel, S., C.L. Montague, S.C. Pennings, S.P. Powers, M. Steinhoff, G. Fricano, C. Schlemme, M. Zhang, J. Oehrig, Z. Nixon, S. Rouhani and J. Michel. 2016. Impacts of the Deepwater Horizon oil spill on salt marsh periwinkles (*Littoraria irrorata*). Environmental Science and Technology 50:643-652.

12

Environmental Toxicology

DAWN K. HOLLIDAY

RUSTY D. DAY

DAVID OWENS

The diamond-backed terrapin (*Malaclemys terrapin*) lives in estuaries along the eastern seaboard of the United States, from Massachusetts to Texas (Ernst and Lovich 2009). These same regions host some of the world's busiest waterways, including Jamaica Bay, New York; Chesapeake Bay and its tributaries in Maryland and Virginia; Charleston Harbor, South Carolina; and the mouth of the Mississippi River, Louisiana. As such, these areas suffer not only from degradation associated with habitat loss and eutrophication but also from significant chemical contamination from urban and agricultural point and nonpoint sources. Until recently, however, environmental toxicology studies rarely focused on terrapins.

In this chapter we introduce the field of environmental toxicology and discuss the current and future concerns facing terrapin populations. We review the available literature on the effects of contaminants on terrapins and the tools used to assess exposure, and where data on terrapins are unavailable, we present data for other turtle species that constitute the closest available surrogate information on toxicology for this taxon.

Contaminants of Concern in Estuaries

The aesthetic, economic, and commercial importance of estuaries has rendered them some of the most heavily degraded habitats. The list of historical, current, and emerging contaminants causing this degradation is extensive, including both organic and inorganic chemicals. Some of these stressors (e.g., metals) are ubiquitous throughout estuaries in the United States, whereas others (e.g., oil spills) are more localized. Here we highlight some of the main contaminants of concern for diamond-backed terrapins.

One toxic metal of significant concern is mercury (Hg) produced from coal combustion, which in the anoxic, bacteria-rich environment of sediments becomes the more bioavailable and toxic form, methylmercury (MeHg) (Landis et al. 2011). Methylmercury is ubiquitous in the brackish water sediments that support terrapins, including important habitats such as the lower Hudson River, New York, and the Patuxent River, Chesapeake Bay (Heyes et al. 2006), and is present, though at highly variable levels, in the estuarine sediments in south Florida (Kannan et al. 1998b). The mercury in the sediments is assimilated by low trophic level organisms that are common prey for terrapins, such as the salt marsh periwinkle snail (*Littoraria irrorata*). The mercury in the snails then efficiently biomagnifies in terrapins through dietary intake (Blanvillain et al. 2007). Unlike many legacy contaminants that have been banned and are declining in the environment, mercury pollution, given the continued reliance on coal for energy, will remain a priority concern for the foreseeable future.

Estuaries such as Chesapeake Bay also receive large inputs of pesticides and herbicides used to treat agricultural crops (Chesapeake Bay Program 2014). Although the use of many of these chemicals in the watershed has been declining (with the exception of pyrethroids, the use of which is increasing), the toxicity of the active ingredients has increased (Hartwell 2011). Exposure of terrapins to pesticides and herbicides has not been examined, but other species, such as the snapping turtle (*Chelydra serpentina*), can take up atrazine from the soil (de Solla and Martin 2011), and this may explain the presence of testicular oocytes in male snapping turtle embryos exposed to atrazine in the laboratory (de Solla et al. 2006).

Many chemicals that have been banned or are out of use are still present in the environment. For example, in spite of the banning of polychlorinated biphenyl (PCB) manufacturing in 1979 (USEPA 2013), PCBs are still present in terrapin habitats, including Chesapeake Bay (Ko and Baker 2004) and the estuaries of New York and New Jersey (Feng et al. 1998; Miller et al. 2011). In addition to these legacy chemicals, newly emerging contaminants

are also potentially affecting terrapin habitat. For example, triclosan, a component of antimicrobial soap, has been identified in estuarine sediments from Boston Harbor, the Hudson River, and Chesapeake Bay (Cantwell et al. 2010). Terrapin tissues sampled from Barnegat Bay, New Jersey, had detectable levels of triclosan and a suite of legacy pollutants, including PCBs, polybrominated diphenyl ethers (PBDEs), and DDT and its metabolites. The presence of triclosan was surprising, given that this chemical is typically associated with wastewater and the New Jersey population was not near a treatment plant (Basile et al. 2011). Another emerging contaminant, perchlorate, a chemical typically associated with contamination at military sites, is also found in some store-bought garden fertilizers (Sridhar et al. 1999) and hence may be in runoff from residential areas. These newer chemicals may prove to be more pervasive than previously thought, once more habitats and biological matrices are examined.

Since 1990, US oil pipelines have leaked more than 415 million liters (110 million gallons) of crude oil and other petroleum products (Frosch and Roberts 2011) into the environment. Accidental releases of oil and its derivatives can also occur from sunken vessels, refinery accidents, faulty storage devices, tanker collisions, and offshore oil rigs. Of 79 oil spill incidents (NOAA 2014), most were within or near possible terrapin habitat. At least three spills have adversely affected terrapin populations.

A spill in Arthur Kill, New Jersey, in the winter of 1990 affected at least 11 adult female terrapins that were found awakened from hibernation and covered in oil. Burger (1994) held the animals for a few weeks in the laboratory and noted a slow righting response, edema, and decreased appetite. In October 2012, Hurricane Sandy caused additional spills in Arthur Kill and surrounding areas, for which no data are currently available.

In April 2000, a pipeline break sent more than 100,000 gallons of crude oil into Swanson Creek, a tributary of the Patuxent River in Chesapeake Bay, Maryland, with a robust population of terrapins studied for more than 20 years by W. M. Roosenburg. As part of a natural resource damage assessment, Wood and colleagues noted more dead embryos from heavily oiled nesting beaches than from reference sites, including fully formed late-stage embryos (Wood and Hales 2001). Based on mortality of all life stages (hatchling, juvenile, adult), Byrd et al. (2002) calculated an estimated injury of approximately 5,245 lost terrapin-years. However, the actual extent of injury may have been substantially larger (W. Roosenburg, Ohio University, pers. comm.). Yet, even Byrd et al.'s (2002) conservative estimate would be a significant blow to the population. One year after the spill, Holliday et al. (2008) collected eggs from known nesting beaches within the spill area. Although many eggs contained polycyclic aromatic hydrocarbons (PAHs), the concentrations did not seem related to the amount of shoreline oiling. Some of the observed hydrocarbons represented background concentrations from the long-term presence of industry and boat traffic, but some may have been from the recent spill.

The Deepwater Horizon accident of 2010 in the Gulf of Mexico became the largest oil spill in US history and contaminated hundreds of miles of shoreline. Based on maps provided by NOAA and personal observations, many terrapin nesting beaches in and around Port Fourchon and Grand Isle, Louisiana, were oiled, and some dead terrapins were collected from these areas by a local landowner. The only published study to date on this topic was unable to relate plasma PAH concentrations to the amount of shoreline oiling or determine the source of the PAHs (Drabeck et al. 2014). Remediation operations also affected the area through increased boat traffic, installation of beach fences, and an increased presence of foot and vehicular traffic from shoreline cleanup and assessment teams. Additionally, as part of the cleanup initiative, large amounts of the chemical dispersant Corexit 9500 were applied to the area. Few data exist on the direct toxicity of this chemical, but it is known to alter the total amount and molecular weight of PAHs available to eggs in the nesting substrate (Rowe et al. 2009). There is much work to be done on Louisiana terrapins, not only in understanding how they respond to spill events at the cellular, organismal, and population levels, but also with regard to long-term effects from living in habitats with recurring oil exposure.

Routes of Exposure

Contaminants in estuaries may be present in the air, water column, sediments, and biota. The manner in which they are introduced into the environment, the environmental compartment into which they partition, and their chemical characteristics determine the route of exposure for terrapins and the tissues or organs that are affected. Dermal exposure and inhalation of contaminants may be common after an oil spill or other catastrophic event. Turtle skin, shell, eyes, and mucous membranes can be damaged by direct contact with petroleum. Oiled turtles were found to exhibit inflammation (Burger 1994) and abnormal development and loss of architecture of the epidermis (Lutcavage et al. 1995). The exposure might occur in the water column or while the turtles are submerged in sediments (see Selman and Baccigalopi 2012, Fig. 1).

A major source of exposure for many contaminant classes is diet. Oil has been detected in the feces of loggerhead sea turtles (*Caretta caretta*) (Lutcavage et al. 1995), suggesting direct consumption of spilled oil. Evidence of the consumption of oil can also be documented by the presence of PAH metabolites in bile (e.g., Barra et al. 2001), assuming the individual is recently deceased. Bioaccumulative compounds such as persistent organic pollutants and heavy metals such as MeHg are efficiently assimilated from the environment by prey and passed on to their predators. Dietary intake is typically the primary source of exposure of wildlife to these contaminants, which for terrapins means exposure through consumption of important prey species such as salt marsh periwinkles, crabs, barnacles, and clams (Tucker et al. 1995; Chapter 11). For example, large female terrapins were found to prefer salt marsh periwinkles (which had higher Hg concentrations), leading to mercury biomagnification (Blanvillain et al. 2007).

Although diet is the most likely route of exposure in most life history stages, exposure of embryos to contaminants can occur in two ways. Terrapins build nests on sandy beaches adjacent to marsh (Chapter 7). Contaminants from the nesting substrate may cross the eggshell to affect the developing embryo. Hydrophilic contaminants such as some pesticides and perchlorate are known to pass through the eggshells of snapping turtles and *Trachemys scripta*, the red-eared slider (de Solla and Martin 2011; Eisenreich et al. 2012). Because of their water solubility, these contaminants are likely to be transient and soon carried away in groundwater. The quantity entering the egg is probably not equivalent to the topical dose, as seen in the exposure of snapping turtle eggs to a metabolite of the pesticide DDT (Portelli et al. 1999). Rowe et al. (2009) replicated oil conditions on a nesting beach and found that low and medium molecular weight PAHs passed from the nest substrate into developing embryos.

The second route of embryonic exposure to contaminants is maternal transfer. During vitellogenesis, the adult female provides yolk to her developing eggs. Lipophilic contaminants stored in her adipose tissues are mobilized and incorporated into the lipid deposited in the egg. The maternal transfer of lipophilic PCBs and PBDEs has been documented in terrapins. Basile et al. (2011) detected both of these chemical families in eggs still within the oviduct of a deceased gravid terrapin. Concentrations were similar in all three oviductal eggs, confirming the long-held expectation that because all eggs are simultaneously provided with yolk, collection of a single terrapin egg is sufficient for assessing the contaminant burden of the entire clutch. Similarly, trace metals that are also essential nutrients can transfer to the developing offspring. Although no data for terrapins exist, studies showed that plasma concentrations of selenium and cadmium in female leatherback sea turtles were positively and significantly correlated with the concentrations of these metals in eggs (Guirlet et al. 2008). The maternal transfer of contaminants represents an important transgenerational effect on the most sensitive stage of development and warrants additional laboratory and field studies.

Terrapins as an Indicator Species

One common method to assess the condition of an ecosystem is to select representative biota from the community and measure the concentrations of environmental contaminants, biomarkers of exposure to contaminants, or other indicators of impaired health. There are many terms associated with variations of this concept, including "sentinel species," "indicator species," and "biomonitoring." Each of these terms reflects a slightly different set of objectives, which in turn lend themselves to a different set of criteria by which to select appropriate species and tissues to collect and analyze. Definitions vary, but sentinel species are those that generally respond to environmental stressors before humans or other components of the ecosystem respond, and serve as an early warning system (Stahl 1997); they are more synonymous with measures of health endpoints. In contrast, indicator species simply detect environmental stressors in a known manner based on scientific observations (Stahl 1997) and are likely to be more resistant to the stressors. Indicator species are often associated with detection of the presence/concentration of environmental contaminants or their health effects. Biomonitoring is the term often applied when an organism is used as a passive or active sampling mechanism to integrate the bioavailable fraction of a chemical contaminant in the environment. Combinations and variations of these approaches and objectives constitute a substantial portion of the published field studies on contaminants. Discrete research efforts have collected and analyzed target compounds in countless species and subsequently declared them to be potential indicator species. However, candidate species for systematic monitoring efforts intended to inform management and policy decisions should be thoroughly vetted to ensure they have suitable life history and physiological characteristics to meet the desired goals. Species whose biology and ecology have been thoroughly studied are the most appropriate candidates, but even well-studied species may require additional experimental work to allow valid interpretation of monitoring data.

Considerations for selecting an appropriate indicator species have been discussed in great detail elsewhere (Luepke 1979). The particular merits of members of the order Testudines in meeting the criteria for effective indicator species have been described, with examples from terrestrial, marine, and freshwater habitats (Meyers-Schone and Walton 1994). The diamond-backed terrapin has characteristics that make it an excellent candidate as an estuarine indicator and has been recommended for this purpose by many investigators (Kannan et al. 1998a, 1999; Burger 2002; Blanvillain et al. 2007; Schwenter 2007; Holliday et al. 2008; Arthur 2009; Green et al. 2010; Basile et al. 2011). Suitable attributes include the terrapin's broad geographic distribution from Cape Cod, Massachusetts, to Corpus Christi, Texas (Carr 1952), a range that includes a wide variety of estuarine environments, including salt marshes, coves, tidal rivers, and mangrove swamps. Attributes also include the terrapin's ability to thrive in a wide range of salinities, even approaching full-strength seawater (Dunson and Mazzotti 1989; Chapter 9). These large ranges in environmental tolerances allow monitoring across a broad geographic range and a variety of estuarine habitats using a single species. The terrapin is also a long-lived species (Mitro 2003), which allows studies on long-term bioaccumulation and chronic health effects (Rowe 2008).

Terrapins also have high site fidelity, with subadults and adults often remaining in a single creek system (Gibbons et al. 2001; Estep 2005), facilitating studies that require spatial resolution at relatively small scales, such as assessing the magnitude and scope of contamination from a point source of pollution. The terrapin has been successfully used to document locally elevated levels of mercury, PCBs, and extractable organohalogens in a salt marsh in the State of Georgia where a chlor-alkali plant once operated (Kannan et al. 1998a, 1999; Blanvillain et al. 2007), despite attempts at remediation. Arthur (2009) used the terrapin to detect higher mercury levels near a coal-fired power plant compared with a reference creek in Chesapeake Bay. As noted above, terrapin eggs collected one year after an oil spill showed PAH concentrations

reflective of persistent background levels in creeks (Holliday et al. 2008). Most high trophic level estuarine vertebrates, such as fish and birds, are migratory or transient residents of estuarine environments, making the resident terrapin a key bioindicator in this habitat.

Another aspect of an indicator species that must be considered is which tissues to use for collection and analysis, and whether these tissues integrate environmental contaminants over the temporal period of interest. Many studies have been performed on aquatic turtles to measure contaminant concentrations in various tissues and thus assess exposure (Meyers-Schone and Walton 1994). Liver, kidney, and muscle from aquatic turtles have been extensively analyzed for heavy metals and persistent organic pollutants (POPs), and the findings are presented throughout this chapter. However, collecting these tissues requires sacrificing the animal or relying on the opportunistic recovery of dead animals, which could bias sampling. Also, because internal organs bioaccumulate contaminants over the life of the organism, age can confound their use for spatial or temporal comparisons.

Noninvasive or nonlethal methods for assessing contaminant exposure have been evaluated for a number of aquatic turtles and are preferable to the use of internal organs, for conservation reasons. Eggs provide data on maternally transferred heavy metals and POPs (Kelly et al. 2008; Basile et al. 2011; Hopkins et al. 2013) and are established biomonitoring units for birds (Vander Pol et al. 2004; Day et al. 2006) and aquatic turtles. As with other amniotes, the turtle egg contains a highly vascularized membrane, known as the chorioallantoic membrane, through which gas exchange and nutrient and waste transport occur. Ismail (2010) studied this membrane as a nonlethal indicator of exposure in terrapins. Because the chorioallantoic membrane is easily collected from hatched eggs, sample procurement has no effect on live animals and can be done relatively quickly. Unfortunately, concentrations of specific PCB congeners in terrapin chorioallantoic membranes did not correlate well with concentrations in liver or in the egg because the less chlorinated PCBs were preferentially partitioned in the membrane, a finding similar to that reported for loggerhead sea turtles (Cobb and Wood 1997).

Other tissues can provide reliable indicators of exposure. Fat can be collected from terrapins by biopsy and may contain lipophilic POPs that partition strongly into this tissue. However, given the invasive nature of the fat biopsy procedure, blood plasma, easily collected, is a preferable tissue, and POP concentrations in blood correlate with those in fat (Basile et al. 2011). The concentrations of heavy metals such as mercury in whole blood or red blood cells have been shown to correlate with internal tissue concentrations in sea turtles (Day et al. 2005), snapping turtles (Golet and Haines 2001; Hopkins et al. 2013), and terrapins (Schwenter 2007). Similar findings have been reported for POPs in the blood of sea turtles (Keller et al. 2004). Thus, blood is a useful tissue for investigating a variety of chemical contaminants.

To further evaluate the use of blood for mercury biomonitoring, Schwenter (2007) performed a methylmercury pharmacokinetic study on diamond-backed terrapins, the only such study ever performed in a reptile. The study provided direct experimental evidence that MeHg peaks in terrapin blood 46 h after ingestion, followed by a rapid decline during redistribution to other tissues (half-time = 41 h) and a long terminal elimination phase (174 days). In a related study, the MeHg concentration in blood was monitored in terrapins during a controlled dosing study of 1.5 years and showed that after regular MeHg ingestion, the concentrations in blood climbed steadily, reflecting long-term bioaccumulation (Schwenter 2007). These studies provided quantitative evidence of how best to use and interpret blood mercury concentrations in terrapins and other turtles captured in the wild. Blood is a dynamic tissue compartment that can respond relatively quickly to changes in dietary intake of MeHg. But the slow elimination rate means there is also a more stable MeHg component in terrapins' blood, so blood MeHg level is representative of the mercury contamination in their local food web under normal conditions.

Despite the merits of using blood for biomonitoring in terrapins, there are some challenges with this approach that must be considered. The small body size of some individuals limits the volume of blood that can be safely collected for analysis. For example, a 1 cc sample of blood would be the approximate maximum advisable from a 300 g turtle, whereas 2 cc could be easily and safely obtained from an adult female. Depending on the blood collection method and the experience of the collector, some lymph may be inadvertently drawn into the tube, diluting and contaminating the blood (Owens, pers. obs.). The studies on mercury described above avoided these problems by spinning down the blood and using only the red blood cell fraction, where nearly all blood mercury is bound (Blanvillain et al. 2007). Also, depending on the field sampling constraints of monitoring efforts, there could be circumstances in which the dynamic nature of blood presents problems. Blanvillain et al. (2007) collected blood from terrapins in the Ashley River, South Carolina, from April to October and found that blood mercury concentrations exhibited seasonal variations. However, in the same study, mercury concentrations in full-depth scute samples from the carapace margin did not vary seasonally. Dermal scutes in turtles consist of layers of inert keratin proteins that integrate the average mercury exposure over the period of scute growth. Therefore, scutes are less sensitive to variability in diet, metabolism, reproductive physiology, or behavior that could cause short-term fluctuations in blood mercury concentrations. The duration of exposure over which a normal terrapin scute integrates contaminants is not known exactly, but the regrowth rate of keratin over previously sampled marginal scutes indicates that this nonlethally sampled tissue may represent a multiyear index.

Use of Biomarkers in Ecotoxicology

What Is a Biomarker?

Huggett et al. (2002) define a biomarker as a physiological, biochemical, or histological indicator of exposure and/or of the effect(s) of a xenobiotic. Other possible biomarkers include behavioral effects that are an organismal manifestation of biochemical changes (e.g., Rowe et al. 2009) and responses studied in the burgeoning field of toxi-cogenomics (Aardema and MacGregor 2002), which will provide a great amount of information about DNA, protein, and epigenetic alterations. Some biomarkers, such as upregulation of mixed function oxidase enzymes and sex reversal, have been widely studied in turtles, whereas others, such as DNA damage, have received little attention. We caution that although some biomarkers may be useful indicators of exposure, their ability to provide information on higher-level effects (organismal, ecosystem) is limited, making their utilization in risk assessment difficult (Forbes et al. 2006).

Biochemical Biomarkers

The metabolic pathways that biotransform many exogenous environmental contaminants are the same as those responsible for the catabolism of en-dogenous steroids. These processes are primarily accomplished by the liver, but the key constituents are also present in gonad, kidney, and other tissues (Ertl and Winston 1998). Metabolism happens in two phases: phase I processes, such as oxidation, reduction, or hydrolysis, convert a parent com-pound by adding or making available a functional group that, during phase II conjugation, becomes a binding site for a ligand, thus tagging the meta-bolite to be recognized for excretion. Phase I oxida-tion reactions are accomplished by a family of cytochrome P450 enzymes (e.g., cytochrome P450 1A, or CYP1A). If baseline enzyme levels are known, one can look for an upregulation of cytochrome P450 as a possible indicator of exposure. During the process of CYP1A induction, heat shock proteins are liberated from the aryl hydrocarbon receptor, which subsequently leads to the production of an aryl hydrocarbon nuclear translocator protein, pro-viding additional biochemical endpoints of inter-est (Whitlock 1999). Alternatively, exposure can be determined indirectly by fluorimetric measure-ment of the rate at which resorufin dye is produced from ethoxyresorufin by the cytochrome-mediated activity of ethoxyresorufin-O-deethylase (EROD) (Stagg and McIntosh 1998). Phase II conjugation enzymes include uridine diphosphate glucurono-syltransferase and glutathione-S-transferase, mak-

ing these additional endpoints of interest (George 1994). These metabolic reactions help the organism eliminate the chemical and, in the process, change the chemical's bioactivity. Catabolism renders some chemical contaminants inactive, whereas other chemicals may be rendered more toxic or carcino-genic than the parent compound.

These biochemical endpoints have been mea-sured in a handful of turtle species. Induction of CYP1A and EROD was found in *C. serpentina* (Bishop et al. 1998) and *Chrysemys picta* (Rie et al. 2000), and induction of EROD and glutathione-S-transferase in *Phrynops geoffroanus* (Venancio et al. 2013), in individuals collected from contami-nated field sites. Richardson et al. (2010) identified the presence of other cytochromes (2K1 and 3A27) in three species of sea turtles, but they did not observe a significant correlation of glutathione-S-transferase activity with total PCB concentrations measured in liver. Laboratory exposures have in-duced CYP1A in *C. picta* (Yawetz et al. 1997, 1998; Schlezinger et al. 2000) and *T. scripta*, as well as in *Mauremys caspica* (Yawetz et al. 1997). (For a com-plete review of phase I and II components mea-sured in reptiles, see Mitchelmore et al. 2005.) There are currently no published studies on any of these endpoints in terrapins. However, hepatic micro-somal EROD activity was induced after a single intracoelomic injection of PCB 126 (20 µg/g) in hatchling terrapins (Holliday, A. Elskus, and W. Roosenburg, unpub. data). These endpoints are clearly relevant for terrapins and can be indicators of exposure; however, they generally require tissue samples from freshly dead individuals.

Metallothioneins, small proteins involved in the detoxification and transport of trace metals, are a potential biomarker of heavy metal exposure (Roesijadi 1994; Carpenè et al. 2007). These proteins bind not only essential trace elements such as zinc but also environmental contaminants, including mercury and cadmium, thus reducing the amount of free metal ions. Metallothioneins have been identified in a number of turtles, including yellow pond turtles (*Mauremys mutica*; Yamamura and Su-zuki 1984), red-eared sliders (Thomas et al. 1994), and loggerhead sea turtles and green sea turtles (*Chelonia mydas*; Andreani et al. 2008). Growth factors, cytokines, and oxidative stress can also induce metallothioneins, making them a powerful marker of oncogenesis (Ruttkay-Nedecky et al. 2013). As such, these cysteine-rich proteins may prove to be a powerful biomarker of more than just exposure to heavy metals.

Plasma Biomarkers

A few studies have examined turtle plasma esterases as a biomarker for exposure to pesticides. Brain acetylcholinesterase activity in the Caspian turtle (*M. caspica*) was inhibited by exposure to metabolites of the pesticide parathion (Yawetz et al. 1983). Clark et al. (2000) measured plasma cholinesterase concentrations in red-eared sliders and found no differences among animals collected from ponds with different levels of pesticide exposure. However, in western pond turtles (*Emys marmorata*), plasma cholinesterase concentrations were significantly lower in populations from areas with a high atmospheric input of pesticides (E. Meyer et al. 2013). Many terrapin populations are likely to be exposed to organophosphate pesticides, but this endpoint has yet to be measured. Because acetylcholinesterase is essential for neurotransmitter degradation in cholinergic synapses, decreased concentrations can have deleterious effects on muscle physiology and transmission of information within the nervous system, with the potential for far-reaching effects at the organismal level.

Vitellogenin is a lipoprotein produced in the liver of mature female turtles as a precursor to egg yolk protein. Because its synthesis is regulated largely by estrogen (Custodia-Lora et al. 2005), males and pre-reproductive individuals do not typically synthesize vitellogenin. Its production can be induced by laboratory exposure to synthetic estrogens in immature sea turtles (Heck et al. 1997) and pond turtles (Tada et al. 2008) and in male red-eared sliders (Palmer and Palmer 1995). In ovo exposure of painted turtles to bisphenol A and 17β-estradiol, however, did not induce liver vitellogenin (Jandegian et al. 2015). Vitellogenin has been successfully measured in a handful of turtle species by both radioimmunoassay (Gapp et al. 1979) and enzyme-linked immunosorbent assay (Palmer

et al. 1998; Herbst et al. 2003). Vitellogenin is also a potent calcium binder and has been implicated in many endocrine-disrupting events in contaminated environments. As such, it can be used as a biomarker for the exposure of oviparous organisms such as turtles to exogenous estrogen mimics. Vitellogenin induction by 17β-estradiol has been measured in only a single laboratory-exposed male terrapin (W. Roosenburg, pers. comm.), so here we summarize some of the literature on other turtle species.

When measuring vitellogenin in field-collected individuals, it is essential to understand not only current and historical concentrations of environmental pollutants but also seasonal variation in the reproductive biology of the species of interest. Male and immature painted turtles sampled from ponds adjacent to cattle fields (Irwin et al. 2001) and from a Cape Cod superfund site (Rie et al. 2005) did not show an induction of vitellogenin. However, the concentration of plasma vitellogenin in female painted turtles was in some cases higher (Irwin et al. 2001) and in other cases lower (Rie et al. 2005; Kitana et al. 2006) than in females collected from reference sites. Plasma vitellogenin concentrations were elevated in runoff-exposed snapping turtle females (J. Meyer et al. 2013). Total vitellogenin measured in female yellow-blotched map turtles (*Graptemys flavimaculata*) collected from a historically affected site was similar to that in females from a reference site, but the monthly pattern differed between sites (Shelby-Walker et al. 2009). Vitellogenin production in juveniles and males and altered vitellogenin production in females may have fitness consequences if associated with impaired reproduction.

Immunotoxicity

Prevalent environmental contaminants such as mercury and POPs have produced immunotoxic effects in laboratory exposures of wildlife (Silkworth et al. 1984; Smialowicz et al. 1989; Harper et al. 1993; Ross et al. 1996; Wu et al. 1999; Segre et al. 2002; Smits et al. 2002). Many field studies have documented correlative relationships between immune function and contaminant exposure in free-ranging individuals (Lahvis et al. 1995; Grasman and Fox 2001; reviewed in Keller et al. 2000). Immunomodulation can be immunosuppression, which is a weakening of the immune response, or hypersensitivity, which includes immune disorders such as autoimmune diseases. Pollution-induced immune suppression can increase the susceptibility of wildlife to disease and has been speculated to be the cause of the increased localized incidence of some turtle diseases, such as the debilitating fibropapillomatosis in sea turtles (Balazs and Pooley 1991).

Although research on turtles is limited, work on loggerhead sea turtles off the coast of the southeastern United States found negative correlations between mercury exposure as measured in blood and both lymphocyte cell counts and ex vivo lymphocyte proliferation of B-cells (Day et al. 2005). Measuring the rate of lymphocyte proliferation after exposure to a mitogen is a common functional assay to assess immune response. Day et al. (2005) corroborated their field observation with an in vitro dosing study and found that lymphocyte proliferation decreased at higher MeHg blood concentrations for B-cells and T-cells. The comparability of the sensitivity of loggerhead and terrapin immune systems to MeHg is unknown, but the immunosuppressive responses observed in loggerheads occurred across blood mercury concentrations comparable to terrapin exposure levels in the wild. This suggests that MeHg is capable of immunomodulation in turtles at environmentally relevant concentrations.

Lymphocyte proliferation has not been assessed in terrapins, but lysozyme activity measured across populations at five sites was found to be lowest in terrapins from Purvis Creek, near the historical site of a chlor-alkali plant where mercury exposure was the highest of all stations sampled (746.2 ng/g; Blanvillain 2005). Lysozymes are bacteriolytic agents secreted by leukocytes (Grinde 1989; Chen et al. 1996), and their levels can be reduced by mercury compounds (Fournier et al. 2000). The Purvis Creek terrapins also exhibited black necrotic lesions on the plastron that were similar to those attributed to bacterial infections of the keratin layer in other turtle species (Lovich et al. 1996; Garner et al. 1997;

Ernst et al. 1999). Shell diseases in the desert tortoise (*Gopherus agassizii*) and the flattened musk turtle (*Sternotherus depressus*) (Dodd 1988; Jacobson et al. 1994) are considered a significant health threat and are probably of comparable concern in terrapins. Polychlorinated biphenyls are also elevated at Purvis Creek, but these were not measured in the study, so their role in the decreased lysozyme activity and shell disease is unknown. These and other organochlorine contaminants also are well-established immunomodulators, and correlations have been reported between total PCB concentrations in plasma and immune parameters in loggerhead sea turtles (Keller et al. 2006). Further studies are required to characterize the exposure and risk of organochlorine immunotoxicity in terrapins.

Overview of Turtle Sex Hormones

Lee (2003) looked closely at testosterone levels in adult male and female terrapins to characterize the general reproductive steroid cycle. She found a distinct postnuptial reproductive cycle, with turtles reinitiating their gonadal growth cycles soon after the summer nesting season (Fig. 12.1). Females completed their ovarian growth in the fall, and testosterone showed a distinct peak in the spring at the time of mating. Although estrogen has not been studied in terrapins, we would expect elevated aromatase in late summer and fall to promote estrogen production, thus reducing testosterone concentrations in females during this time. Males demonstrated testicular growth in the fall, coincident with peak testosterone production. In Charleston, South Carolina, limited courtship and mating were observed in October (Estep 2005), while the peak of copulatory behavior was clearly in the spring, in April and May, well after testosterone began to decline for males.

Other than Lee's (2003) observations of testosterone, there have been no studies of sex steroids in terrapins. Although mammalian or other reptilian studies may be instructive in developing a model, these systems may not be representative of the physiology of the unique, estuarine-dwelling terrapin. However, estrogen is of particular importance

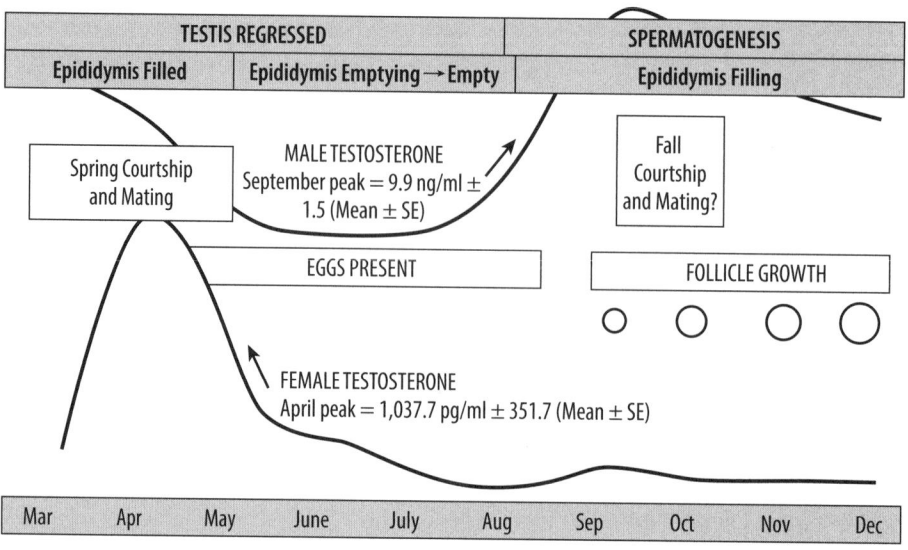

Fig. 12.1. Reproductive and testosterone cycles of diamond-backed terrapins in Charleston, South Carolina, and surrounding areas. Terrapins have a reproductive cycle best described as postnuptial (Saint Girons 1963), in which turtles mate and nest in spring and then, after a very active summer foraging period, immediately show gonadal recrudescence (both testis and ovary) before hibernation. Adapted from Lee (2003), with permission

in all turtles due to its clear link to calcium mobilization from the liver and bones, particularly during eggshell formation. Progesterone is secreted in elevated titers in this critical physiological sequence, as it moves into the bloodstream at high concentrations coincident with ovulation. Corpora lutea, which are left behind in the ovary when oocytes move into the oviduct, are the site of progesterone production (Owens 1997; Hamann et al. 2003). The dramatic surge in progesterone is likely to play an important role at this time, first in albumin secretion around the fertilized ovum (in the distal oviduct), then in secretion of the critical fibrous protein matrix around the albumin (in the medial oviduct), and finally (possibly with estrogen) in secretion of the calcium crystal shelling (in the basal oviduct).

Estrogen also is critical to the full cellular maturation of the oviduct, both developmentally and annually during the recrudescent reproductive season (for a review, see Norris 2007). In 3- to 4-year-old green sea turtles, both progesterone and testosterone injected intramuscularly induced minimal and abnormal oviduct growth, whereas estrogen produced apparently normal oviduct development in these 40 to 50 kg immature animals

(Owens 1976). In nesting female sea turtles, estrogen is elevated after ovulation during the internesting period in these multi-clutch species. We would expect the situation to be the same in terrapins, which also lay multiple clutches, but this has not been tested as far as we are aware.

Although terrapin endocrinology remains largely uninvestigated, other species of turtles, and particularly sea turtles, have received some useful and relevant attention. As an example, Licht, Papkoff, and others studied pituitary tissues from the green sea turtle, obtained from the commercial turtle farm on Grand Cayman Island (Licht et al. 1979; Licht 1984). They purified both follicle-stimulating hormone and luteinizing hormone and determined apparently distinct functions for these molecules (Licht 1984), although functional questions in turtle species are still not fully resolved (Norris 2007). These two turtle glycoproteins, as well as thyrotropin from the same gene family, have amino acid sequences similar, but not identical, to those of the mammalian hormones, and they have physiological and phylogenetic relatedness (conserved molecular structures) to similar glycoproteins in other amniotic tetrapods (Norris 2007).

Endocrine Disruptors

The presence of endocrine disruptors (EDs) in an animal presents serious concerns for both individual survival and reproductive success. Reptiles are good models for studying endocrine disruption because of their varying modes of sex determination (Crain and Guilette 1998). In terms of regulatory biology, turtles use the expected suite of steroid and other hormones, including testosterone, estradiol, progesterone, corticosterone (CORT), aldosterone, and thyroid hormones (T3 and T4) (Norris 2007). Other potentially important molecules, less well studied, include melatonin, leptin, somatotropic peptides, and many others, each with the potential to be involved in endocrine disruption. There are many similarities between mammalian and reptilian regulatory systems, which argues for turtle researchers to take advantage of the strong mammalian literature in identifying and formatting appropriate questions (Boggs et al. 2011). For example, the steroid-like molecule diethylstilbestrol, infamously used in human reproductive medicine (Norris 2007), was shown to cause abnormal male and female gonadal development in the green sea turtle (Owens et al. 1979).

By definition, EDs affect the way that cells communicate. They might be agonists or antagonists of receptors, or act as modulators to change the receptor environment. Regardless of the exact mechanism, endogenous hormone signaling is likely to be affected. It is difficult to use plasma hormone concentrations as a biomarker of exposure because hormones in both sexes vary seasonally with changes in reproduction and may vary across sites, irrespective of environmental estrogens or antiandrogens. No consistent pattern of sex steroid production in contaminated vs. reference populations is reported in the literature. Turtles collected from affected sites exhibited higher concentrations (J. Meyer et al. 2013), lower concentrations (Rie et al. 2005), or statistically similar concentrations (de Solla et al. 1998) of testosterone or estradiol. In a controlled laboratory experiment, Kitana et al. (2006) challenged male and female painted turtles from two sites with ovine follicle-stimulating hormone. Female and male turtles from the reference site increased estradiol and testosterone production, whereas only females from the Cape Cod superfund site were not affected by the exposure.

We can readily assess some effects of xenoestrogens by examining hatchling sex ratios. Turtles are oviparous organisms in which the sex of the offspring is determined by the incubation temperature of the egg (Bull 1980; Chapter 10). Many studies have shown that environmental contaminants with endocrine-disrupting effects can override the effect of temperature on gonadal differentiation (e.g., Raynaud and Pieau 1985; Wibbels et al. 1993).

In these sex reversal studies, chemicals are dissolved in a vehicle (e.g., ethanol, acetone, or dimethyl sulfoxide) and topically applied to an egg during the temperature-sensitive period. Gonads are then harvested during development or at hatching and sectioned for histological identification of sex. Many studies use 17β-estradiol as a positive control. Topical application of 17β-estradiol to terrapin eggs produced female turtles at male-producing temperatures, whereas application of an aromatase inhibitor (4-OH androstenedione) did not cause sex reversal (Jeyasuria et al. 1994). Ethinyl estradiol, a synthetic derivative of 17β-estradiol commonly found in aquatic environments, is often used as a positive control in mammalian and fish studies and should be considered an acceptable positive control in future turtle studies.

Despite the lack of sex reversal data after contaminant exposure in terrapins, such a reversal has been shown many times in other aquatic species. For example, red-eared slider eggs incubated at male-producing temperatures produced 50% or 100% female clutches when exposed, respectively, to 100 μg per egg of the PCBs 2′,3′,4′,5′-tetrachloro-4-biphenylol or 2′,4′,6′-trichloro-4-biphenylol (Bergeron et al. 1994). Simultaneous exposure to both PCBs produced interactive effects, with 35% female clutches seen at even lower doses. Similarly, Willingham and Crews (1999) produced 20% more females at male-producing temperatures by exposing red-eared slider eggs to trans-nonachlor, p,p′-DDE, and chlordane, and 50% more females when the eggs were exposed to Aroclor 1242. Furthermore, 30% of painted turtles exposed in ovo to just 0.05 ng per egg of bisphenol A exhibited disruption of sexual differentiation, with degradation of testicular tubules and ovarian-like expansion of the cortex (Jandegian et al. 2015).

Sex reversal does not result from application of all EDs. Willingham et al. (2000) exposed red-eared slider eggs to the pesticides toxaphene, dieldrin, or p,p′-DDD and did not find any sex reversal at male-producing temperatures. Similarly, snapping turtle and green sea turtle eggs incubated at male-producing temperatures and spotted with as much as 0.1 mg/kg (Portelli et al. 1999) and 66.5 µg/µL of DDE, respectively, also did not result in female hatchlings. Sex is typically confirmed by gonadal histology. Although sex reversal may not have occurred in these few examples, exposure to EDs may have resulted in changes in gonadal ultrastructure. Contaminants that alter the architecture of the testis without identifiable feminization might still produce deleterious reproductive effects, and this area warrants further study.

Hypothalamic-Pituitary-Adrenal Axis and Thyroid Hormones

The hypothalamic-pituitary-adrenal axis has been well studied in reptiles, but only to a limited degree in turtles, with most observations devoted to understanding the stress axis and corticosterone secretion, especially during reproduction (Tokarz and Summers 2011). An interesting question in reptiles, including turtles, is the possibility that the normal axis is modulated (Moore and Jessop 2003) or inhibited during demanding events such as reproduction. Valverde et al. (1999) and Jessop et al. (2000) provided supportive evidence in sea turtles that actively nesting females show an inhibitory modulation (presumably at the central nervous system level) that would be expected to enable a female to complete her nesting sequence under added stress, even from approaching predators.

In diamond-backed terrapins, CORT concentrations have been measured as an indicator of stress under various conditions (Harden 2013; Winters 2013). Harden (2013) found that terrapins held for extended periods in natural enclosures did not exhibit increased CORT concentrations, which she interpreted as indicating that the animals were not being stressed in their anthropogenic enclosures. In sea turtles, handling of animals of all ages will cause sharp increases of CORT concentrations within 30 min (Valverde et al. 1999). Winters (2013) studied adult nesting females in New Jersey. Just as in sea turtles, the females did not show a CORT stress response 5 min after capture, but they did show a response after 30 or 60 min later. They did not seem to be stressed (i.e., CORT concentrations did not increase) when they encountered bulkhead beach structures as they emerged to nest (Winters 2013).

Thyroid hormones have not been examined in terrapins and have not been very carefully studied in turtles. The few studies available leave a rather confusing picture. Although clearly of major metabolic importance, the thyroid regulatory system and its physiological roles are complex in seasonally inactive poikilotherms, including many turtles (Norris 2007). In mammals, PCBs are known to decrease plasma concentrations of T4 (thyroxine) (van Birgelen et al. 1992), and metabolites of some polybrominated flame retardants may competitively bind to T4 receptors (de Wit 2002). Because turtles generally show higher concentrations of T4 than T3 (triiodothyronine) (Moon et al. 1999), exposure to these contaminants may have far-reaching metabolic consequences and could be a mechanism underlying the decreases in metabolic rates seen in PCB-exposed terrapins (Holliday et al. 2009). Although Eisenreich et al. (2009) did not find a significant effect of PCBs on snapping turtles' thyrosomatic index, there was a trend in which individuals from contaminated sites given contaminated food had smaller thyroid glands. However, Eisenreich et al. (2012) demonstrated that in ovo exposure to perchlorate in red-eared sliders and snapping turtles reduced glandular thyroxine concentrations. In many turtle species—depending on age, reproductive condition, nutritional state, captive vs. wild status, and ambient temperatures—thyroid hormone changes can range from pronounced to negligible (Licht et al. 1985; Moon et al. 1999). In fact, the diamond-backed terrapin would be an excellent choice for a more careful and controlled set of observational experiments. These experiments could then be the basis

Bone and Calcium Physiology

Mammals have a well-documented three-hormone system of calcium regulation, with parathyroid hormone from the parathyroid glands promoting hypercalcemia, calcitonin from the ultimobranchial glands promoting hypocalcemia, and vitamin D (1,25-dihydroxycholecalciferol) in the gut facilitating calcium absorption. It is generally thought that reptiles conform to this pattern, although detailed work has not been undertaken for most groups (Norris 2007). Clark pioneered work on calcium metabolism in turtles (Clark and Laverty 1985), finding that freshwater turtles seem to regulate calcium differently from terrestrial turtles and other reptiles. She found that parathyroidectomies in freshwater turtles caused little change in circulating calcium but caused tetany and hypocalcemia in nearly all other animals studied, including tortoises. The role of these hormones in normal physiology needs to be clarified before we can begin to understand how these processes may be disrupted with exposure to EDs.

In addition to defense and support, turtle bone is a source of physiological buffer against acidosis during diving (Jackson 2000) and a source of calcium for the shelling of eggs (de Buffrénil and Francillon-Vieillot 2001). These physiological processes are under endocrine control, so a disruption of hormone signaling could compromise some of these functions and have serious consequences for fitness. Much research is needed to further elucidate the patterns associated with normal bone development and ossification in juvenile turtles so that we can better understand the mechanisms underlying disruption. Holliday and Holliday (2012) used computed tomography and histology to examine bone structure in terrapins exposed to PCB 126, a known ED. Exposed animals had less bone in the cortical region and greater total amounts of low-density bone in femurs, but not in ribs or plastrons (Holliday and Holliday 2012). The PCBs may have affected bone deposition via osteoblasts and calcitonin, or may have affected bone resorption via osteoclasts and parathyroid hormone, or operated through some other, unidentified mechanism. Additional studies need to examine why some bony elements seem to be more affected than others and whether these differences might manifest in bony carapacial elements, providing another biomarker of exposure.

Organismal Effects

Effects of contaminant exposure at the organismal level include mortality and sublethal endpoints such as anomalies in morphology, growth, metabolic rate, reproductive function, and behavior. These endpoints are not considered biomarkers, because it is difficult to tease apart the effects of multiple variables at the organismal level. Smaller animals collected from exposed natural populations may be indicators of contaminant exposure or may reflect the complex interplay of a number of variables of which the presence of a specific chemical is only one piece of the puzzle.

Environmental contaminants have been suspected of causing developmental and morphological anomalies in field-collected snapping turtles in the Great Lakes area (Bishop et al. 1998) and in individual turtles exposed to pesticides in the laboratory (de Solla and Martin 2011). Similarly, wild-caught terrapins from Jamaica Bay, New York, had a higher incidence of deformities than those collected at other sites (Horn and Burke 2008). These anomalies included minor scute irregularities and major deformities such as a recessed mandible or malformed tail and limbs, which could directly affect fitness.

To examine mortality, Eisenreich et al. (2009) used a factorial design to test for the effects of maternally derived contaminants (i.e., using contaminated or reference collection sites) and of exposure through diet. Survival over the first few months was equivalent across the different treatments. However, at 8 months post-hatching, after an overwintering period, mortality significantly increased in the maternally exposed individuals: there was a lag period between exposure and effect. This same delay in mortality was seen in terrapins exposed to PCB 126 in the laboratory. For the first 150 days after intraperitoneal injection of PCB 126, survival was similar in the control and treatment groups. After that point, however, there was a precipitous decline in survival in the treatment group and a further divergence in growth (Ford 2005). These two studies clearly demonstrate that had the researchers followed growth and survival for only a few months, these patterns would have been missed and the effects of PCBs underestimated in turtles. In terrapins exposed to MeHg in their food, it took nearly a year to show effects on food consumption, mortality, and righting response (Schwenter 2007). Even then, not all turtles showed the negative responses, demonstrating how individual variability can make detecting organismal endpoints difficult (Schwenter 2007).

Juvenile size has important fitness consequences because it can affect (1) future adult body size, thus constraining clutch size; (2) overwintering success; and (3) vulnerability to predation. Many studies have shown contaminant-exposed turtles to be smaller (Bishop et al. 1994). Hatchling terrapins exposed in the laboratory to PCB 126 were smaller and exhibited very different growth rates than control animals (Holliday et al. 2009). These same PCB-exposed animals also had depressed metabolic rates. Neither growth nor metabolic rate differed from that of control animals until at least 90 days after exposure, again suggesting a latent period between exposure and effect. As stated earlier, a possible explanation for this pattern of delayed and reduced growth is that exposure to PCBs causes a depletion of thyroid hormone (van Birgelen et al. 1992). If this were the case, one would expect these same individuals to have a reduced metabolic rate and decreased bone density, as these are also associated with hypothyroidism. As predicted, PCB 126–exposed terrapins showed reduced carbon dioxide production at 90 days after exposure (Holliday et al. 2009) and decreased apparent bone mineral density (Holliday and Holliday 2012). Eisenreich et al. (2012) further supported this idea by showing that snapping turtles exposed in ovo to perchlorate had decreased thyroxine concentrations and depressed metabolic rates. Although these animals were producing less thyroxine, their thyroid was enlarged, suggesting at least some part of the endocrine feedback loop was functioning properly.

Endocrine disruptors and neurotoxins can also affect behavior. The most common behavioral measure in turtles is the righting response, because it is easily measured and may be indicative of decreased physical strength, reduced motivation, or other general health concerns. Schwenter (2007) found a significant delay in righting response in terrapins after approximately 1 year of receiving regular dietary doses of MeHg, with females exhibiting a stronger impairment than males. Burger (1994) measured a delayed righting ability in terrapins collected after the 1990 Arthur Kill oil spill. However, in a laboratory experiment, Neuman-Lee and Janzen (2011) exposed *Graptemys ouachitensis* and *G. pseudogeographica* eggs to atrazine-contaminated sands and did not find a significant difference in righting, swimming, or foraging performance. Similarly, Rowe et al. (2009) found no effect of hydrocarbons on snapping turtles' righting response, and they found no effect on predator avoidance during a simulated attack. One behavioral endpoint that has not received attention is how contaminants, especially those with estrogenic or antiandrogenic properties, might affect reproductive behavior. Admittedly, this would be difficult to measure because it would require sexually mature and reproductively active animals to be observed under laboratory conditions. However, if any contaminant exposure produces male-male pairings similar to those caused by atrazine exposure in amphibians (Hayes et al. 2010), the population level effects could be significant.

Population Effects

Biomarkers and organismal endpoints such as those discussed above provide the toxicological foundation for understanding mechanisms, sensitivity, and other measurable biological outcomes associated with exposure to environmental contaminants. This chapter focuses on our current understanding of these key elements, but to effectively bridge the gap between toxicology, risk assessment, and conservation, researchers should provide information to managers on how these effects translate into population level effects. The merits and challenges associated with implementing this strategy for reptiles are discussed in more detail elsewhere (Selcer 2006). Confounding factors such as trophic interactions, predator-prey dynamics, habitat degradation, fisheries mortality, and climate and other abiotic factors can have a profound effect on terrapin populations, making it difficult to assess the effects of contaminants. The long life span and slow reproduction of terrapins also delay the ability to detect population changes. Despite these challenges, future toxicological research should target endpoints that can be more directly translated into changes in fecundity, growth, survivorship, and/or mortality to improve population level risk assessments.

Conclusions

Environmental toxicology in turtles is a relatively young field. Throughout this chapter we have noted the need for additional studies of specific toxicological endpoints. We have also stressed the need for studies of "typical" physiology so that, knowing how the system functions in the absence of EDs and other exogenous chemicals, we can better understand the effects of environmental perturbation. Many of the studies presented here measured effects without an understanding of underlying mechanisms. These mechanisms may be conserved across taxa, and turning to the literature on other, more extensively studied species (e.g., fish, mammals, birds), we may enhance our understanding of these physiological processes in turtles. Our key recommendations for future studies are summarized in Table 12.1.

We know little about how different contaminants affect the subcellular, cellular, and organismal processes of turtles, let alone terrapins specifically. We know even less about how suites of stressors might bring about physiological change and endocrine disruption. Like all organisms, terrapins regularly experience multiple stressors. These stressors are varied and include not only the contaminants addressed in this chapter but also fluctuations in temperature, salinity, and other ecological and physiological variables. Multiple stressors encountered simultaneously may produce effects greater than the combined sum of the single ef-

Table 12.1. Key recommendations for future terrapin environmental toxicology studies

Broad goals	Specific needs
Investigate underlying mechanisms	Endocrinology studies Pharmacokinetics in adults and eggs
Develop new nonlethal techniques	New biomarkers or tissues Behavioral endpoints
Determine suitability of surrogate species	Comparative studies with other emydids
Strive for ecological realism	Lab studies with multiple stressors Mimic routes of exposure and concentrations
Translate to the population level	Long-term monitoring of populations Endpoints usable in risk assessment

fects. For example, Zaga et al. (1998) reported synergistic deleterious effects in larval development when frogs were exposed to carbamate (an insecticide) and ultraviolet light. However, interactions can also be antagonistic or nonexistent. When Holliday et al. (2009) exposed terrapins to PCB 126 and a range of natural salinities, no interactive effects were observed. Nevertheless, salinity, pH, and temperature are likely to be of increasing concern given predictions by climate change scientists, so these variables will be of increasing importance in the coming years. Many studies in the past, perhaps due to practical limitations, typically examined a particular contaminant, or a specific mixture, or a set of ecological variables. Clearly, there is a need for multiple-stressor experiments, and such research will increase ecological realism.

We encourage researchers conducting laboratory and field studies on terrapins to strive for ecological realism by including additional variables when possible. For example, Rowe et al. (2009) exposed snapping turtle eggs to hydrocarbon mixtures that had been physically dispersed by percolation through sand or mixtures chemically dispersed using a commercially available dispersant (Corexit 9500). The inclusion of percolated oil provides information on how nesting substrate could affect exposure of eggs within a nest. In the experiment, the low molecular weight PAHs were trapped in the sand, thus exposing the eggs to more of the medium and high molecular weight hydrocarbons. Furthermore, the addition of a chemical dispersant such as Corexit mimics what the turtles would experience during the response to an oil spill. The dispersant altered the exposure of eggs to different PAH congeners but had no direct effects on any of the endpoints measured. These data enhance our ability to quantify damage in the case of oil spills or other such accidents and may provide additional insight into the relationship between exposure and effects.

Globally, turtle populations are in decline, and researchers must be careful not to reduce population size. Many environmental toxicology studies, by necessity, euthanize individuals; consequently, research has largely focused on the effects of contaminants on the younger age classes. When possible, terrapin researchers should consider using a surrogate species. The suitability of different species for this role has been much debated. Based on current knowledge, we suggest using painted turtles or red-eared sliders as surrogates, given our better understanding of their biology and their commercial availability. We also encourage all researchers to employ nondestructive techniques when possible. We support Hopkins et al. (2001) in their argument for the development of additional nonlethal sampling methods. Some techniques are already being implemented, such as the use of scutes for assessing mercury load (Blanvillain et al. 2007), blood samples for measuring vitellogenin concentrations (Irwin et al. 2001), and genotoxicity studies (Matson et al. 2005), as well as ultrasound for reproductive studies in adults (Lee 2003).

The tidal creek habitat of terrapins has been labeled a "sentinel habitat" for our coastal environment, partly because it receives direct runoff of anthropogenic pollutants from upland areas (Holland et al. 2004). The evidence presented in this chapter suggests that terrapins are an effective indicator species for assessing the degree of environmental contamination in their estuarine habitat. But what about assessing the risk of toxicological effects on terrapins? Are terrapins sensitive enough to toxicants to make them a sentinel species in this senti-

nel habitat? And how much risk do contaminants pose to terrapin populations, compared with habitat loss and fisheries bycatch (discussed in Chapters 14 and 16)? The threshold of toxicity for various contaminants can vary broadly among taxa, and limited data currently exist for Testudines, or even Reptilia. Additional work on terrapin toxicology is needed to allow better risk assessments and epidemiological studies for terrapin populations. Field studies can provide some of the needed health information through nonlethal means and minimal effects on populations. This includes defining normal ranges of basic blood cell counts and plasma chemistries and performing correlative toxicological studies by comparing contaminant exposure with clinical parameters and biomarkers of immunotoxicity, genotoxicity, endocrine disruption, and neurotoxicity. However, additional experimental laboratory studies may be needed to definitively characterize toxicology and to further develop the terrapin as an indicator species. This will include better defining key endpoints that link cellular and biochemical responses to organismal and population effects. Diamond-backed terrapins are a hardy species that survives and reproduces in captivity, are of a manageable size, and in most areas are not currently threatened or endangered. Given these factors, additional studies on terrapin toxicology would further develop this species as a valuable estuarine sentinel and a chelonian toxicological model and would allow a more effective assessment of the risk of contaminants to populations of terrapins and other reptiles.

ACKNOWLEDGMENTS

We thank M. Lee, R. Estep, J. Schwenter, C. Arthur, and G. Blanvillain for providing access to their studies of reproduction, behavior, and mercury in terrapins. We also thank the anonymous reviewers and the book editors for their constructive comments.

REFERENCES

Aardema, M.J. and J.T. MacGregor. 2002. Toxicology and genetic toxicology in the new era of "toxicogenomics":

impact of "-omics" technologies. Mutation Research 499:13-25.

Andreani, G., M. Santoro, S. Cottignoli, M. Fabbri, E. Carpenè and G. Isani. 2008. Metal distribution and metallothionein in loggerhead (*Caretta caretta*) and green (*Chelonia mydas*) sea turtles. Science of the Total Environment 390:287-294.

Arthur, C. 2009. Mercury contamination along the eastern coast of the United States: assessment of the diamondback terrapin, *Malaclemys terrapin*, as an indicator species. MS thesis, College of Charleston, Charleston, SC. 60 pages.

Balazs, G.H. and S.G. Pooley. 1991. Research plan for marine turtle fibropapilloma. NOAA Technical Memorandum NMFS no. NOAA-TM-NMFS-SWFSC-156. National Oceanographic and Atmospheric Administration, National Marine Fisheries Service, Honolulu. http://www.pifsc.noaa.gov/tech/NOAA_Tech_Memo_156.pdf.

Barra, R., J.C. Sanchez-Hernandez, R. Orrego, O. Parra and J.F. Gavilan. 2001. Bioavailability of PAHs in the Biobio River (Chile): MFO activity and biliary fluorescence in juvenile *Oncorhynchus mykiss*. Chemosphere 45:439-444.

Basile, E.R., H.W. Avery, W.F. Bien and J.M. Keller. 2011. Diamondback terrapins as indicator species of persistent organic pollutants: using Barnegat Bay, New Jersey as a case study. Chemosphere 82:137-144.

Bergeron, J.M., D. Crews and J.A. McLachlan. 1994. PCBs as environmental estrogens: turtle sex determination as a biomarker of environmental contamination. Environmental Health Perspectives 102:780-781.

Bishop, C.A., G.P. Brown, R.J. Brooks, D.R.S. Lean and J.H. Carey. 1994. Organochlorine contaminant concentrations in eggs and their relationship to body size, and clutch characteristics of the female common snapping turtle (*Chelydra serpentina serpentina*) in Lake Ontario, Canada. Archives of Environmental Contamination and Toxicology 27:82-87.

Bishop, C.A., P. Ng, K.E. Pettit, S.W. Kennedy, J.J. Stegeman, R.J. Norstom and R.J. Brooks. 1998. Environmental contamination and developmental abnormalities in eggs and hatchlings of the common snapping turtle (*Chelydra serpentina serpentina*) from the Great Lakes-St. Lawrence River basin (1989-91). Environmental Pollution 101:143-156.

Blanvillain, G. 2005. Using diamondback terrapins (*Malaclemys terrapin*) as sentinel species for mercury monitoring in estuaries along the southeast coast of the US. MS thesis, College of Charleston, Charleston, SC. 146 pages.

Blanvillain, G., J.A. Schwenter, R.D. Day, D. Point, S.J. Christopher, W.A. Roumillat and D.W. Owens. 2007. Diamondback terrapins, *Malaclemys terrapin*, as a sentinel species for monitoring mercury pollution of estuarine systems in South Carolina and Georgia, USA. Environmental Toxicology and Chemistry 26:1441-1450.

Boggs, A.S.P., N. Botteri, H. Hamlin and L.J. Guillette Jr. 2011. Endocrine disruption in reptiles. Pages 373-396 in D.O. Norris and K.H. Lopez (eds.), Hormones and Reproduction of Vertebrates. Volume 3. Reptiles. Elsevier, San Diego, CA.

Bull, J.J. 1980. Sex determination in reptiles. Quarterly Review of Biology 55:3-21.

Burger, J. 1994. Immediate effects of oil spills on organisms in the Arthur Kill. Pages 115-129 in J. Burger (ed.), Before and After an Oil Spill: The Arthur Kill. Rutgers University Press, New Brunswick, NJ.

Burger, J. 2002. Metals in tissues of diamondback terrapin from New Jersey. Environmental Monitoring and Assessment 77:255-263.

Byrd, H., E. English, R. Greer, H. Hinkelday, W. Kicklighter, N. Meade, J. Michel, T. Tomasi and R. Wood. 2002. Estimate of total injury to diamondback terrapins from the Chalk Point oil spill. National Oceanic and Atmospheric Administration (NOAA) Damage Assessment, Remediation, and Restoration Program. 16 pages. https://casedocuments.darp.noaa.gov/northeast/chalk_point/pdf/cpar2038.pdf.

Cantwell, M.G., B.A. Wilson, J. Zhu, G.T. Wallace, J.W. King, C.R. Olsen, R.M. Burgess and J.P. Smith. 2010. Temporal trends of triclosan contamination in dated sediment cores from four urbanized estuaries: evidence of preservation and accumulation. Chemosphere 78:347-352.

Carpenè, E., G. Andreani and G. Isani. 2007. Metallothionein functions and structural characteristics. Journal of Trace Elements in Medicine and Biology 21:35-39.

Carr, A. 1952. Handbook of Turtles. Cornell University Press, Ithaca, NY. 542 pages.

Chen, S.C., T. Yoshida, A. Adams, K.D. Thompson and R.H. Richards. 1996. Immune responses of rainbow trout to extracellular products of *Mycobacterium* sp. Journal of Aquatic Animal Health 8:216-222.

Chesapeake Bay Program. 2014. Agriculture. http://www.chesapeakebay.net/issues/issue/agriculture.

Clark, D.R. Jr., J.W. Bickham, D.L. Baker and D.F. Cowman. 2000. Environmental contaminants in Texas, USA, wetland reptiles: evaluation using blood samples. Environmental Toxicology and Chemistry 19:2259-2265.

Clark, N.B. and G. Laverty. 1985. Role of parathyroid hormone in regulation of calcium in reptiles. Pages 843-845 in B. Lofts and W.N. Holmes (eds.), Current Trends in Comparative Endocrinology. Volume 2. Hong Kong University Press, Hong Kong.

Cobb, G.P. and P.D. Wood. 1997. PCB concentrations in eggs and chorioallantoic membranes of loggerhead sea turtles (*Caretta caretta*) from the Cape Romain National Wildlife Refuge. Chemosphere 34:539-549.

Crain, D.A. and L.J. Guilette Jr. 1998. Reptiles as models of contaminant-induced endocrine disruption. Animal Reproductive Science 53:77-86.

Custodia-Lora, N., A. Novillo and I.P. Callard. 2005. Synergistic role for pituitary growth hormone in the regulation of hepatic estrogen and progesterone receptors and vitellogenesis in female freshwater turtles, *Chrysemys picta*. General and Comparative Endocrinology 140:25-32.

Day, R.D., S.J. Christopher, P.R. Becker and D.W. Whitaker. 2005. Monitoring mercury in the logger-head sea turtle, *Caretta caretta*. Environmental Science and Technology 39:437-446.

Day, R.D., S.S. Vander Pol, S.J. Christopher, W.C. Davis, R.S. Pugh, K.S. Simac, D.G. Roseneau and P.R. Becker. 2006. Murre eggs (*Uria aalge* and *Uria lomvia*) as indicators of mercury contamination in the Alaskan marine environment. Environmental Science and Technology 40:659-665.

de Buffrénil, V. and H. Francillon-Vieillot. 2001. Ontogenetic changes in bone compactness in male and female Nile monitors (*Varanus niloticus*). Journal of Zoology 254:539-546.

de Solla, S.R., C.A. Bishop, G. van der Kraak and R.J. Brooks. 1998. Impact of organochlorine contamination on levels of sex hormones and external morphology of common snapping turtles (*Chelydra serpentina serpentina*) in Ontario, Canada. Environmental Health Perspectives 106:253-260.

de Solla, S.R. and P.A. Martin. 2011. Absorption of current use pesticides by snapping turtle (*Chelydra serpentina*) eggs in treated soil. Chemosphere 85:820-825.

de Solla, S.R., P.A. Martin, K.J. Fernie, B.J. Park and G. Mayne. 2006. Effects of environmentally relevant concentrations of atrazine on gonadal development of snapping turtles (*Chelydra serpentina*). Environmental Toxicology and Chemistry 25:520-526.

de Wit, C.A. 2002. An overview of brominated flame retardants. Chemosphere 46:583-624.

Dodd, C.K. Jr. 1988. Disease and population declines in the flattened musk turtle *Sternotherus depressus*. American Midland Naturalist 119:394-401.

Drabeck, D.H., M.W.H. Chatfield and C.L. Richards-Zawacki. 2014. The status of Louisiana's diamondback terrapin (*Malaclemys terrapin*) populations in the wake of the Deepwater Horizon oil spill: insights from population genetic and contaminant analyses. Journal of Herpetology 48:125-136.

Dunson, W.A. and F.J. Mazzotti. 1989. Salinity as a limiting factor in the distribution of reptiles in Florida Bay: a theory for the estuarine origin of marine snakes and turtles. Bulletin of Marine Science 44:229-244.

Eisenreich, K.M., K.M. Dean, M.A. Ottinger and C.L. Rowe. 2012. Comparative effects of *in ovo* exposure to sodium perchlorate on development, growth, metabolism and thyroid function in the common snapping turtle (*Chelydra serpentina*) and red-eared slider (*Trachemys scripta elegans*). Comparative Biochemistry and Physiology 156C:166-170.

Eisenreich, K.M., S.M. Kelly and C.L. Rowe. 2009. Latent mortality of juvenile snapping turtles from the upper Hudson River, New York, exposed maternally and via the diet to polychlorinated biphenyls (PCBs). Environmental Science and Technology 43:6052-6057.

Ernst, C.H., T.S.B. Akre, J.C. Wilgenbusch, T.P. Wilson and K. Mills. 1999. Shell disease in turtles in the Rappahannock River, Virginia. Herpetological Review 30:214-215.

Ernst, C.H. and J.E. Lovich. 2009. Turtles of the United States and Canada. 2nd edition. Johns Hopkins University Press, Baltimore, MD. 827 pages.

Ertl, R.P. and G.W. Winston. 1998. The microsomal mixed function oxidase system of amphibians and reptiles: components, activities and induction. Comparative Biochemistry and Physiology 121C:85-105.

Estep, R.L. 2005. Seasonal movement and habitat use patterns of a diamondback terrapin (*Malaclemys terrapin*) population. MS thesis, Grice Marine Laboratory, College of Charleston, Charleston, SC. 113 pages.

Feng, H., J.K. Cochran, H. Lwiza, B.J. Brownawell and D.J. Hirschberg. 1998. Distribution of heavy metal and PCB contaminants in the sediments of an urban estuary: the Hudson River. Marine Environmental Research 45:69-88.

Forbes, V.E., A. Palmqvist and L. Bach. 2006. The use and misuse of biomarkers in ecotoxicology. Environmental Toxicology and Chemistry 25:272-280.

Ford, D. 2005. Sublethal effects of stressors on physiological and morphological parameters in the diamondback terrapin, *Malaclemys terrapin*. PhD dissertation, Ohio University, Athens, OH. 136 pages.

Fournier, M., D. Cyr, B. Blakely, H. Boermans and P. Brousseau. 2000. Phagocytosis as a biomarker of immunotoxicity in wildlife species exposed to environmental xenobiotics. American Zoologist 40:412-420.

Frosch, D. and J. Roberts. 2011. Pipeline spills put safeguards under scrutiny. New York Times, September 9. http://www.nytimes.com/2011/09/10/business/energy-environment/agency-struggles-to-safeguard-pipeline-system.html.

Gapp, D.A., S.M. Ho and I.P. Callard. 1979. Plasma levels of vitellogenin in *Chrysemys picta* during the annual gonadal cycle: measurement by specific radio-immunoassay. Endocrinology 104:784-790.

Garner, M.M., R. Herrington, E.W. Howerth, B.L. Homer, V.F. Nettles, R. Isaza, E.B. Shotts Jr. and E.R. Jacobson. 1997. Shell disease in river cooters (*Pseudemys concinna*) and yellow-bellied turtles (*Trachemys scripta*) in a Georgia (USA) lake. Journal of Wildlife Diseases 33:78-86.

George, S.G. 1994. Enzymology and molecular biology of phase II xenobiotic-conjugating enzymes in fish. Pages 37-86 in D.C. Malins and G.K. Ostrander (eds.), Aquatic Toxicology: Molecular, Biochemical, and Cellular Perspectives. CRC Press/Lewis Publishers, Boca Raton, FL.

Gibbons, W., J.E. Lovich, A.D. Tucker, N.N. FitzSimmons and J.L. Greene. 2001. Demographic and ecological factors affecting conservation and management of the diamondback terrapin (*Malaclemys terrapin*). Chelonian Conservation Biology 4:66-74.

Golet, W.J. and T.A. Haines. 2001. Snapping turtles as monitors of chemical contaminants in the environment. Reviews in Environmental Contamination and Toxicology 135:93-153.

Grasman, K.A. and G.A. Fox. 2001. Associations between altered immune function and organochlorine contamination in young Caspian terns (*Sterna caspia*) from Lake Huron, 1997-1999. Ecotoxicology 10:101-114.

Green, A.D., K.A. Buhlmann, C. Hagen, C. Romanek and J.W. Gibbons. 2010. Tissue distribution and maternal transfer of mercury in diamondback terrapins with implications for human health. In Proceedings of the 2010 South Carolina Water Resources Conference, October 13-14, 2010, Columbia Metropolitan Convention Center, Columbia, SC. http://media.clemson.edu/public/restoration/scwrc/2010/manuscripts/t5/greena_10scwrcpaper.pdf.

Grinde, B. 1989. Lysozyme from rainbow trout, *Salmo gairdneri* Richardson, as an antibacterial agent against pathogens. Journal of Fish Disease 12:95-104.

Guirlet, E., K. Das and M. Girondot. 2008. Maternal transfer of trace elements in leatherback turtles (*Dermochelys coriacea*) of French Guiana. Aquatic Toxicology 88:267-276.

Hamann, M., C. Limpus and D. Owens. 2003. Reproductive cycles of males and females. Pages 135-161 in P. Lutz, J. Musick and J. Wyneken (eds.), The Biology of Sea Turtles. Volume 2. CRC Press, Boca Raton, FL.

Harden, L. 2013. Seasonal variation in ecology and physiology of diamondback terrapins (*Malaclemys terrapin*) in North Carolina. PhD dissertation, University of North Carolina at Wilmington, Wilmington, NC. 141 pages.

Harper, N., K. Connor and S. Safe. 1993. Immunotoxic potencies of polychlorinated biphenyl (PCB), dibenzofuran (PCDF) and dibenzo-*p*-dioxin (PCDD) congeners in C57BL/6 and DBA/2 in mice. Toxicology 80:217-227.

Hartwell, S.I. 2011. Chesapeake Bay watershed pesticides use declines but toxicity increases. Environmental Toxicology and Chemistry 30:1223-1231.

Hayes, T.B., V. Khoury, A. Narayan, M. Nazir, A. Park, T. Brown, L. Adame, E. Chan, D. Bucholz, T. Stueve and S. Gallipeau. 2010. Atrazine induces complete feminization and chemical castration in male African clawed frogs (*Xenopus laevis*). Proceedings of the National Academy of Sciences 107:4612-4617.

Heck, J., D. MacKenzie, D. Rostal, K. Medler and D. Owens. 1997. Estrogen induction of plasma vitellogenin in the Kemp's ridley sea turtle (*Lepidochelys kempi*). General and Comparative Endocrinology 107:280-288.

Herbst, L.H., L. Siconolfi-Baez, J.H. Torelli, P.A. Klein, M.J. Kerben and I.M. Schumacher. 2003. Induction of vitellogenesis by estradiol-17β and development of enzyme-linked immunosorbant assays to quantify plasma vitellogenin levels in green turtles (*Chelonia mydas*). Comparative Biochemistry and Physiology 135B:551-563.

Heyes, A., R.P. Mason, E.-H. Kim and E. Sunderland. 2006. Mercury methylation in estuaries: insights from using measuring rates using stable mercury isotopes. Marine Chemistry 102:134-147.

Holland, A.F., D.M. Sanger, C.P. Gawle, S.B. Lerberg, M.S. Santiago, G.H.M. Riekerk, L.E. Zimmerman and G.I. Scott. 2004. Linkages between tidal creek ecosystems and the landscape and demographic attributes of their watersheds. Journal of Experimental Marine Biology and Ecology 298:151-178.

Holliday, D.K., A.A. Elskus and W.M. Roosenburg. 2009. Impacts of multiple stressors on growth and metabolic rate of *Malaclemys terrapin*. Environmental Toxicology and Chemistry 28:338-345.

Holliday, D.K. and C.M. Holliday. 2012. The effects of the organopollutant PCB 126 on bone density in juvenile diamondback terrapins (*Malaclemys terrapin*). Aquatic Toxicology 109:228-233.

Holliday, D.K., W.M. Roosenburg and A.A. Elskus. 2008. Spatial variation in polycyclic aromatic hydrocarbon concentrations in eggs of diamondback terrapins, *Malaclemys terrapin*, from the Patuxent River, Maryland. Bulletin of Environmental Contamination and Toxicology 80:119-122.

Hopkins, B.C., M.J. Hepner and W.A. Hopkins. 2013. Non-destructive techniques for biomonitoring spatial, temporal, and demographic patterns of mercury bioaccumulation and maternal transfer in turtles. Environmental Pollution 177:164-170.

Hopkins, W.A., J.H. Roe, J.W. Snodgrass, B.P. Jackson, D.E. Kling, C.L. Rowe and J.D. Congdon. 2001. Nondestructive indices of trace element exposure in squamate reptiles. Environmental Pollution 115:1-7.

Horn, E. and R. Burke. 2008. Possible effects of endocrine disrupting chemicals on diamondback terrapin (*Malaclemys terrapin*) from Jamaica Bay, NY. Section VII, pages 1-23, in C.A. McGlynn and J.R. Waldman (eds.), Final Reports of the Tibor T. Polgar Fellowship Program, 2007. Hudson River Foundation, New York, NY.

Huggett, R.J., R.A. Kimerle, P.M. Mehrle Jr. and H.L. Bergman. 2002. Biomarkers: Biochemical, Physiological, and Histological Markers of Anthropogenic Stress. Lewis Publishers, Boca Raton, FL.

Irwin, L.K., S. Gray and E. Oberdorster. 2001. Vitellogenin induction in painted turtle, *Chrysemys picta*, as a biomarker of exposure to environmental levels of estradiol. Aquatic Toxicology 55:49-60.

Ismail, N. 2010. Bioaccumulation of polychlorinated biphenyls in the northern diamondback terrapin (*Malaclemys terrapin terrapin*). MS thesis, Temple University, Philadelphia, PA. 112 pages.

Jackson, D.C. 2000. Living without oxygen: lessons from the freshwater turtle. Comparative Biochemistry and Physiology 125A:299-315.

Jacobson, E.R., T.J. Wronski, J. Schumacher, C. Reggiardo and K.H. Berry. 1994. Cutaneous dyskeratosis in free-ranging desert tortoises, *Gopherus agassizii*, in the Colorado desert of Southern California. Journal of Zoo and Wildlife Medicine 25:68-81.

Jandegian, C.M., S.L. Deem, R.K. Bhandari, C.M. Holliday, D. Nicks, C.S. Rosenfeld, K.W. Selcer, D.E. Tillitt, F.S. vom Saal, V. Vélez-Rivera, Y. Yang and D.K. Holliday. 2015. Developmental exposure to bisphenol A (BPA) alters sexual differentiation in painted turtles (*Chrysemys picta*). General and Comparative Endocrinology 216:77-85.

Jessop, T.S., M. Hamann, M.A. Read and C.J. Limpus. 2000. Evidence for a hormonal tactic maximizing green turtle reproduction in response to a pervasive ecological stressor. General and Comparative Endocrinology 118:407-417.

Jeyasuria, P., W.M. Roosenburg and A.R. Place. 1994. Role of P-450 aromatase in sex determination of the diamondback terrapin, *Malaclemys terrapin*. Journal of Experimental Zoology 270:95-111.

Kannan, K., M. Kawano, Y. Kashima, M. Matsui and J.P. Giesy. 1999. Extractable organohalogens (EOX) in sediment and biota at an estuarine marsh near a former chloralkali facility. Environmental Science and Technology 33:1004-1008.

Kannan, K., H. Nakata, R. Stafford, G.R. Masson, S. Tanabe and J.P. Giesy. 1998a. Bioaccumulation and toxic potential of extremely hydrophobic polychlorinated biphenyl congeners in biota collected at a superfund site contaminated with Aroclor 1268. Environmental Science and Technology 32:1214-1221.

Kannan, K., R.G. Smith, R.F. Lee, H.L. Windom, P.T. Heitmuller, J.M. Macauley and J.K. Summers. 1998b. Distribution of total mercury and methyl mercury in water, sediment, and fish from south Florida estuaries. Archives of Environmental Contamination and Toxicology 34:109-118.

Keller, J.M., J.R. Kucklick, M.A. Stamper, C.A. Harms and P.D. McClellan-Green. 2004. Associations between organochlorine contaminant concentrations and clinical health parameters in loggerhead sea turtles from North Carolina, USA. Environmental Health Perspectives 112:1074-1079.

Keller, J.M., P.D. McClellan-Green, J.R. Kucklick, D.E. Keil and M.M. Peden-Adams. 2006. Effects of organochlorine contaminants on loggerhead sea turtle immunity: comparison of a correlative field study and in vitro exposure experiments. Environmental Health Perspectives 114:70-76.

Keller, J.M., J.N. Meyer, M. Mattie, T. Augsperger, M. Rau, J. Dong and E.D. Lewin. 2000. Assessment of immunotoxicity in wild populations: review and recommendations. Reviews in Toxicology 3:167-212.

Kelly, S.M., K.A. Eisenreich, J.E. Baker and C.L. Rowe. 2008. Accumulation and maternal transfer of polychlorinated biphenyls (PCBs) in snapping turtles of the upper Hudson River, New York, USA. Environmental Toxicology and Chemistry 27:2565-2574.

Kitana, N., W. Khonsue, S.J. Won, V.A. Lance and I.P. Callard. 2006. Gonadotropin and estrogen responses in freshwater turtle (*Chrysemys picta*) from Cape Cod, Massachusetts. General and Comparative Endocrinology 149:49-57.

Ko, F. and J.E. Baker. 2004. Seasonal and annual loads of hydrophobic organic contaminants from the Susquehanna River basin to the Chesapeake Bay. Marine Pollution Bulletin 48:840-851.

Lahvis, G.P., R.S. Wells, D.W. Kuehl, J.L. Stewart, H.L. Rhinehart and C.S. Via. 1995. Decreased lymphocyte responses in free-ranging bottlenose dolphins (*Tursiops truncatus*) are associated with increased concentrations of PCBs and DDT in peripheral blood. Environmental Health Perspectives 103:67-71.

Landis, W.G., R.M. Sofield and M.-H. Yu. 2011. Introduction to Environmental Toxicology: Molecular Substructures to Ecological Landscapes. 4th edition. CRC Press, Boca Raton, FL. 514 pages.

Lee, M.A. 2003. Reproductive biology and seasonal testosterone patterns of the diamondback terrapin, *Malaclemys terrapin*, in the estuaries of Charleston, South Carolina. MS thesis, University of Charleston, Charleston, SC. 110 pages.

Licht, P. 1984. Reptiles. Pages 206-282 in G.E. Lamming (ed.), Marshall's Physiology of Reproduction. Churchill Livingstone, Edinburgh.

Licht, P., G.L. Breitenbach and J.D. Congdon. 1985. Seasonal cycles in testicular activity, gonadotropin, and thyroxin in the painted turtle, *Chrysemys picta*, under natural conditions. General and Comparative Endocrinology 59:130-139.

Licht, P., J. Wood, D.W. Owens and F. Wood. 1979. Serum gonadotropins and steroids associated with breeding activities in the green sea turtle *Chelonia mydas*. I. Captive animals. General and Comparative Endocrinology 39:274-289.

Lovich, J.E., S.W. Gotte, C.H. Ernst, J.C. Harshbarger, A.F. Laemmerzahl and J.W. Gibbons. 1996. Prevalence and histopathology of shell disease in turtles from Lake Blackshear, Georgia. Journal of Wildlife Diseases 32:259-265.

Luepke, N.P. 1979. Monitoring Environmental Materials and Specimen Banking: Proceeding of the International Workshop, Berlin (West), 23-28 October 1978. Martinus Hijhoff Publishers, The Hague, Netherlands. 541 pages.

Lutcavage, M.E., P.L. Lutz, G.D. Bossart and D.M. Hudson. 1995. Physiologic and clinicopathologic effects of crude oil on loggerhead sea turtles. Archives of Environmental Contamination and Toxicology 28:417-422.

Matson, C.W., G. Palatnikov, A. Islamzadeh, T.J. McDonald, R.L. Autenrieth, K.C. Donnelly and J.W. Bickham. 2005. Chromosomal damage in two species of aquatic turtles (*Emys orbicularis* and *Mauremys caspica*) inhabiting contaminated sites in Azerbaijan. Ecotoxicology 14:513-525.

Meyer, E., D. Sparling, and S. Blumenshine. 2013. Regional inhibition of cholinesterase in free-ranging western pond turtles (*Emys marmorata*) occupying California mountain streams. Environmental Toxicology and Chemistry 32:692-698.

Meyer, J.L., S. Rogers-Burch, J.K. Leet, D.L. Villeneuve, G.T. Ankley and M.S. Sepulveda. 2013. Reproductive physiology in eastern snapping turtles (*Chelydra serpentina*) exposed to runoff from a concentrated animal feeding operation. Journal of Wildlife Diseases 49:996-999.

Meyers-Schone, L. and B.T. Walton. 1994. Turtles as monitors of chemical contaminants in the environment. Pages 93-153 in G.W. Ware (ed.), Reviews of Environmental Contamination and Toxicology. Springer, New York, NY.

Miller, R.E.L., K.J. Farley, J.R. Wands, R. Santore, A.D. Redman and N.B. Kim. 2011. Fate and transport modeling of sediment contaminants in the New York/New Jersey harbor estuary. Urban Habitats 6. http://www.urbanhabitats.org/v06n01/fateandtransport _full.html.

Mitchelmore, C.L., C.L. Rowe and A.R. Place. 2005. Tools for assessing contaminant exposure and effects in reptiles. Pages 63-122 in S. Gardner and E. Oberdorster (eds.), Toxicology of Reptiles. CRC Press, Boca Raton, FL.

Mitro, M. 2003. Demography and viability of a diamondback terrapin population. Canadian Journal of Zoology 81:716-726.

Moon, D.-Y., D.W. Owens and D.S. MacKenzie. 1999. The effects of fasting and increased feeding on plasma thyroid hormones, glucose, and total protein in sea turtles. Zoological Science 16:579-586.

Moore, I.T. and T.S. Jessop. 2003. Stress, reproduction and adrenocortical modulation in amphibians and reptiles. Hormones and Behavior 43:39-47.

Neuman-Lee, L.A. and F.J. Janzen. 2011. Atrazine exposure impacts behavior and survivorship of neonatal turtles. Herpetologica 67:23-31.

NOAA (National Oceanic and Atmospheric Administration). 2014. Incident News, Emergency Response Division, Office of Response and Restoration. http://www.incidentnews.noaa.gov/map?id=8577.

Norris, D.O. 2007. Vertebrate Endocrinology. 4th edition. Elsevier Academic Press, New York, NY. 550 pages.

Owens, D.W. 1976. The endocrine control of reproduction and growth in the green sea turtle *Chelonia mydas*. PhD dissertation, University of Arizona, Tucson, AZ. 108 pages.

Owens, D.W. 1997. Hormones in the life history of sea turtles. Pages 315-341 in P. Lutz and J. Musick (eds.), The Biology of Sea Turtles. CRC Press, Boca Raton, FL.

Owens, D.W., J.R. Hendrickson and D. Endres. 1979. Somatic and immune responses to bovine growth hormone, bovine prolactin and diethylstilbestrol in the green sea turtle. General and Comparative Endocrinology 38:53-61.

Palmer, B.D., L. Huth, D.L. Pieto and K.W. Selcer. 1998. Vitellogenin as a biomarker for xenobiotic exposure in an amphibian model system. Environmental Toxicology and Chemistry 17:30-36.

Palmer, B.D. and S.K. Palmer. 1995. Vitellogenin induction by xenobiotic estrogens in the red-eared turtle and African clawed frog. Environmental Health Perspectives 103:19-25.

Portelli, M.J., S.R. de Solla, R.J. Brooks and C.A. Bishop. 1999. Effect of dichlorodiphenyltrichloroethane on sex determination of the common snapping turtle (*Chelydra serpentina serpentina*). Ecotoxicology and Environmental Safety 43:284-291.

Raynaud, A. and C. Pieau. 1985. Embryonic development of the genital system. Pages 149-300 in C. Gans and F. Billett (eds.), Biology of the Reptilia. Volume 15. Development B. John Wiley and Sons, New York, NY.

Richardson, K.L., M.L. Castro, S.C. Gardner and D. Schlenk. 2010. Polychlorinated biphenyls and biotransformation enzymes in three species of sea turtles from the Baja California Peninsula of Mexico.

Archives of Environmental Contamination and Toxicology 58:183-193.

Rie, M.T., N. Kitana, K.A. Lendas, S.J. Won and I.P. Callard. 2005. Reproductive endocrine disruption in a sentinel species (*Chrysemys picta*) on Cape Cod, Massachusetts. Archives of Environmental Contamination and Toxicology 48:217-224.

Rie, M.T., K.A. Lendas, B.R. Woodin, J.J. Stegeman and I.P. Callard. 2000. Hepatic biotransformation enzymes in a sentinel species, the painted turtle (*Chrysemys picta*), from Cape Cod, Massachusetts: seasonal-, sex- and location related differences. Biomarkers 5:382-394.

Roesijadi, G. 1994. Metallothionein induction as a measure of response to metal exposure in aquatic animals. Environmental Health Perspectives 102(Suppl. 12):91-95.

Ross, P., R. De Swart, R. Addison, H. Van Loveren, J. Vos and A. Osterhaus. 1996. Contaminant-induced immunotoxicity in harbour seals: wildlife at risk? Toxicology 112:157-169.

Rowe, C.L. 2008. "The calamity of so long life": life histories, contaminants, and potential emerging threats to long-lived vertebrates. BioScience 58:623-631.

Rowe, C.L., C.L. Mitchelmore and J.E. Baker. 2009. Lack of biological effects of water accommodated fractions of chemically- and physically-dispersed oil on molecular, physiological and behavioral traits of juvenile snapping turtles following embryonic exposure. Science of the Total Environment 407:5344-5355.

Ruttkay-Nedecky, B., L. Nejdl, J. Gumulec, O. Zitka, M. Masarik, T. Eckschlager, M. Stiborova, V. Adam and R. Kizek. 2013. The role of metallothionein in oxidative stress. International Journal of Molecular Science 14:6044-6066.

Saint-Girons, H. 1963. Spermatogénèse et evolution cyclique des caractères sexuels secondaires chez les Squamata. Annales des Sciences Naturelles–Zoologie et Biologie Animale, 12e ser. 5:461-478.

Schlezinger, J.J., J. Keller, L.A. Verbugge and J.J. Stegeman. 2000. 3,3′4,4′-Tetrachlorobiphenyl oxidation in fish, bird and reptile species: relationship to cytochrome P450 1A inactivation and reactive oxygen production. Comparative Biochemistry and Physiology 125C:273-286.

Schwenter, J.A. 2007. Monitoring mercury in the diamondback terrapin (*Malaclemys terrapin*): kinetics and accumulation of an emerging global contaminant. MS thesis, Grice Marine Biological Laboratory, College of Charleston, Charleston, SC. 225 pages.

Segre, M., S.M. Arena, E.H. Greeley, M.J. Melancon, D.A. Graham and J.B. French Jr. 2002. Immunological and physiological effects of chronic exposure of *Peromyscus leucopus* to Aroclor 1254 at a concentration similar to that found at contaminated sites. Toxicology 174:163-172.

Selcer, K.W. 2006. Reptile ecotoxicology: studying the effects of contaminants on populations. Pages 267-297 in S.C. Gardner and E. Oberdorster (eds.), Toxicology of Reptiles. CRC Press, Boca Raton, FL.

Selman, W. and B. Baccigalopi. 2012. Effectively sampling Louisiana diamondback terrapin (*Malaclemys terrapin*) populations, with description of a new capture technique. Herpetological Review 43:583-588.

Shelby-Walker, J.A., C.K. Ward and M.T. Mendonça. 2009. Reproductive parameters in female yellow-blotched map turtles (*Graptemys flavimaculata*) from a historically contaminated site vs. a reference site. Comparative Biochemistry and Physiology 154A:401-408.

Silkworth, J.B., L. Antrim and L.S. Kaminsky. 1984. Correlations between polychlorinated biphenyl immunotoxicity, the aromatic hydrocarbon locus, and liver microsomal enzyme induction in C57BL/6 and DBA/2 mice. Toxicology and Applied Pharmacology 75:156-165.

Smialowicz, R.J., J.E. Andrews, M.M. Riddle, R.R. Rogers, R.W. Luebke and C.B. Copeland. 1989. Evaluation of the immunotoxicity of low level PCB exposure in the rat. Toxicology 56:197-211.

Smits, J.E., K.J. Fernie, G.R. Bortolotti and T.A. Marchant. 2002. Thyroid hormone suppression and cell-mediated immunomodulation in American kestrels (*Falco sparverius*) exposed to PCBs. Archives of Environmental Contamination and Toxicology 43:338-344.

Sridhar, S., T.W. Collette, A.W. Garrison, N.L. Wolfe and S.C. McCutcheon. 1999. Perchlorate identification in fertilizers. Environmental Science and Technology 33:3469-3472.

Stagg, R. and A. McIntosh. 1998. Biological effects of contaminants: determination of CYP1A-dependent monooxygenase activity in dab by fluorimetric measurement of EROD activity. ICES Techniques in Marine Environmental Sciences, no. 23. International Council for the Exploration of the Sea, Copenhagen. 16 pages.

Stahl, J.R.G. 1997. Can mammalian and non-mammalian "sentinel species" data be used to evaluate the human health implications of environmental contaminants? Human and Ecological Risk Assessment 3:329-335.

Tada, N., A. Nakao, H. Hoshi, M. Saka and Y. Kamata. 2008. Vitellogenin, a biomarker for environmental estrogenic pollution of Reeves' pond turtles: analysis of similarity for its amino acid sequence and cognate mRNA expression after exposure to estrogen. Journal of Veterinary Medical Science 70:227-234.

Thomas, P., K.N. Baer and R.B. White. 1994. Isolation and partial characterization of metallothionein in the liver of the red-eared turtle (*Trachemys scripta*) following interperitoneal administration of cadmium. Comparative Biochemistry and Physiology 107C:221-226.

Tokarz, R.R. and C.H. Summers. 2011. Stress and reproduction in reptiles. Pages 169-213 in D.O. Norris and K.H. Lopez (eds.), Hormones and Reproduction of Vertebrates. Volume 3. Elsevier, San Diego, CA.

Tucker, A.D., N.N. FitzSimmons and J.W. Gibbons. 1995. Resource partitioning by the estuarine turtle *Malaclemys terrapin*: trophic, spatial, and temporal foraging constraints. Herpetologica 51:167-181.

USEPA (United States Environmental Protection Agency). 2013. Polychlorinated biphenyls (PCBs). http://www .epa.gov/epawaste/hazard/tsd/pcbs/about.htm.

Valverde, R.A., D.W. Owens, D.S. MacKenzie and M.S. Amoss. 1999. Basal and stress-induced corticosterone levels in olive ridley sea turtles (*Lepidochelys olivacea*) in relation to their mass nesting behavior. Journal of Experimental Zoology 284:652-662.

van Birgelen, A., J. van der Kolk, H. Poiger, M. van den Berg and A. Brouwer. 1992. Interactive effects of 2,2′,4, 4′,5,5′-hexachlorobiphenyl and 2,3,7,8-tetrachlorodibenzo-*p*-dioxin on thyroid hormone, vitamin A, and vitamin K metabolism in the rat. Chemosphere 25:1239-1244.

Vander Pol, S.S., P.R. Becker, J.R. Kucklick, R.S. Pugh, D.G. Roseneau and K.S. Simac. 2004. Persistent organic pollutants in Alaskan murre (*Uria* spp.) eggs: geographical, species, and temporal comparisons. Environmental Science and Technology 38:1305-1312.

Venancio, L.P.R., M.I.A. Silva, T.L. da Silva, V.A.G. Moschetta, D.A.P. de Campos Zuccari, E.A. Almeida and C.R. Bonini-Domingos. 2013. Pollution-induced metabolic responses in hypoxia-tolerant freshwater turtles. Ecotoxicology and Environmental Safety 97:1-9.

Whitlock, J.P. Jr. 1999. Induction of cytochrome P4501A1. Annual Review of Pharmacology and Toxicology 39:103-125.

Wibbels, T., P. Gideon, J.J. Bull and D. Crews. 1993. Estrogen- and temperature-induced medullary cord regression during gonadal differentiation in a turtle. Differentiation 53:149-154.

Willingham, E. and D. Crews. 1999. Sex reversal effects of environmentally relevant xenobiotic concentrations on the red-eared slider turtle, a species with temperature-dependent sex determination. Annual Review of Pharmacology and Toxicology 113:429-435.

Willingham, E., T. Rhen, J.T. Sakata and D. Crews. 2000. Embryonic treatment with xenobiotics disrupts steroid hormone profiles in hatchling red-eared slider turtles (*Trachemys scripta elegans*). Environmental Health Perspectives 108:329-332.

Winters, J.M. 2013. The effects of bulkheading on diamondback terrapin nesting in Barnegat Bay, New Jersey. PhD dissertation, Drexel University, Philadelphia, PA. 100 pages.

Wood, R.C. and L.S. Hales Jr. 2001. Comparison of northern diamondback terrapin (*Malaclemys terrapin terrapin*) hatching success among variably oiled nesting sites along the Patuxent River following the Chalk Point Oil Spill of April 7, 2000. http://www .dtwg.org/Bibliography/Gray/Wood%20and%20 Hales%202001.pdf.

Wu, P.J., E.H. Greeley, L.G. Hansen and M. Segre. 1999. Immunological, hematological, and biochemical responses in immature white-footed mice following maternal Aroclor 1254 exposure: a possible bioindicator. Archives of Environmental Contamination and Toxicology 36:469-476.

Yamamura, M. and K.T. Suzuki. 1984. Isolation and characterization of metallothionein from the tortoise *Clemmys mutica*. Comparative Biochemistry and Physiology 79C:63-69.

Yawetz, A., M. Benedek-Segal and B. Woodin. 1997. Cytochrome P4501A immunoassay in freshwater turtles and exposure to PCBs and environmental pollutants. Environmental Toxicology and Chemistry 16:1802-1806.

Yawetz, A., I. Sidis and A. Gasith. 1983. Metabolism of parathion and brain cholinesterase inhibition in Aroclor 1254-treated and untreated Caspian terrapin (*Mauremys caspica rivulata*, Emydidae, Chelonia) in comparison with two species of wild birds. Comparative Biochemistry and Physiology 75C:377-382.

Yawetz, A., B. Woodin and J.J. Stegeman. 1998. Cytochromes P450 in liver of the turtle *Chrysemys picta picta* and the induction and partial purification of CYP1A-like proteins. Biochimica et Biophysica Acta 1381:12-26.

Zaga, A., E.E. Little, C.F. Rabeni and M.R. Ellersieck. 1998. Photoenhanced toxicity of a carbamate insecticide to early life stage anuran amphibians. Environmental Toxicology and Chemistry 17:2543-2553.

PART II FISHERIES AND CONSERVATION CHALLENGES

13

History of Commercial Fisheries and Artificial Propagation

VICTOR S. KENNEDY

There are many Tortoyses both of lande and sea kinde, their backes & bellies are shelled very thicke; their head, feete, and taile, which are in appearance, seeme ougly as though they were members of a serpent or venemous: but notwithstanding they are very good meate, as also their egges.
—Hariot (1588)

I have caught with mine angle, Pike, Carpe, Eele, Perches of six severall kindes, Creafish, and the Torope or little turtle, besides many smaller kinds.
—Alexander Whitaker (1624)

Turtles in colonial America were mentioned in an early account by English explorer Thomas Hariot (1588), and Alexander Whitaker (1624) may have provided the first written mention in the Chesapeake Bay region of the diamond-backed terrapin, *Malaclemys terrapin* ("Torope" is thought to be a Virginian Algonquin word; see Chamberlain 1902). Hariot's statement about turtles as food reflects the reality that turtles have been eaten by humans worldwide (Thorbjarnarson et al. 2000). Terrapins are no exception. Their remains have been found in Indian kitchen middens along the US Atlantic coast (e.g., Sutherland 1974; Scarry and Scarry 1997; Handley 2001), and Whitaker (1624) listed them with the food fish and crayfish he caught while angling.

The history of the terrapin fishery on the US Atlantic Coast reveals terrapins to be numerous in the early nineteenth century, but apparently with little or no market value. Laffan (1877) reported that elderly residents of Maryland's Eastern Shore could remember when terrapins were fed to pigs. Before commercial demand for terrapins began to grow, the animals were undoubtedly a local and readily harvested food item for people living along estuarine shores. For example, "Bob the Sea Cook" (1881), a *New York Times* food columnist, wrote that in 1847, a resident of an unidentified small cove in Chesapeake Bay, with his son, could catch "one, two, and three dozen" terrapins in the early morning, sending what they did not eat to Baltimore or Norfolk for sale. A plantation owner living about 40 miles north of Cape Charles, Virginia, lamented around the same time that a nearby salt marsh teemed with terrapins and that "we get kind of sick of them here. At about this time of year, August, it's terrapin all the time." Again, various newspaper stories (e.g., New York Times 1886; New York Sun 1904) told of how former US Secretary of

State John M. Clayton (1796-1856) would occasionally spend $1 or $2 for an oxcart load of terrapins that were shoveled alive into his cellar in Delaware, as a stock for eating over time.

Gradually, terrapins became a commercial food and, by the nineteenth century, an important item on some hotel menus. Thus, Laffan (1877) wrote that a nineteenth-century "plain winter dinner in Maryland" included terrapin à la Maryland along with oysters, canvasback ducks, a crab and lettuce salad, baked Irish potatoes, fried hominy cakes, and plain celery. As demand rose and harvests fell during the nineteenth century, there was a variably successful focus on artificial propagation. In the twentieth century, as propagation efforts waned and as fishery harvests continued, natural abundances continued to decline. In recent years, most states harboring terrapins have either banned their harvest or put conservation regulations in place.

The Fishery

Although terrapins occur from Massachusetts to Texas, their early commercial exploitation was mainly centered in Chesapeake Bay and North Carolina (Table 13.1) (Coker 1906). Chesapeake animals grew larger than those in North Carolina (Coker 1906) and had a better reputation for taste, with those from the Potomac River reputed to be the finest in North America (Washington Post 1880; Washington Star 1884)—although another source touted Chester River, Maryland, animals (Baltimore Sun 1887). Notably, President Lincoln's second inaugural ball in Washington, DC, in March 1865 included terrapin stew in the buffet (Gambino 2013). My search of the New York Public Library's collection of hotel and steamship menus revealed 1,816 separate dishes that included terrapin in some form between 1854, when a terrapin dish was first included on a menu in the collection, and 1935, when the last menu mentioning terrapins occurred.

Until the late 1800s, there was no regular effort to estimate either natural abundances or catch statistics for commercial aquatic species, including terrapins. To remedy this shortcoming, the US Fish

Table 13.1. Estimated weight of diamond-backed terrapins landed in 1880 along the Atlantic Coast, by state

State	Pounds landed	Kilograms	Estimated terrapin numbers
New York	1,800	816	1,059
New Jersey	9,000	4,082	5,294
Delaware	30,708	13,929	18,064
Maryland	30,000	13,608	17,647
Virginia	165,600	75,115	97,412
North Carolina	123,000	55,792	72,353
South Carolina	23,400	10,614	13,765
Georgia	19,800	8,981	11,647
Florida, east coast	3,000	1,361	1,765
Total	406,308	184,298	239,006

Source: Data (weight in pounds) as reported by True (1887).
Note: Weights are converted to kilograms. Numbers of animals captured are estimated by dividing pounds landed by 1.7 lb, the average weight of a terrapin as calculated by Velema and Speir (2007).

Commission was established in 1871, morphing into the US Fish and Fisheries Commission in 1881 and the US Bureau of Fisheries in 1903 (NOAA 2008). Thus, I cite here early anecdotal reports of catches and prices for terrapins as reported primarily in newspaper articles until Fish Commission statistics became available, as well as some scattered efforts by others to assemble data on harvests.

One such effort was that of McCauley (1945), who used earlier literature reports and other sources to track the rise and fall of terrapin harvests in Maryland, estimating that the weight of terrapins harvested between 1880 and 1936 ranged from a peak of ~40,437 kg (40.4 metric tons) in 1891 to a low of ~373 kg (0.4 metric tons) in 1920 (Fig. 13.1). McCauley (1945) cautioned that the estimates were general and may well have been underestimates. On the other hand, because the estimates included data compiled from dealers' records and because some dealers sold southern terrapins, the data might overestimate Maryland's landings. Note also that, even where commercial landings were reported, some terrapins were being captured for personal use (e.g., "Bob the Sea Cook" 1881) and so were not included in landings. Regardless of the absolute accuracy of the numbers, the picture is one of high landings in the late nineteenth century,

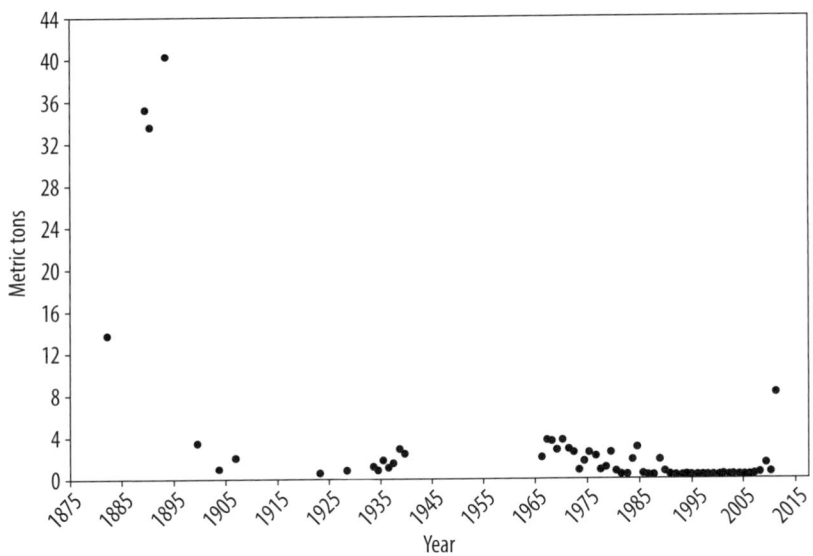

Fig. 13.1. Diamond-backed terrapin landings in Maryland from 1880 to 2006. Data for 1880-1936 from McCauley (1945); data for 1962-2006 from Velema and Speir (2007)

followed by a rapid decline at the start of the twentieth century and a low level of harvest over the first four decades of that century.

Maryland landings continued to be low from 1962 to 2006 (Fig. 13.1). Beginning in 2002, the terrapin harvest in Maryland began to increase (Velema and Speir 2007), reaching a maximum in 2006, when 10,278 animals estimated to weigh 7.9 metric tons were landed. This catch was correlated with an increasing demand for turtle flesh from Asian markets (Rein 2007; Roosenburg et al. 2008), and its magnitude stimulated Maryland's harvest ban in 2007 (Roosenburg et al. 2008).

Another pattern of rapid decline in landings in the twentieth century is seen in data for all US states as assembled by the federal government. There was a steep decline from 1950 to very low landings through the present, presumably due in part to some states' bans on harvesting (Fig. 13.2).

A variety of devices have been used to capture terrapins (True 1887; Chapter 2). In cold seasons, some terrapins are buried in marshes. Knowledgeable hunters would thrust a pole into the center of a telltale mound and pry out any terrapin encountered, placing it in a bag before continuing (Washington Post 1880; True 1887; Coker 1906). Uncommonly, near Beaufort and Morehead City, North Carolina, hunters set the marsh on fire in winter, heating the marsh surface and apparently warming buried terrapins, which subsequently appeared on the surface (Earll 1887; True 1887).

Dredges captured buried terrapins overwintering in water too deep to wade through. Modified oyster dredges (True 1887) or "drags" (Brooks 1893) captured terrapins in Somerset County, Maryland, as early as 1837 (Ducatel 1837; type of dredge not specified). The winter terrapin fishery in Pamlico and Roanoke sounds of North Carolina began in 1845 when William Midgett modified an oyster dredge (True 1887). A typical terrapin dredge was about 1 to 1.3 m wide and included a wooden upper bar and an iron lower bar, held apart by 36 cm diameter hoops at each end of the bars. A bag 1.3 m long, with meshes large enough to allow mud and debris to pass through, was attached to the dredge to hold terrapins dug out of the mud by the teeth on the lower bar as the dredge was pulled over the bottom. The teeth were typically ~8 cm long and located ~5 cm apart. Dredging occurred from small sailboats, with a dredge over each side and a third deployed from the stern. In recent years, scrapes modified after the design used to capture molting blue crabs (*Callinectes sapidus*) have been used in winter to capture dormant terrapins for population assessments (Haramis et al. 2011).

In warm weather, when terrapins feed, baited hooks could attract a hungry animal. Dip nets captured inquisitive individuals that popped their

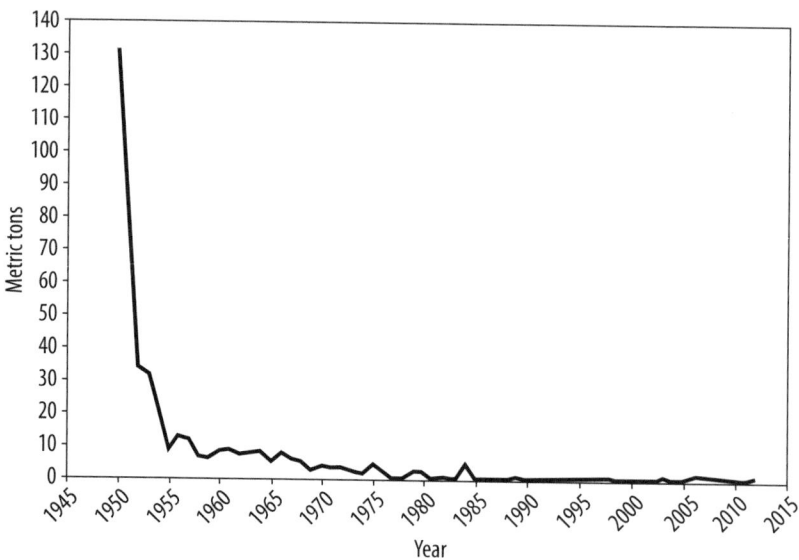

Fig. 13.2. Diamond-backed terrapin landings in the United States from 1950 to 2012. Data from NOAA, Office of Science and Technology, National Marine Fisheries Service, www.st .nmfs.noaa.gov/commercial -fisheries/commercial -landings/annual-landings /index

heads out of the water when the side of a boat was rapped sharply (New York Times 1886; Thom 1898). Seines of 10 to 12 cm mesh, 120 to 180 m long and 6 to 7 m deep, were pulled in shallow creeks south of Cape Fear, North Carolina (Earll 1887; True 1887). In this region, seiners again exploited terrapins' curiosity by rapping on the boat, or "bucking," to entice terrapins to leave the bottom and become trapped by the seine (True 1887). Or, seiners would "gouge the bottom and stir and beat the bottom to scare terrapins into the net" (Coker 1906). There was an account of hundreds of terrapins sometimes taken in a single seine haul before the 1880s (New York Sun 1904; location not named). Traps or fyke nets used to capture terrapins had four or five nested 1 m hoops linked to form a 1.3 m long device (True 1887; Thom 1898). Each hoop supported a funnel-shaped net that could be baited with fish (although baiting is not necessary; see Chapter 2). A trap was tied to a pole and had a portion that remained above water to allow captured terrapins to breathe.

Females were preferred, both because they grew larger than males and because they often contained eggs, considered a delicacy. Most terrapins found on land are females seeking to lay eggs, so locals in some regions of North Carolina used dogs to follow the scent of individuals that had come onshore (Earll 1887; True 1887). Other females seeking nest-

ing sites were sometimes captured by arranging boards on end along the water's edge on nesting beaches. The boards steered the females toward boxes sunk at intervals in the sand, from which trapped females would be retrieved (New York Sun 1904).

The start of commercial markets for terrapins in cities is not clear, but terrapins were available in the early nineteenth century. Recall that "Bob the Sea Cook" (1881) wrote that a Chesapeake Bay resident sold terrapins beyond his own needs in Norfolk and Baltimore as early as 1847. True (1887) reported that Captain John B. Etheridge dredged 2,150 terrapins from Roanoke Sound waters in 1849 and sold them in Norfolk for $400 (~19 cents each). Encouraged by his success, he immediately dredged 1,900 more animals and sold them in Baltimore for $350 (~18 cents each). Other North Carolina residents copied his efforts, stocking some ponds in summer to provide ~10,000 animals for cold-weather sales; one-sixth of the harvest went to Baltimore, one-third each went to Philadelphia and New York, and the remainder were sold in Virginia.

By 1877, Baltimore was becoming a center for distribution of captured terrapins, both domestically and overseas, from September to March (Baltimore Gazette 1877). This trend continued so that, although a larger quantity of terrapins was landed in Virginia and North Carolina than in

Maryland in 1880 (Table 13.1), by 1886 Baltimore had become "the greatest terrapin market in the world," with catches in the Bay and its tributaries in 1886 estimated to be worth $1,500,000 in trade annually (New York Times 1886). (See Fig. 13.1 for the rapid increase in the late nineteenth century in Maryland landings alone.)

For commercial purposes, terrapins were sorted into three size categories as measured along the plastron (True 1887; Coker 1920). Bulls (usually males but also small females) were smaller than 5″ (12.7 cm); heifers or half counts ranged from 5″ to 6″ (12.7 to 15.2 cm); and counts were larger than 6″ (15. 2 cm). Larger animals were worth more.

Although terrapins could be cooked and the meat canned, live animals were preferred. Because the greatest demand was in January and February, animals captured earlier would be kept in pens in cool, dark places such as cellars (Washington Post 1880). Terrapins do not need to eat in cold conditions so their care and maintenance was minimal, although sometimes in autumn months they were fed soft crabs or oysters, abundant and inexpensive in the nineteenth century (Baltimore Gazette 1877; Washington Post 1880; New York Times 1888).

As domestic markets expanded, processors would pack live terrapins, surrounded by "seagrass," in a barrel or wire basket before shipping the animals without food or ice (Washington Post 1880); most animals were delivered to the purchaser generally unscathed. Overseas shipments in winter went to England, France, and Germany (Baltimore Gazette 1877; Baltimore Sun 1889); nearly all would survive. Demand for terrapins by French gourmands encouraged the Jardin des Plantes in Paris to determine whether terrapins could be propagated in France. The Jardin had some animals shipped from Baltimore in 1878; they arrived in excellent condition (Baltimore American 1878). I found no further information on this effort in France. Naturalist Charles Lucien Jules Laurent Bonaparte, Prince de Canino, imported terrapins into Italy, as reported by De Kay (1842); De Kay did not know how successful this effort was. The absence of established terrapin populations in Europe suggests limited success.

Over time, the price for terrapins rose as demand grew and abundances fell. In 1877, animals measuring ≥18 cm sold for $25 to $36 a dozen in Baltimore, whereas 30 years earlier they had sold for $6 a dozen (Laffan 1877). By 1888, large animals sold for $36 to $42 a dozen in Baltimore, more if they were females containing eggs (New York Times 1888). Prices continued to rise, and in 1894, despite terrapins being plentiful in Baltimore markets, the largest (18-20 cm) cost about $100 a dozen, with 16 cm animals fetching $70 a dozen (Baltimore Sun 1894). Prices reached $120 a dozen by 1904 (New York Sun 1904) and $96 to $125 a dozen for the largest and highest quality animals in 1906 (Coker 1906). Use of an inflation calculator (http://www.westegg.com/inflation) reveals that $6 in 1860 was equivalent to ~$6 in 1906, so the increased prices were driven by the market rather than inflation. As market demand dropped, so did prices. A Maryland dealer told McCauley (1945) that diminished demand in 1938 had depressed prices to $2 for 15 cm animals and $3 for 18 cm animals ($24 and $36 a dozen, respectively).

The prices described above for the nineteenth century were for fresh-caught animals. As mentioned, terrapins were often held captive until market conditions were right. Animals held in cellars with hard floors tried to dig through the floor, wearing away their toes (Lowry 1888) or developing corns on their feet and scratches on their plastron. Such evidence of captivity lowered their price during the period when animals remained abundant (New York Times 1888).

Because few people could identify real terrapin meat in a dish, scarcity and associated increased cost led to "terrapin" dishes being adulterated with meat of other turtles, including less-desirable terrapins from North Carolina (Coker 1906) as well as freshwater turtles (Baltimore Sun 1887). Veal and chicken chunks might also be added (Baltimore Sun 1887). Pigeon eggs (Laffan 1877) or freshwater turtle eggs (New York Times 1888) were sometimes substituted for terrapin eggs, and faux eggs might be manufactured (Baltimore Sun 1887), including from chicken eggs (Hammond 1918).

There was one apparently limited use for terrapins beyond that of food. Stevenson (1903) wrote that oil extracted from Chesapeake Bay terrapins was thought to have medicinal qualities, especially for rheumatism, but that most of the oil was bottled and used in local homes rather than having market value. Oil from turtles was substituted for cod-liver oil in the Seychelles and Jamaica (Brannt 1896), and a turtle-egg oil in Brazil was used like butter (Reese 1919), but terrapin oil never built a commercial appeal.

Artificial Propagation or Farming

Over time, fears grew that terrapin populations were becoming disturbingly depleted. Based on True's (1887) description, an estimated 115,059 terrapins were harvested in 1880 from Chesapeake Bay alone (Virginia + Maryland; Table 13.1), whereas 24 years later the yield along the Atlantic Coast was not expected to be more than 12,000, with only 3,600 from Chesapeake Bay (New York Sun 1904). In 1898, a Baltimore dealer who had been accustomed to sales of $25,000 in earlier years sold just $4,000 worth of animals, even though prices remained high (New York Times 1899). Truitt (1939) illustrated the decline of the terrapin industry in Maryland by reporting that there were 57 terrapin dealers in 1904, 39 in 1916, and 7 in 1937. In addition to harvesting the adults, hunters collected eggs from nests as food (New York Times 1891; Baltimore Sun 1897), thereby hindering population recoveries. Attention turned to artificial propagation (or farming, as it was called then) to supplement natural production.

Early propagation efforts began simply, using pens that held terrapins until market conditions improved. For example, Ducatel (1837) writes of Somerset County residents "parking" terrapins in a square ditch excavated to allow tidal movement and surrounded by 1 to 1.3 m high plank walls. Terrapins held in their thousands in these pounds were fed clams, crabs, and other inexpensive (at the time) food until they could be sold for higher prices in winter. The Commissioners of Fisheries of Maryland (1876), concerned with the continuing decline of terrapin numbers, encouraged entrepreneurs to not just pen terrapins but farm them by providing pounds with nesting beaches where eggs could be laid.

One successful farm was at Cedar Point, about 50 km south of Mobile, Alabama. It covered 1.2 hectares, with the pound enclosed by pilings and with two canals supplying saltwater through sluices to a number of ditches, 3.3 m wide by 33 m long (Mobile Register 1881). Terrapins laid their eggs on the sand banks lining the ditches. The owner purchased up to 8,000 animals annually from local hunters for $3 a dozen, and the farm held an estimated 20,000 to 25,000 terrapins. Crabs and fish were fed to the terrapins at an annual cost of $1 per dozen reptiles, which were later sold to New York vendors for $8 to $18 a dozen. Shipments (up to 12,000 a season) began on October 1 and ended about May 10. Lowry (1888) gave a short report of his visit to that farm in 1883, noting that it had been in operation for 12 to 15 years.

Elsewhere, True (1887) described a 1.6 hectare enclosure at Roanoke Island, inside the Outer Banks of North Carolina, into which the tide ebbed and flowed through a brick channel closed at the sound end by wire mesh. This pound held 3,000 to 6,000 animals, with nesting sand beaches available for egg laying. Likewise, Charles Lewis's farm on Hog Island in the Potomac River reportedly had thousands of terrapins in stock after 5 years (Vallandigham 1894).

Coker (1906) undertook a study of terrapins on behalf of a joint US–North Carolina research operation (see below), beginning by visiting terrapin pounds that had been established near Charleston, South Carolina, and Crisfield, Maryland. A. B. Riggin & Co. and A. T. LaVallette & Co. farmed terrapins in Crisfield (Baltimore Sun 1897; Wilson 1977). LaVallette moved from Philadelphia in 1887 to live on Hammock Point near Crisfield and rear terrapins in pounds he constructed on a 7 acre property (Forest and Stream 1899; Meyer 2005; Fincham 2008). Initially he was very successful, buying terrapins from local watermen, holding them in the pounds (see photographs in Rhodes 2006), where they were fed local crab waste, then selling them to high-end restaurants in cities such as Philadelphia and New York. He provided those restaurants with recipes for terrapin dishes and signed sole-supplier contracts with them, thus cornering the market for a time. His business thrived. In 1890,

LaVallette provided 28 barrels of terrapins worth about $4,700 for a banquet in New York's Delmonico's restaurant, hosted by Jay Gould (Forest and Stream 1899), and in 1896 filled an order worth $3,000 for a dinner for a Chinese diplomat visiting New York (Baltimore Sun 1897). His pounds were constructed so that nesting could occur, and eventually they held an estimated 15,000 animals of all sizes (Forest and Stream 1899). Although LaVallette initially benefited from increased prices as terrapin numbers declined, his business ended in the early 1900s, apparently because local terrapin abundances continued to decline, as well as for personal reasons (Meyer 2005).

A very successful farm was established by A. M. Barbee at the Isle of Hope outside Savannah, Georgia, in 1889 (New York Times 1912). It continued for many years and was a tourist destination until at least 1940 (Georgia Board of Education 1940), written about entertainingly by Mitchell (1939). The farm included a roofed, open-sided building 50 m long by 20 m wide holding 18 pens (Reese 1919). Male and female animals were held in separate pens except during breeding season. In warmer months, estuarine water was pumped into the pens three times a week while the animals were fed shrimp, fish, and garden vegetables. Outside the feeding times, a freshwater trough in the center of all pens was kept filled. The wooden sides of the pens were sunk a meter or more below ground to keep the terrapins from digging out, and sand over boards was used as flooring (Moulton 1914; Reese 1919). Dogs guarded the pens at night to protect against mammalian predators and human poachers.

From June through August, females were provided with a hillock of sand for egg laying. The eggs were retrieved and placed in one of 30 incubators, marked with the expected hatching date (Moulton 1914; Reese 1919). These incubators were shallow wooden trays, 1 m long and 0.5 m wide, filled with screened sand and humus; the trays were held on tables, with legs standing in buckets of water to keep ants from feeding on hatchlings (Moulton 1914). Hatchlings remained buried in sand until spring, whereupon they were moved to marshes enclosed by wire mesh and guarded by

paid workers. Losses were few. Barbee claimed to have about 1,800 adult animals available as broodstock, with about 7,000 young hatching between July 1 and the end of January, of which 95% would survive (New York Times 1912). Carr (1952) reported that Barbee told him that the price of a dozen farmed terrapins in 1921 was $90. The firm also manufactured cans of Barbee's Diamond Back Terrapin Soup, with a label promising a $1,000 forfeit if the can did not contain "pure diamond back terrapin" (see photograph in Cooper 2002, p. 20).

Barbee built a portable leather case with an air vent in one end as an incubator, which held a compartmented wooden box filled with sand and fertile terrapin eggs. Maintenance consisted of a weekly sprinkling of water. On a 1912 visit to New York City, he transported 100 eggs in his incubator, and two hatchlings emerged during the train trip to Philadelphia (New York Times 1912). Another 48 hatched in that city, followed by more in New York. Barbee's efforts attracted continued attention in newspapers (e.g., New York Tribune 1914) and magazines (e.g., Moulton 1914), with headlines predicting a solid future for terrapin farming.

Although terrapin farms attracted much attention and some laudatory headlines, not all succeeded. For example, Lowry (1888), convinced of the feasibility of terrapin farming by his 1883 visit to Cedar Point farm, started a farm near Easton, Maryland, in 1884. The site had a 0.6 hectare pond connected by an inlet to Chesapeake Bay and fed by a small freshwater stream. He enclosed the pond and a small amount of dry land with a board fence. In May 1885 he added about 1,000 terrapins to the farm and fed them chopped blue crabs as well as the occasional fish. He did not provide any more information about the success of the farm, other than noting that hatchlings grew very slowly. A later report (New York Sun 1904) of a terrapin pond below Easton with a tidal stream connecting it to the Bay noted that the pond had been provided with 1,000 animals (year not specified), but after 6 months few animals could be found. This may have been Lowry's farm.

Other, similarly unsuccessful farming efforts were described (New York Sun 1904). In one, a Fulton Market fish dealer placed 5,000 young animals in an enclosed cave adjoining Pleasure Bay, New Jersey, but only about 100 were found a few months later. In another, around 1899, about 10,000 young animals were held in an enclosure near the mouth of the Potomac River, but only about 500 remained after two years. Terrapins are escape artists. They can burrow below shallow fences or climb over them (Hay 1906). In addition, birds such as crows, gulls, and herons prey on terrapin eggs and hatchlings, and foxes, skunks, rats, and raccoons will take all terrapin life cycle stages (MDTTF 2001) unless protective netting is spread over a nesting beach or an enclosed hatchling nursery is provided (True 1887; Coker 1906). Clearly, farming terrapins was a hit-or-miss endeavor.

In spite of variable results of farming efforts, interest continued as harvests declined. Ultimately the federal government was asked to provide guidance because of the increasing scarcity of the animal in most Atlantic coastal states (Smith 1895). Once laid in an appropriate and protected habitat, terrapin eggs seemed very robust, needing no maintenance until they hatched. The commission understood, however, that the main hindrances to terrapin farming occurred post-hatching, including the animal's very slow growth rate. There was also a lack of understanding by farmers of culture requirements (water conditions, availability of a nesting beach, food type and quantity) that would support successful reproduction and faster growth.

The Fish Commission eventually began a terrapin culture study in Beaufort in 1902 in conjunction with the North Carolina Geologic and Economic Survey (Coker 1906). However, the initial observation pound was too small for satisfactory breeding, so work at Beaufort ceased in 1903 when the (now) Bureau of Fisheries established a larger program at Lloyds, Maryland, beginning in 1904 (Hay 1906). This study persisted for five years and demonstrated that female terrapins would lay eggs even when captive for two seasons or more; most eggs were fertile and hatched. Hay (1906) found that food should be abundant, primarily crabs but also fish and snails. Coker (1906) recommended that farm pounds should be supplied with regularly changed water to flush away wastes that otherwise

accumulated rapidly, fouling the habitat. Similarly, terrapins should not be crowded, and dead animals should be removed promptly.

The study in Lloyds ended in 1909, and the research was transferred back to Beaufort, where large concrete tanks were built and where research continued until 1949 (Coker 1951; Wolfe 2000). In 1924, a cooperative agreement between the Bureau of Fisheries and the North Carolina Fisheries Board supported further terrapin studies (Wolfe 2000). Accounts of the research were prepared by Hay and Aller (1913), Hay (1917), Barney (1922), and Hildebrand and Hatsel (1926), and findings were summarized in a major review by Hildebrand (1929). Results of studies of life history characteristics such as growth, reproduction, and sex ratios, including those reported by Hildebrand (1929), are summarized in Chapter 6. Here I focus on the general purposes of the cooperative experiments and describe the facilities as they finally evolved.

The general goals of the Beaufort study included (Hildebrand 1929):

- Determining space requirements for holding adults and raising young
- Determining space required for egg laying
- Estimating the natural sex ratio of the experimental animals and the ratio required for successful breeding
- Evaluating the effects of the overwinter feeding of young on their growth, maturity, and survival
- Learning about disease in young hatchlings (yearlings and older did not seem to have diseases)
- Performing selective breeding experiments
- Crossbreeding Texas and Carolina terrapins to attain higher growth rates

Hildebrand and Hatsel (1926) provided a number of figures showing the facilities at Beaufort, as well as construction plans and details. The facilities included concrete tanks for holding adult broodstock outdoors (Fig. 13.3, A, B) and an enclosed building for rearing terrapin hatchlings (Fig. 13.3, C, D). The tanks were built by the water's edge to allow exchange of water by the tides. Tanks were protected from poaching by a barbed-wire fence located

Fig. 13.3. US Bureau of Fisheries' terrapin study facilities in Beaufort, North Carolina. (A) Adult terrapins were kept in outdoor concrete tanks as breeding stock, with tanks flushed by every tide and a sand beach available for egg laying. (B) Charles Hatsel fed chopped fish to terrapins in outdoor tanks twice a week. (C) Aquarium and terrapin rearing house. (D) Charles Hatsel feeding baby terrapins in tanks inside the rearing house. All images from NOAA Central Library Historical Fisheries Collection.

about 7 to 8 m from the tanks (Hildebrand and Hatsel 1926). Sandy beaches were provided for egg laying, with boards inserted in the sand after laying ended to retain the young. The beaches also were used as a base for feeding chopped fish to the terrapins (Fig. 13.3, B), which rapidly learned to come on a particular signal (a call, handclap, knock on the tank wall, sound of food being chopped, etc.). The Beaufort experiments kept hatchlings warm and fed over winter in a separate building (Fig. 13.3, C, D), accelerating growth under these artificial conditions (Hildebrand 1929). Boards were placed as an overhang along the top of the rearing tanks to prevent the agile young from escaping from the wooden tanks (Fig. 13.3, D).

Terrapin rearing at Beaufort continued after 1909 until the program ended in 1949 (Wolfe 2000). A new building (Fig. 13.3, C), constructed in 1939 to 1940, expanded the rearing space. Experiments under the care of Charles Hatsel were very productive, with perhaps 17,000 hatchlings produced in the peak year of 1941 (or 1942; accounts are not clear). Animals were used for experimental purposes and to provide broodstock and hatchlings to North Carolina, the federal government's collaborator in this effort, as well as to Virginia, South Carolina, and Georgia. Some were even shipped to California in an unsuccessful effort to establish terrapins in San Francisco Bay (Brown 1971). Coker (1951) reported that, over the course of the program at Beaufort, more than 249,000 young animals were hatched and distributed by August 1949. The Beaufort Laboratory released most of their broodstock after Hatsel retired in 1947 and federal appropriations ceased in 1948. In 1954, Maryland State biologists retrieved the few remaining adults when the

Beaufort Laboratory finally stopped feeding the holdovers from the defunct program (Wolfe 2000), but I found no account of what happened to the adults removed to Maryland.

The success of the Bureau of Fisheries' program apparently encouraged a company led by Dr. Charles Duncan to establish the Beaufort Terrapin Farm at nearby Town Creek, North Carolina, in 1913 (Hildebrand 1929; Wolfe 2000). Concrete ponds and a nursery were built (Wolfe 2000), and more than 3,000 terrapins were purchased as broodstock. Over time, as many as 25,000 hatchlings were held over winter at this facility (Coker 1920). However, despite this success in rearing hatchlings, the project was abandoned after 1918 because the cost of local labor had tripled with the onset of World War I and the war depressed markets for luxuries (Hildebrand 1929). Coker (1951) added that the nursery was heated by anthracite coal, which was not available for that purpose during the war.

In spite of the success of the Bureau of Fisheries' experiments and of some terrapin farms, demand for terrapins diminished in the early twentieth century (Carr 1952). As noted above, one proposed cause was that the war suppressed high living, and another was that the onset of Prohibition in 1920 removed the alcohol supplements (e.g., sherry and madeira; Baltimore Sun 1887) that enlivened the flavor of the dish (Coker 1920). However, Carr (1952) concluded that it was more likely that the keen demand for terrapins was a food fad that had eventually faded away, a suggestion supported by Coker's (1920) report of a terrapin dealer who lamented, "It is the old buyers only that come to me for diamond-backs." This hints at a decline in younger people's familiarity with the reptile as a food item as abundances fell. Nevertheless, a market persisted in New York City's Fulton Market, where seafood firm Walter T. Smith sold terrapins as a medicinal resource to elderly Chinese residents (Mitchell 1939).

Despite the mixed results of earlier endeavors, terrapin farming continues in Maryland and Louisiana. The market strategy has shifted from supplying adults for food to producing hatchlings for overseas markets, primarily for the Asian pet trade and the stocking of emerging turtle farms. Eliminating the need for US farmers to rear terrapins past the hatchling stage has resulted in US profit margins that make the endeavor financially worthwhile (T. Leuteritz, USFWS, pers. comm. to W. Roosenburg). Furthermore, numerous farms in China are now rearing terrapins for commercial sale (B. Horne, Wildlife Conservation Society, pers. comm. to Roosenburg) with broodstock of unknown origin, but clearly from the United States. Perhaps the greatest threat from farming is the strain on wild US populations to provide broodstock for large-scale propagation. The move to close the New Jersey fishery in 2016 was stimulated by the capture of more than 3,600 individuals that were sold as broodstock to a terrapin farm in Maryland in 2013.

Managing the Resource

In addition to declines in abundances due to the overharvesting of adults and eggs, other factors can adversely affect terrapin populations (MDTTF 2001). One involves loss of the intertidal habitat essential for nesting. Habitat can be altered by erosion control measures such as installing bulkheads and stone revetments (Chapter 14) and by land development that disturbs beaches or places impediments between nesting beaches and the water. If land development separates the water from a nesting beach by a road, roadkill of females moving toward the beach to lay eggs may occur (Chapter 14). Boats can kill terrapins (Chapter 15), and terrapins may suffocate in crab pots that have no escape mechanisms (Chapter 16).

Concerns about managing the terrapin fishery while conserving populations have led to various legal restrictions on the fishery. The State of Maryland, a once important source of commercial terrapins, serves as an example of such conservation actions (Velema and Speir 2007). Its first law concerning terrapins was passed in 1878, setting an open season from November 2 to March 31 and a minimum plastron size limit of 5″ and banning possession, destruction, or disturbance of eggs. About two decades later, the Maryland Game and Fish Protective Association examined the state's laws on terrapin harvest, finding them "good, but they are irregular and they are not enforced" (Thom 1898). The association recommended a $10 fine for possession of terrapins between April 1 and November 1, to prevent holding terrapins in pounds; a similar fine for selling animals smaller than 8 cm in plastron length; and regulations limiting terrapin sales to the county where they were harvested (Thom 1898). Such a bill was enacted in the Maryland legislature in 1900 (New York Times 1900). For 129 years, Maryland allowed commercial capture of terrapins with modest restrictions, including open seasons that varied over time from 3 to 9 months in length. Ultimately, in 2007, the state banned the commercial capture of terrapins, although it allows farming under a permit. Recreational crab pots must be fitted with a turtle excluder device to keep trapped terrapins from drowning (Chapter 16).

All 16 states in which the species occurs now regulate the harvest or possession of terrapins (Table 13.2). Eleven ban harvest, and eight allow possession of a specific number of animals or with a permit. Aquaculture (farming) is allowed under permit by five states, but most other states do not mention terrapin culture in their laws; in most instances, aquaculture is not possible because possession of terrapins is banned. Various additional

Table 13.2. State regulations applied to terrapins as of 2018

State	Harvest restrictions*	Terrapin aquaculture permitted?
Massachusetts	Listed as "Threatened." No harvest allowed. No possession without a permit.	Not explicitly stated.
Rhode Island	No harvest allowed. No possession without a permit.	Not explicitly stated. Egg possession needs a permit.
Connecticut	No harvest allowed.	Not explicitly stated.
New York	No harvest allowed. Eggs and nests are protected.	Not explicitly stated.
New Jersey	No harvest allowed.	Not explicitly stated.
Delaware	Open season Sep 1–Nov 15. Catch limit 4 per day. Eggs are protected. One individual can be possessed without a permit.	Terrapins can be reared in private ponds, but requires a permit.
Maryland	No harvest allowed. Noncommercial possession limited to 3 individuals legally obtained from outside the state or >4" length.	Yes, with a permit.
Virginia	No harvest or possession allowed.	Not explicitly stated.
North Carolina	Species of "Special Concern," so cannot be harvested or possessed.	Not explicitly stated.
South Carolina	Commercial harvest banned. Two animals may be kept for noncommercial purpose.	Commercial sale is prohibited.
Georgia	Listed as "Unusual." Some freshwater turtles can be harvested; no mention of terrapins. Terrapins cannot be held as pets.	Turtle farms are permitted, but terrapins are not mentioned among the turtle species that can be held.
Florida	Noncommercial capture of 1 animal per day allowed; limit of 2 in possession. Gear includes hand, dip net, baited hook, minnow seine.	Yes, with a permit, but taking eggs is prohibited.
Alabama	No harvest allowed.	Not explicitly stated.
Mississippi	Species of "Special Concern." Noncommercial possession of 4 animals allowed. Eggs are protected.	Yes, with a permit, but breeding stock limited to 8 wild-caught animals.
Louisiana	Commercial season June 16–Apr 14; license required. Noncommercial capture allowed with a fishing license. Cannot be trapped. Minimum length 6". Egg harvest banned.	Permitted.
Texas	No harvest allowed.	No person may possess a diamond-backed terrapin at any time except under a scientific, educational, or zoological permit.

*Size refers to plastron length.

conservation efforts are underway to protect the species in many states (Chapters 14-18).

ACKNOWLEDGMENTS

Information about state regulations was provided by A. Baxter, R. Boettcher, J. Boundy, B. Brennessel, R. Chambers, R. Clay, J. Crawford, G. Guillen, K. Marion, A. Morris, H. Niederriter, T. Norton, K. Shelton, C. Sornborger, R. Stanley, W. Turner, M. Whilden, J. Wiebe, and J. Wnek. The comments of three referees, Willem Roosenburg, and Deborah Coffin Kennedy were helpful in improving the chapter and are appreciated.

REFERENCES

Baltimore American. 1878. Chesapeake terrapin for France. Reprinted in New York Times, February 2.

Baltimore Gazette. 1877. Fish and fowl. How oysters, terrapin, fish, and game are shipped. Reprinted in New York Times, February 17.

Baltimore Sun. 1887. Maryland terrapin. How they are caught and cooked—tricks of caterers. Reprinted in New York Times, December 12.

Baltimore Sun. 1889. Terrapin. Reprinted in New York Times, January 6.

Baltimore Sun. 1894. How terrapin prices vary. Reprinted in New York Times, February 25.

Baltimore Sun. 1897. Terrapin culture. A Maryland expert talks entertainingly about it—a profitable industry. Reprinted in New York Times, December 26.

Barney, R.L. 1922. Further notes on the natural history and artificial propagation of the diamond-back terrapin. Bulletin of the US Bureau of Fisheries 37:91-111.

"Bob the Sea Cook." 1881. The old Maryland coast. The way they cooked terrapin forty years ago. New York Times, August 14.

Brannt, W.T. 1896. A Practical Treatise on Animal and Vegetable Fats and Oils. 2nd edition. Volume II. Henry Carey Baird and Co., Philadelphia. 753 pages.

Brooks, W.K. 1893. The diamond-back terrapin. Pages 261-263 in Maryland: Its Resources, Industries, and Institutions. Prepared for the Board of World's Fair Managers of Maryland by Members of Johns Hopkins University and Others. Baltimore, MD.

Brown, P.R. 1971. The story of California diamondbacks. Herpetology 5(2):37-38.

Carr, A. 1952. Handbook of Turtles. Cornell University Press, Ithaca, NY. 542 pages.

Chamberlain, A.F. 1902. Algonkian words in American English: a study in the contact of the white man and the Indian. Journal of American Folk-lore 15:240-267.

Coker, R.E. 1906. The natural history and cultivation of the diamond-back terrapin with notes on other forms of turtles. North Carolina Geological Survey Bulletin 14:1-67.

Coker, R.E. 1920. The diamond-back terrapin: past, present and future. Scientific Monthly 11(2):171-186.

Coker, R.E. 1951. The diamond-back terrapin in North Carolina. Pages 219-230 in H.F. Taylor (ed.), Survey of Marine Fisheries of North Carolina. University of North Carolina Press, Chapel Hill, NC.

Commissioners of Fisheries of Maryland. 1876. Report to the General Assembly, January 1. John F. Wiley, Annapolis, MD. 66 pages.

Cooper, P.W. 2002. Isle of Hope: Wormsloe and Bethesda. Arcadia Publishing, Charleston, SC. 128 pages.

De Kay, J.E. 1842. Zoology of New York, or the New-York Fauna. Part III. Reptiles and Amphibia. W. & A. White & J. Visscher, Albany, NY. 98 pages.

Ducatel, J.T. 1837. Outlines of the physical geography of Maryland, embracing the prominent geological features. Transactions of the Maryland Academy of Science and Literature 1:24-53.

Earll, R.E. 1887. Part XIII. The fisheries of South Carolina and Georgia. Pages 501-518 in G.B. Goode (ed.), The Fisheries and Fishery Industries of the United States. Section II. A Geographical Review of the Fisheries Industries and Fishing Communities for the Year 1880. US Bureau of Fisheries, Washington, DC.

Fincham, M.W. 2008. The men who would be kings. Chesapeake Quarterly 7(4):10-14.

Forest and Stream. 1899. Terrapin farming. Volume 53, page 191.

Gambino, M. 2013. Document deep dive: the menu from President Lincoln's Second Inaugural Ball. Smithsonian.com, January 15. http://www.smithsonianmag .com/history/document-deep-dive-the-menu-from -president-lincolns-second-inaugural-ball-1510874.

Georgia Board of Education. 1940. Georgia: A Guide to Its Towns and Countryside. University of Georgia Press, Athens, GA. 559 pages.

Hammond, M.M.E. 1918. The Swedish, French, and American Cook Book. Mrs. Ericsson Hammond, New York. 480 pages.

Handley, B.M. 2001. The Blue Goose midden (8IR15): a Malabar II occupation on the Indian River Lagoon. Florida Anthropologist 54(3-4):103-121.

Haramis, G.M., P.F.P. Henry and D.D. Day. 2011. Using scrape fishing to document terrapins in hibernacula in Chesapeake Bay. Herpetological Review 42:170-177.

Hariot, T. 1588. A Briefe and True Report of the New Found Land of Virginia. Privately printed, London.

42 pages. https://archive.org/stream/cu31924028784571 #page/n17/mode/thumb.

Hay, W.P. 1906. Cultivation of the diamond-back terrapin at Lloyds, Maryland. Chapter V in R.E. Coker, The natural history and cultivation of the diamond-back terrapin with notes on other forms of turtles. North Carolina Geological Survey Bulletin 14:35-37.

Hay, W.P. 1917. Artificial propagation of the diamond-back terrapin. US Bureau of Fisheries Economic Circular 5(revised):1-21.

Hay, W.P. and H.D. Aller. 1913. Artificial propagation of the diamond-back terrapin. US Bureau of Fisheries Economic Circular 5:1-14.

Hildebrand, S.F. 1929. Review of experiments on artificial culture of diamond-back terrapin. Bulletin of the US Bureau of Fisheries 45:25-70.

Hildebrand, S.F. and C. Hatsel. 1926. Diamond-back terrapin culture at Beaufort, N.C. US Bureau of Fisheries Economic Circular 60:1-20.

Laffan, W.M. 1877. Canvas-back and terrapin. Scribner's Monthly 15(1):1-13.

Lowry, R.C. 1888. Terrapin culture. Letter to the editor. Forest and Stream, December 13, page 31. http://eshore.vcdh.virginia.edu/node/1889.

McCauley, R.H. Jr. 1945. The Reptiles of Maryland and the District of Columbia. Privately published, Hagerstown, MD. 194 pages.

MDTTF (Maryland Diamondback Terrapin Task Force). 2001. Findings and recommendations. Final Report to the Secretary of the Maryland Department of Natural Resources. Annapolis, MD. 10 pages.

Meyer, E.L. 2005. Easy come, easy go. Chesapeake Bay Magazine 34(April):74-79, 104-107.

Mitchell, J. 1939. The same as monkey glands. Pages 314-324 in Up in the Old Hotel, and Other Stories. Original story in the New Yorker. Published in book of collected stories, Random House, 1993. Reissued by Vintage Books, New York, 2008.

Mobile Register. 1881. A Mobile terrapin farm. Reprinted in New York Times, February 4.

Moulton, R.H. 1914. Raising terrapin in incubators. Technical World Magazine, October, pages 246-249.

New York Sun. 1904. How to save the terrapin. They are decreasing so fast that they won't last long. January 3.

New York Times. 1886. The terrapin industry. How the palatable reptile is captured. Its favorite haunts in Chesapeake Bay and how terrapin farming is carried on. November 21.

New York Times. 1888. Canvas-back duck trust. No monopoly in terrapins possible. February 5.

New York Times. 1891. Our tables will suffer. Terrapin and canvas-back surely disappearing. February 15.

New York Times. 1899. Terrapin growing scarcer. Receipts at Baltimore alarm dealers. Prices soaring upward. November 12.

New York Times. 1900. Bill to protect terrapin. March 7.

New York Times. 1912. Hatches terrapin in Broadway hotel. Largest shipper from Georgia carries his sand incubator with him. They thrive in his bathtub. October 15.

New York Tribune. 1914. Diamond backed crop of world's only terrapin farm. July 26.

NOAA (National Oceanic and Atmospheric Administration). 2008. United States Fish Commission Annual Reports. NOAA Central Library. ftp://ftp.library.noaa.gov/docs.lib/htdocs/rescue/cof/data_rescue_fish_commission_annual_reports.html.

Reese, A.M. 1919. Outlines of Economic Zoology. P. Blackiston's Son and Co., Philadelphia, PA. 316 pages.

Rein, L. 2007. Md. officials aim to halt harvest of prized terrapins. Washington Post, March 6.

Rhodes, J. 2006. Crisfield: The First Century. Arcadia Publishing, Charleston, SC. 128 pages.

Roosenburg, W.M., J. Cover and P.P. van Dijk. 2008. Legislative closure of the Maryland terrapin fishery: perspectives on a historical accomplishment. Turtle and Tortoise Newsletter 12:27-30.

Scarry, J.F and C.M. Scarry. 1997. Subsistence remains from prehistoric North Carolina archaeological sites. Report Prepared for North Carolina Division of Archives and History. On file at the North Carolina Office of State Archaeology, Raleigh, NC.

Smith, H.M. 1895. Report of the division of statistics and methods of the fisheries. US Commission of Fish and Fisheries Report of the Commissioner for the Year Ending June 30, 1893. Volume 19, pages 52-77.

Stevenson, C.H. 1903. Aquatic products in arts and industries: fish oils, fats, and waxes. Fertilizers from aquatic products. US Fish Commission Report for 1902. Pages 177-279 + 16 plates. Washington, DC.

Sutherland, D.R. 1974. Excavations at the Spanish Mount shell midden Edisto Island, South Carolina. South Carolina Antiquities 6(1):185-195.

Thom, W.H.DeC.W. 1898. The diamond-back terrapin. Forest and Stream 50(8):151.

Thorbjarnarson, J., C.J. Lagueux, D. Bolze, M.W. Klemens and A.B. Meylan. 2000. Human use of turtles: a worldwide perspective. Pages 33-84 in M.W. Klemens (ed.), Turtle Conservation. Smithsonian Institution Press, Washington, DC.

True, F.W. 1887. The turtle and terrapin fisheries. 2. The terrapin fishery. Pages 499-503 in G.B. Goode (ed.), The Fisheries and Fishery Industries of the United States. Section V. History and Methods of the Fisheries. Volume II. Part XIX. US Commission of Fish and Fisheries, Washington, DC.

Truitt, R.V. 1939. Diamondback terrapin. Pages 93-96 in R.V. Truitt, Our Water Resources and Their Conservation. Contribution no. 27. Chesapeake Biological Laboratory, Solomons, MD.

Vallandigham, E.N. 1894. He raises terrapin. Jamestown [North Dakota] Weekly Alert, November 29, and Lewiston [Maine] Evening Journal, November 30.

Velema, G.J. and H. Speir. 2007. The 2006 diamondback terrapin fishery in Maryland's Chesapeake Bay. Maryland Department of Natural Resources Fisheries Technical Memorandum 35:1-11.

Washington Post. 1880. A talk about terrapins. How Washington's favorite delicacy is obtained and stewed. Reprinted in New York Times, December 5.

Washington Star. 1884. Terrapin and terrapin-eaters. Reprinted in New York Times, May 4.

Whitaker, A. 1624. Good News from Virginia, Sent from James His Towne This Present Moneth of March, 1623. Reprinted as pages 579-588 in A. Brown, The Genesis of the United States. Volume II. Riverside Press, Cambridge, MA, 1890.

Wilson, W.T. 1977. History of Crisfield and Surrounding Areas on Maryland's Eastern Shore. Gateway Press, Baltimore, MD. 405 pages.

Wolfe, D.A. 2000. A History of the Federal Biological Laboratory at Beaufort, North Carolina 1899-1999. National Oceanic and Atmospheric Administration, Beaufort, NC. 312 pages.

14

Conservation in Terrestrial Habitats: Mitigating Habitat Loss, Road Mortality, and Subsidized Predators

JOHN C. MAERZ
RICHARD A. SEIGEL
BRIAN A. CRAWFORD

The majority of animals have complex life cycles that require the use of multiple, complementary habitats (Wilbur 1980; Dunning et al. 1992; Moran 1994). As a result, the management of many species can be challenging because it requires assessing and targeting interventions toward multiple, concurrent threats that affect different habitats, or the same habitat, or the ability of animals to move among habitats. Like most aquatic turtles, diamond-backed terrapins (*Malaclemys terrapin*) spend the vast majority of their lives in aquatic habitats; however, their shelled eggs require a terrestrial life stage, and terrestrial conditions regulate many critical vital rates, including nest and hatchling survival and sex determination. Moreover, a suite of anthropogenically altered terrestrial threats can have large effects on terrapin population dynamics disproportionate to the time spent on land.

We summarize here the terrestrial habits of terrapins and discuss proximate terrestrial threats and management options. Chapter 16 addresses the effects and management of terrapin bycatch in commercial and recreational crab pots, which is the primary anthropogenic threat to terrapins in aquatic habitats. However, bycatch interacts with a suite of terrestrial threats to affect terrapin populations, and terrestrial management needs to be included as part of a comprehensive terrestrial-aquatic management strategy for terrapin populations (Hart 2005; Crawford et al. 2014a; Gilliand et al. 2014). A failure to account for threats to both aquatic and terrestrial life stages may render management activities targeted at a subset of threats or life stages ineffective. We discuss how complementary terrestrial management creates flexibility and opportunities for compensatory actions when certain threats to terrapin populations cannot be fully remediated. We acknowledge that coastal development is the ultimate threat to terrapin populations and the cause of many of the proximate threats addressed here. However, many coastal areas are already well developed and there is a need for guidance on proximate threats and management actions. The broader issue of coastal development is beyond the scope of this chapter. We also do not directly address threats or management activities related to climate change and rising sea levels, which are important aspects of coastal environmental change that warrant attention.

Terrestrial Habits, Life Cycle, and Demography

Like many freshwater turtles, terrapins primarily live in aquatic environments, with terrestrial habits limited to females making brief annual excursions onto land to nest and hatchlings spending some period of their first year on land before dispersing to aquatic habitats. Terrestrial movements of juvenile (>1 year) and adult male terrapins are believed to be rare, but individuals have occasionally been encountered (<2% of observations) during road surveys (Szerlag and McRobert 2006; Crawford 2011). Terrapins show similar nesting behavior throughout the species' range, though patterns can vary geographically and are associated with changes in coastal environments and movement of the species into more inland, brackish habitats (Roosenburg 1994). Generally, females nest from late spring through midsummer and may make multiple nesting excursions within a single breeding season.

Diel nesting patterns may vary geographically, potentially in relation to predation risk and timing of high tide. Several studies showed daylight nesting of terrapins concentrated around diurnal high tide (Fig. 14.1) (Burger and Montevecchi 1975; Zimmerman 1992; Goodwin 1994; Butler et al. 2006; Szerlag and McRobert 2006; Crawford et al. 2014b). Nesting during the high tide may reduce the time and distance female terrapins must travel and may be related to higher survival among nests laid above the high-tide line. In localities where tidal influences are <1 m, daytime nesting occurs in late morning through midafternoon

(Seigel 1980). Females nest throughout the day and night, with small peaks at dusk and dawn and a major peak between 10:00 and 14:00, on the Patuxent River in Maryland (W. Roosenburg, Ohio University, pers. comm.), and approximately half of females nest at night in Sandy Neck, Massachusetts (Auger and Giovannone 1979). As in most freshwater turtles (Steen et al. 2012), the terrestrial portions of terrapins' nesting routes are generally short, <50 m, but can be as long as 1,600 m in Massachusetts (Auger and Giovannone 1979). Nesting duration ranges from 15 to 120 min, with the majority of time allocated to nest site selection (Roosenburg 1994). As is true of many species of turtles, females will frequently abandon nesting, often in response to disturbance by predators or humans.

Female terrapins and other turtles often exhibit bias in their selection of nesting habitats and microhabitats that is related to embryonic survival, development rate, and hatchling sex ratio (Wilson 1998; Refsnider and Janzen 2010). Mean incubation time varies with latitude and microhabitat, ranging from as short as 66 to 69 days in central and northern Florida (Seigel 1980; Butler et al. 2004) to 74 days in New Jersey and 77 days in New York (Scholz 2007). However, local factors such as vegetation that shades nests can slow development and create greater local variation of incubation time than is seen over wide latitudes (Goodwin 1994; Wnek et al. 2013). Generally, turtles are more likely to nest in open habitats that are warmer, which results in faster development and shorter times to hatching.

Nesting typically takes place a few meters above the spring high tide (Burger and Montevecchi 1975; Butler et al. 2004) and generally in areas of sandy soil. Terrapins have been reported nesting in a variety of habitats, including areas of bare sand, exposed and vegetated dunes, short or tall dense grasses, and dense shrubs. Attributes of sandy soils and vegetation that may be important in selection of nesting habitat include access, thermal and moisture conditions, soil composition and texture, disturbance, and predation risk (Burger and Montevecchi 1975; Goodwin 1994; Roosenburg 1994; Roosenburg and Place 1995; Feinberg and Burke 2003; Ner 2003; Scholz 2007; Hackney 2010; Grosse et al. 2015). Roosenburg (1996) used data on

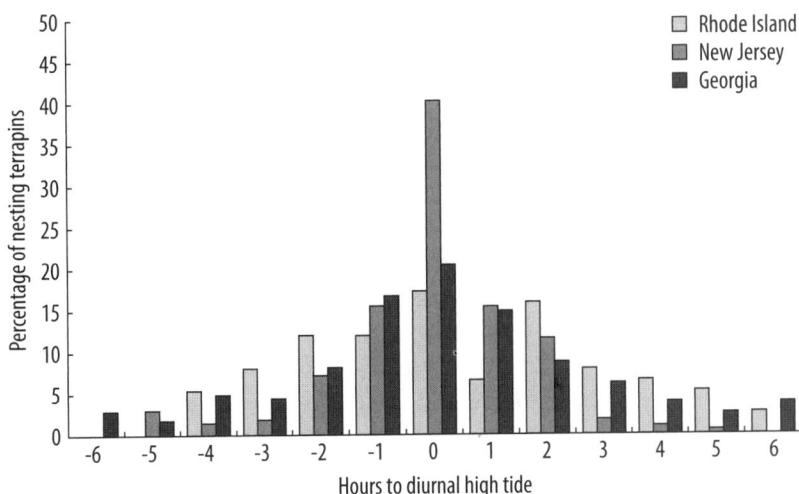

Fig. 14.1. Proportion of nesting female terrapins observed as a function of proximity to daytime high tide for three sites across the Atlantic portion of the species' range. Data sources: Rhode Island, Goodwin (1994); New Jersey, Burger and Montevecchi (1975); Georgia, Crawford et al. (2014b)

egg size to infer that females are more selective about nest sites when they have larger eggs. There may also be "cultural" maternal effects related to nest site selection and fidelity that can affect population responses to alterations in habitat (Freedberg and Wade 2001).

Female terrapins exhibit nest site fidelity within and across years (Goodwin 1994; Sheridan et al. 2010; Crawford 2011), though nest site fidelity and selectivity may vary geographically (Roosenburg 1994). Crawford (2011) reported capturing 50% and 80% of nesting female terrapins within 50 m and 200 m, respectively, of their capture location during the previous nesting season. Sheridan et al. (2010) used six highly polymorphic loci to estimate relative relatedness among nesting female terrapins at seven sites distributed over the 70 km long Barnegat Bay Estuary, New Jersey. Using multilocus autocorrelation coefficients, they found positive correlations among genotypes of nesting females at distances as proximate as 5 m and up to distances of 150 m, with the strongest positive correlations among females nesting within 50 m of each other. In other words, females nesting in close proximity within nesting beaches are more closely related, suggesting a tendency toward natal philopatry among terrapins. Assuming that nest fidelity emerges from natal imprinting of nesting location, the potential for local inheritance of maternal nest sites may be important for understanding nesting patterns in relation to historical and cur-

rent habitat conditions, as well as for predicting the effects of terrestrial threats and management actions.

Hatching occurs in midsummer through late summer, but hatchlings may remain in the nest for some period and emerge in the late summer, fall, or the following spring, after which they appear to disperse into high marsh environments (Lovich et al. 1991; Muldoon and Burke 2012). Hatchling turtles generally exhibit orientation and rapid movement toward preferred habitats to avoid predation and dehydration and, depending on species, will orient toward or away from silhouettes of tall vegetation (reviewed by Coleman et al. 2014). Burger (1976) reported that hatchling terrapins in New Jersey moved quickly downslope into grassy vegetation, which would be consistent with bias toward rapid dispersal from elevated nesting areas into marsh vegetation. Coleman et al. (2014) showed that <1-week-old, 1- to 8-week-old, and 1-year-old Mississippi diamond-backed terrapins (*M. terrapin pileata*) hatched in captivity oriented toward marsh vegetation and away from open water. Muldoon and Burke (2012) reported that hatchlings that emerged in the fall in New York moved upward in elevation (rather than down toward more aquatic habitats) until spring, when the behavior reversed. They also reported finding hatchlings on land up to 9 months after emergence from the nest, which may indicate prolonged terrestrial habits for hatchlings in some environments.

Once in the high marsh, hatchlings appear to avoid open water and tend to stay close to vegetation, often burrowing in the mud or under dense wrack (Lovich et al. 1991; Coleman et al. 2014). Avoidance of open water may be related to reducing predation risk or difficulty in swimming and maneuvering in strong tidal currents, and movement into densely vegetated high marsh may reduce predation and dehydration risk (Coleman et al. 2014). Laboratory studies suggested that hatchling and juvenile terrapins may also use high marsh and terrestrial habitats to reduce osmotic stress that can inhibit growth (Dunson 1985; Kinneary 2008; Holliday et al. 2009; Chapter 9).

Demographic rates associated with the terrapin life cycle are similar to those of most aquatic turtles (Congdon et al. 1993; Heppell 1998; Ernst and Lovich 2009). Generally, terrapin life history is characterized by (1) high annual adult survival; (2) potentially low but highly variable egg survival; and (3) slow growth and low juvenile survival (Table 14.1). Females generally mature at between 6 and 8 years, though some smaller subspecies may mature at 4 years. Females lay 1 to 3 clutches of 5 to 15 eggs per year. Hatchling sex ratio is environmentally determined, with warmer temperatures producing female hatchlings and incubation temperature and egg size positively affecting juvenile growth and time of maturity (Roosenburg and Place 1995; Roosenburg and Kelley 1996; Chapter 10). There are few estimates of adult annual survival, despite the existence of several multiyear population studies across the species' range (e.g., Mitro 2003; Dorcas et al. 2007). Mitro (2003) estimated adult annual survival between 0.94 and 0.96, which is consistent with estimates of adult annual survival for freshwater turtles in the family Emydidae (Heppell 1998). There are few direct or indirect estimates of hatchling or juvenile annual survival (aptly referred to as the "missing years"). Mitro (2003) estimated juvenile annual survival between 0.45 and 0.57, which is lower than estimates for other emydids (Heppell 1998); however, Mitro did not distinguish hatchling (<1 year of age) from juvenile (>1 year of age). As with other species, hatchling survival is likely to be lower than juvenile annual survival (Heppell 1998). Draud and

Zimnavoda (2004) documented 33% mortality in hatchling terrapins over a 29 day period, though this resulted from predation by introduced Norway rats. Crawford et al. (2014a) estimated hatchling annual survival at ~25%, consistent with rates estimated for other emydids (Heppell 1998).

There is little opportunity to evaluate geographic variation for most terrapin demographic rates. We do know that terrapin clutch size increases significantly with latitude, with a small concurrent increase in mean age at maturity, from 6 to 8 years (Table 14.1; Chapter 6). Based on patterns among turtles and other reptiles, one would expect higher adult terrapin survival rates related to delayed maturity at higher latitudes (Shine and Iverson 1995). Nest or egg survival rates to hatching are highly variable among sites and years, and no clear geographic trends are evident. Rather, variation in nest and egg survival is governed by local conditions such as nest and predator abundance and vegetation (Table 14.1; Chapter 6).

All demographic parameters (rates) influence population growth, but not all demographic parameters are sufficiently variable or malleable to be used as effective targets for management (de Kroon et al. 2000; Spencer and Janzen 2010). As with models of population growth in other turtles (Heppell 1998; Enneson and Litzgus 2008), model estimates of terrapin population growth rates are most sensitive to variation in adult female survival (Mitro 2003; Crawford et al. 2014a; however, see Gilliand et al. 2014). The models indicate that 3% to 5% additive reductions in female survival can cause population declines. However, rapid evolution and negative density-dependent effects on adult fecundity and juvenile growth and survival may be greater for turtles than previously expected and thus may have some capacity to offset small reductions in adult survival (Spencer and Janzen 2010). Models suggest that terrapin population growth is least sensitive to nest or egg survival (Mitro 2003; Crawford et al. 2014a). However, nest or egg survival is both naturally and anthropogenically highly variable (Table 14.1). In addition, terrapins have temperature-dependent sex determination (Chapter 10), which creates the potential for sex-specific nest or egg survival rates. Therefore, given

Table 14.1. Stage-specific estimates of diamond-backed terrapin survival and reproductive rates in the presence and absence of identified threats

Stage parameter	Estimate or change in estimate (negative change indicated by −)	State	Identified threat	Comment	Reference
Nest survival	0.828	FL	None	Following raccoon removal	Munscher et al. 2012
	0.552 to 0.853	NJ	None	Open, loamy soils with no dredge materials	Wnek et al. 2013
	0.626 to 0.721	NY	None	Open, grassy habitat; predator exclusion devices used	Scholz 2007
	0.060 to 0.700	MD	Raccoon predation	Artificial nests; survival varied among beaches and years	Roosenburg & Place 1995
	0.078	NY	Raccoon predation	—	Feinberg & Burke 2003
	0.083 to 0.161	MD	Raccoon predation	—	Roosenburg 1991
	0.128	RI	Raccoon predation	—	Goodwin 1994
	0.169 to 0.228	FL	Raccoon predation	5%–16% of nests lost to storm inundation	Butler et al. 2004
	0.156	GA	Raccoon predation	Simulated nests with chicken eggs; shrub vegetation	Grosse et al. 2015
	0.205	VA	Raccoon predation	—	Ruzicka 2006
	0.233	FL	Raccoon predation	Cessation of raccoon removal	Munscher et al. 2012
	0.250 to 0.840	NJ	Raccoon predation	Reported higher predation, specifically by mammals, in area of higher tree and shrub cover	Burger 1977
	0.340	NJ	Raccoon predation	—	Burger 1976
	0.381	GA	Raccoon predation	Estimate from monitored nests only	Crawford et al. 2014b
	0.475 to 0.491	FL	Raccoon predation	Estimate from monitored nests only	Munscher et al. 2012
	0.548	GA	Raccoon predation	Simulated nests with chicken eggs; open grassy vegetation	Grosse et al. 2015
	0.674	NY	Raccoon predation	An additional 14.7% of eggs in surviving nests failed to hatch	Ner 2003
	0.000 to 0.594	NJ	Dredge soils	Eggs fail to develop in fresh dredge soils; attributed to high salt concentrations in dredge soils; survival increases 1 year post-dredge	Wnek et al. 2013
	0.111 to 0.634	NJ	Shade	Shaded, loamy soils with no dredge materials	Wnek et al. 2013
Hatchling annual survival	0.253	GA	None	Estimated from population model with other known demographic rates	Crawford et al. 2014a
Juvenile annual survival	0.446 to 0.565	RI	—	Estimated from population model with other known demographic rates; included hatchling year in juvenile survival estimate	Mitro 2003
Adult annual female survival	0.952	RI	—	—	Mitro 2003
	0.908	MD	—	—	W. Roosenburg & W. Kendall, Ohio University, pers. comm.
	0.099 to −0.448	MD	Bycatch	—	Roosenburg et al. 1997

(continued)

Table 14.1. (continued)

Stage parameter	Estimate or change in estimate (negative change indicated by −)	State	Identified threat	Comment	Reference
	−0.060 to −0.110	SC	Bycatch	—	Hoyle & Gibbons 2000
	−0.088	NJ	Road mortality	—	Szerlag & McRobert 2006
	−0.044 to −0.164	GA	Road Mortality	—	Crawford et al. 2014a
Mean clutch size	16	RI	—	—	Goodwin 1994
	13	MD	—	—	Roosenburg 1991
	12 to 13	NY	—	Estimated from protected nests	Ner 2003
	12	MA	—	—	Zimmerman 1992
	11	NY	—	—	Feinberg & Burke 2003
	10	NJ	—	—	Burger & Montevecchi 1975
	7	VA	—	—	Reid 1955
	7	SC	—	—	Zimmerman 1992
	7	FL	—	—	Seigel 1980
Mean clutch frequency per year	≥2	NY	—	—	Feinberg & Burke 2003
	1 to 2	RI	—	—	Goodwin 1994
	1 to 3	MD	—	—	Roosenburg 1991

our knowledge about factors that affect nest survival and sex ratios of hatchlings, the nest stage is more valuable as a management target than is reflected by its estimated relative effects on population growth.

Terrestrial Threats to Populations

Though terrapins spend only a brief portion of their lives in terrestrial habitats, a suite of terrestrial threats can have a disproportionately large effect on terrapin populations. Nesting terrapins confronted with modified access and alterations to nesting habitats by bulkheads or coastal roads can experience altered nest success or high mortality rates (Roosenburg 1994; Winters 2013; Crawford et al. 2014a). Additionally, in developed areas, mesopredators (e.g., raccoons) often increase in abundance due to food subsidies from trash and the absence of apex predators. Higher abundances of these mesopredators can increase the mortality of both adults and eggs (Table 14.1). When examined in isolation, specific terrestrial threats can appear to have only localized or small effects. However,

most specific terrestrial threats are part of a larger, cumulative set of terrestrial threats, all of which are related to coastal development. As a result, specific terrestrial threats are often spatially correlated and may interact within and across life stages to have severe negative effects on terrapin populations (Fahrig and Rytwinski 2009; Crawford et al. 2014a).

Alterations to Nesting Habitats

Though not consistently identified among the threats to terrapin populations, intensifying coastal land use and associated changes in land cover are emerging as important factors influencing local and regional terrapin abundance (Seigel and Gibbons 1995). For example, Winters (2013) and Wnek et al. (2013) reported that 38% of the shoreline of Barnegat Bay, New Jersey, has been altered in ways detrimental to terrapin nesting habitat. We know of only one study that has examined the effects of terrestrial land use on terrapin populations at large spatial scales. Isdell (2014) found that terrestrial land use, notably agricultural land use, adjacent to marsh habitat was negatively correlated with adult terrapin occupancy of tidal creeks within the lower Chesapeake Bay, Virginia. How agricultural practices affect terrapin populations is unknown. A study of wood turtles (*Glyptemys insculpta*) in an agricultural landscape estimated that as many as 20% of adults were killed by agricultural heavy machinery (Saumure et al. 2007). In addition, agricultural practices probably affect nesting habitats through soil disturbance and the planting of dense vegetation.

Our knowledge about anthropogenic effects on nesting habitats is limited to estimates of local effects on terrapin behavior or nest success. Roosenburg (1991) highlighted that increasing coastal development and the common practice of constructing bulkheads result in the loss of intact shoreline. Bulkheads can block terrapins' access to proximate, preferred nesting sites, which can increase females' nesting migration distances, or cause females to nest in suboptimal habitats, or concentrate nests into smaller areas (Roosenburg 1994; Roosenburg et al. 2014). Winters (2013) used telemetry to measure the effects of barriers on terrapin nesting migrations in Barnegat Bay, New Jersey. She found that barriers could increase total travel distance up to 500%. When displaced females spatially concentrate nests in remaining habitats, this can lead to higher nest predation rates (Roosenburg and Place 1995). In addition, anthropogenic structures can shade terrapin nesting areas, resulting in longer development times and increased male-biased sex ratios in hatchlings (Wnek et al. 2013).

Coastal development practices that alter soils may also be detrimental to terrapin nesting (Wnek et al. 2013). Terrapins nest in sandy soils, which have properties such as large particle sizes that improve gas diffusion and increases water potential that reduces hydric constraints on developing embryos (Roosenburg 1994; Wnek et al. 2013). The modification of soils, including the addition of organic materials or compaction, may affect embryonic survival. The common practice of dredging to clear shipping channels, fill uplands, and construct causeways can adversely affect terrapin nesting habitats (Wnek et al. 2013; however, see Roosenburg et al. 2014). In highly developed coastal areas,

dredge fill areas and causeways may be the only habitat available for terrapin nesting (Wood and Herlands 1997; Wnek et al. 2013). Dredge soils contain high concentrations of organic solids, fine sediments, and salts that reduce gas exchange and increase hydric constraints on terrapin eggs. Wnek et al. (2013) demonstrated that terrapin embryos are incapable of developing in freshly dredged soil, and they hypothesized that high salt content causes embryos to desiccate. Terrapin embryos could develop in aged dredge fill that had leached some of the salts and other contaminants, and we should note that dredge soils have been used to restore terrapin nesting habitat within Chesapeake Bay, Maryland (Roosenburg et al. 2014). Also, our understanding of the effects of dredge soils on terrapin nesting success is relevant to the risks posed by sea level rise. Rising sea level and more frequent inundation of nesting areas due to storm surges will increase the salinity of low-lying terrestrial habitats (Michener et al. 1995). If terrapin embryos are sensitive to high soil salinity, then frequent seawater inundation of nesting areas might reduce terrapins' nesting success.

Increasing coastal development also leads to shifts in vegetation that affect nesting success directly by interacting with other factors to influence nest survival and hatchling sex ratio. Terrapins prefer to nest in areas of patchy, short vegetation, where embryos develop faster and tend to produce a higher proportion of female hatchlings (Burger and Montevecchi 1975; Goodwin 1994; Roosenburg 1994; Feinberg and Burke 2003; Ner 2003; Scholz 2007; Hackney 2010; Grosse et al. 2015). The dense planting of dune grasses to control erosion or dense areas of invasive plants can increase terrapin nest failure (Roosenburg 1991; Wnek et al. 2013). Some plant roots will infiltrate and kill terrapin eggs (Lazell and Auger 1981), and dense grasses can reduce soil water potential, resulting in higher egg failure rates. Dense herbaceous cover or the succession or deliberate planting of trees and shrubs can also reduce terrapin nest survival and skew hatchling sex ratios. Dense vegetation shades nests, lowering temperature and resulting in longer development times, male-biased hatchling sex ratios, and potentially higher egg mortality (Wnek et al. 2013).

Vegetation can also be an important determinant of nest predation rates. Roosenburg and Place (1995) found that dense herbaceous cover (grasses) reduced the rate of nest predation, presumably because the dense vegetation hindered predator foraging; however, the herbaceous cover also shaded nests, resulting in 100% male hatchlings. In contrast, woody vegetation such as shrubs and trees may increase predation risk while also shading terrapin nests. Burger (1977) reported high mammalian nest predation in wooded, shrub, and edge habitats, and Hackney (2010) found that terrapin nests in shrub or edge habitats closer to marshes had a higher probability of being preyed upon than nests in open, sandy areas farther from the marsh. Grosse et al. (2015) demonstrated experimentally that shaded nests in shrub habitats had lower survival rates as a result of higher raccoon predation, and surviving nests in these habitats produced 100% male hatchlings. In contrast, open, grassy habitats had the highest nest survival rates and produced 100% female hatchlings. Collectively, these studies show that altered terrestrial vegetation can directly and indirectly influence nest survival and hatchling sex ratios.

Terrestrial Predators

Predation on turtle nests is spatially and temporally variable but is increasingly a threat to population persistence in developed landscapes. Terrestrial mortality of terrapins and other turtles, particularly nest and hatchling mortality, is clearly related to predator abundance and human activities that concentrate turtle nesting (Table 14.1). Known terrestrial predators of adult terrapins are raccoons and foxes, while terrestrial nest and hatchling predators include ghost crabs, grackles, crows, willets, kingsnakes, foxes, raccoons, armadillos, introduced Norway rats, and red imported fire ants (Burger 1977; Goodwin 1994; Roosenburg and Place 1995; Feinberg and Burke 2003; Butler et al. 2004; Draud and Zimnavoda 2004; Hackney 2010; Muldoon and Burke 2012; Munscher et al. 2012; Roosenburg et al. 2014). The abundance of mesopredators (e.g., raccoon) has increased, particularly in developed areas, because of the loss of apex pred-

ators and the presence of anthropogenic resource subsidies. Many studies report predation by raccoons on adult female terrapins during oviposition; however, the effect on population number and structure has not been documented. Seigel (1980) and Feinberg and Burke (2003) documented a relatively high number of adult female terrapins killed by raccoons over a 2 year period in Florida and New York, respectively, which indicates that adult mortality could be high during nesting in areas of high predator abundance. Burger and Montevecchi (1975) and Roosenburg (1994) suggest that high mammalian predation risk may be a selective pressure promoting diurnal nesting habits by female terrapins.

Terrapin nest and juvenile survival rates can be high in areas with few mammalian predators but are often extremely low in areas with a high density of predatory mammals (Table 14.1). Raccoons routinely depredate up to 95% of nests at sites with high raccoon densities (references in Table 14.1). Several studies have noted that mammalian predation of nests generally occurs within 48 h of nesting (Burger 1977; Roosenburg 1991; Goodwin 1994; Munscher et al. 2012; Crawford et al. 2014b), and as mentioned above, nests in shrubby or wooded habitats may be more vulnerable to mammalian predators. Several studies have also documented that predators are significant sources of mortality for hatchlings both in the nest and after emergence (Feinberg and Burke 2003; Butler et al. 2006). A study in Oyster Bay Harbor, New York, found that introduced rats depredated 67% of terrapin hatchlings within 30 days of emerging from nests (Draud and Zimnavoda 2004).

Roads

Roads have become a pervasive fixture of most landscapes, with conservation implications for many species. Along the East and Gulf coasts of the states that encompass the range of the diamond-backed terrapin, 60% to 80% of all land area is within 400 m of a road, and these regions are experiencing rapid annual increases in traffic (Riitters and Wickham 2003; Riitters et al. 2003; Baird 2009). Empirical data show spatial and temporal changes in turtle populations associated with increasing road cover (Marchand and Litvaitis 2004; Steen and Gibbs 2004; Aresco 2005a; Gibbs and Steen 2005), and models predict that road mortality can cause substantial declines of turtle populations at regional scales (Fahrig et al. 1995; Gibbs and Shriver 2002; Litvaitis and Tash 2008; Langen et al. 2009; Beaudry et al. 2010).

Roads can affect terrapin populations through the permanent loss of habitat, through increased mortality from vehicle strikes, and by creating a barrier to movement between terrestrial and aquatic habitats (for reviews of the effects of roads on various taxa, see Forman and Alexander 1998; Trombulak and Frissell 2000; Fahrig and Rytwinski 2009). In general, turtles use mixed habitat networks for migrating, mate searching, nesting, and hibernating that often take them across roadways (Gibbs and Shriver 2002; Aresco 2005b; Brennessel 2006; Beaudry et al. 2008; Langen et al. 2009; Steen et al. 2012). Road mortality is particularly high among turtles because they are slow moving, do not avoid roads, and instinctively respond to oncoming cars by freezing and retracting into their shells (Gibbs and Shriver 2002; van Langevelde and Jaarsma 2004; Fahrig and Rytwinski 2009; Lima et al. 2015). Moreover, because road mortality of aquatic turtles is generally related to movements for breeding and nesting (Wood and Herlands 1997; Gibbs and Shriver 2002; Beaudry et al. 2010), it disproportionately affects mature females (Steen et al. 2006; Grosse et al. 2011; Crawford et al. 2014a). Among studies that have monitored terrapin activity on roads, >98% of individuals observed or struck by cars were adult females (Wood and Herlands 1997; Hoden and Able 2003; Szerlag and McRobert 2006; Crawford et al. 2014b).

Because terrapin populations are strongly governed by high adult female survival, even minor, additive mortality of females killed on roads is predicted to lead to population declines (Crawford et al. 2014a; Gilliland et al. 2014). However, few studies have evaluated the regional or local effects of roads on terrapins or identified factors that cause terrapin mortality to vary on roads. Grosse et al. (2011) found that, at the regional scale, proximity to roads did not correlate with terrapin abundance or

sex ratio along the coast of Georgia. However, they did find that terrapin abundance was notably low adjacent to causeways that bisect marshes to barrier islands. This finding is consistent with several studies that report high numbers of terrapins killed along causeways that bisect marshes. For example, Hoden and Able (2003) documented 450 terrapins killed during the nesting season on 8 km of Great Bay Boulevard near Tuckerton, New Jersey, between 1999 and 2002; Wood and Herlands (1997) documented 4,020 terrapins killed during the nesting season on 28 km of causeways to the Cape May Peninsula between 1989 and 1995. Between 2009 and 2013, Crawford et al. (2014b) recorded 757 female terrapins killed during nesting seasons along 9 km of the causeway to Jekyll Island, Georgia. They also directly estimated adult female terrapin mortality over 3 years along the causeway and found that mean annual adult female mortality from vehicle collisions during nesting migrations was 11% (range 4.4%–16.4%). They estimated that an additive mortality of 3.1% from vehicle collisions was sufficient to cause negative population growth along the causeway. Consistent with the high rate of female mortality, terrapin populations in the creeks adjacent to the Jekyll Island causeway had a 4:1 male-biased sex ratio (Maerz et al., unpub. data), compared with M:F sex ratios of 0.6:1 to 1:1 in other tidal creeks in the region (Grosse et al. 2011). Similarly, Avissar (2006) reported that the M:F sex ratio of the terrapin population along the Garden State Parkway in New Jersey, where terrapin road mortality is frequent, increased from 0.6:1 to 6:1 between 1989 and 2003. Mean terrapin carapace length declined over the same 14 year period, which is consistent with a loss of older adult females.

Causeways constructed through marshes create "ecological traps" for female terrapins (Schlaepfer et al. 2002). With increasing coastal development, elevated causeways through the marsh become the only available nesting habitat for terrapins. Females are naturally attracted to nest in open, elevated habitat near the marsh (Szerlag and McRobert 2006). Portions of the causeway nearer the marsh are often planted or overgrown with shrubs, while areas near the road are mowed to maintain short

Fig. 14.2. Adult female terrapins, during nesting movements, emerging onto (A) Great Bay Boulevard causeway near Tuckerton, New Jersey, and (B) the Downing-Musgrove Causeway to Jekyll Island, Georgia. The circles on the road in (A) indicate female terrapins from the perspective of a driver. In addition to the causeways' representing elevated areas above and immediately proximate to marsh, we draw attention (A, B, and C) to the shrub and tree lines most proximate to the marsh line, in contrast to shorter, mowed vegetation adjacent to the road edge. Image sources: (A) G. Sakowicz; (B) B. A. Crawford; (C) J. C. Maerz

vegetation (Fig. 14.2), making them the most attractive parts of causeways for nesting (Wood and Herlands 1997; Crawford et al. 2014b). Traffic volumes on coastal roads peak between May and August (Baird 2009), overlapping with terrapins' nesting season throughout the range of the species (Seigel 1980; Zimmerman 1992; Wood and Herlands 1997; Feinberg and Burke 2003; Crawford et al. 2014b). Because female terrapins often nest around the diurnal high tide, they are likely to be active on roads during the day, when traffic volume is highest.

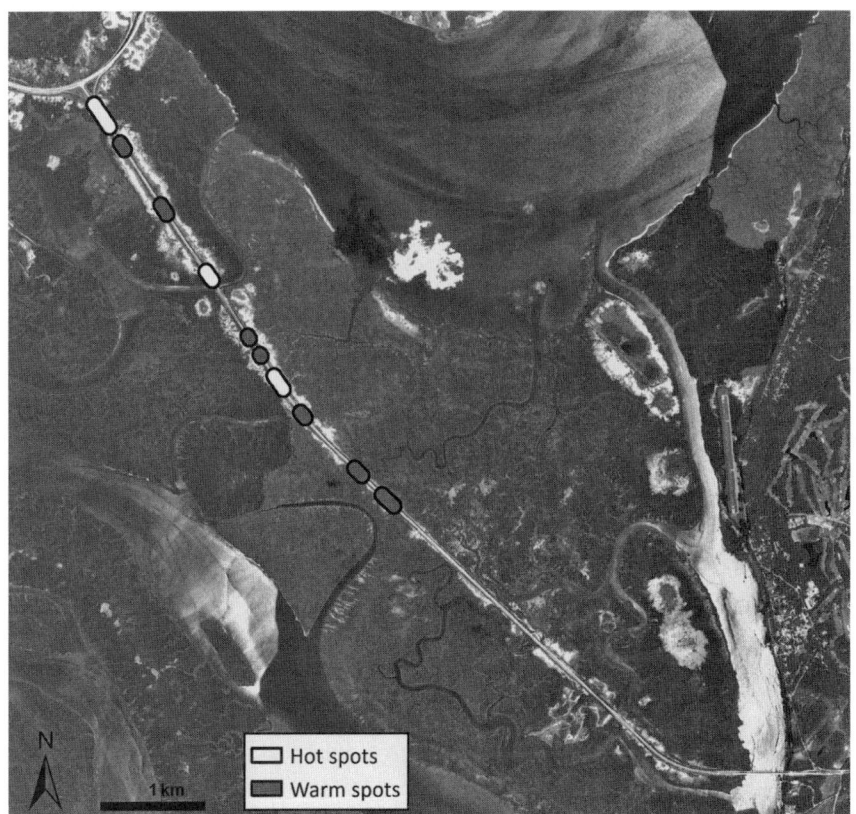

Fig. 14.3. Spatial patterns of adult female terrapin activity over 2 years on the Downing-Musgrove Causeway to Jekyll Island, Georgia. In this study, 29.2% of females were observed within the three hot spots spanning 803 m of the causeway (9.2% of total length), and 51.9% of females were observed within all hot spots and warm spots spanning 1,859 m (21.3%) of the causeway. Note that the three hot spots are in an area with extensive unvegetated high marsh (light gray), but the specific locations of the hot spots are proximate to local points where vegetated high marsh (dark gray) extends up to the causeway. From Crawford et al. (2014b)

At finer spatial scales, a portion of terrapin mortality along roads can be spatially or temporally clustered. Areas of concentrated activity or mortality on roads are known as threat "hot spots," and periods of concentrated activity are known as threat "hot moments" (Beaudry et al. 2008, 2010). Crawford et al. (2014b) found that 30% of terrapin road mortality occurring over a 2 year period was distributed among three hot spots representing only 0.8 km (9%) of the causeway to Jekyll Island, Georgia (Fig. 14.3). These hot spots were spatially stable across years, probably reflecting high interannual nest site fidelity among females. Crawford et al. (2014b) found no clear relationship between terrapin road mortality and proximity to tidal creeks or shrub or grass cover above the high marsh; however, they did find that female terrapins nested more frequently along the roads near areas of well-vegetated high marsh. More concentrated nesting along densely vegetated areas of marsh may be related to migrating

females' minimization of exposure to high temperatures or predators or to higher survival of emerging hatchlings. Crawford et al. (2014b) also demonstrated that 52% of terrapin road mortality was concentrated within a 3 h window around the diurnal high tide, when females generally prefer to nest (Figs. 14.1, 14.4), and that within that 3 h window, the probability that a terrapin would be encountered by a vehicle on the road was nearly double (0.6–0.8) during the first 30 days of the nesting season compared with the last 30 days (0.2–0.4; Fig. 14.4). Other studies also reported more concentrated peaks in terrapin nesting during the first 20 to 30 days of the nesting season (Burger and Montevecchi 1975; Feinberg and Burke 2003). By integrating information on the spatial and temporal predictability of activity, there is the potential to develop management strategies that focus on hot spots or hot moments so as to reduce terrapin mortality on coastal roadways.

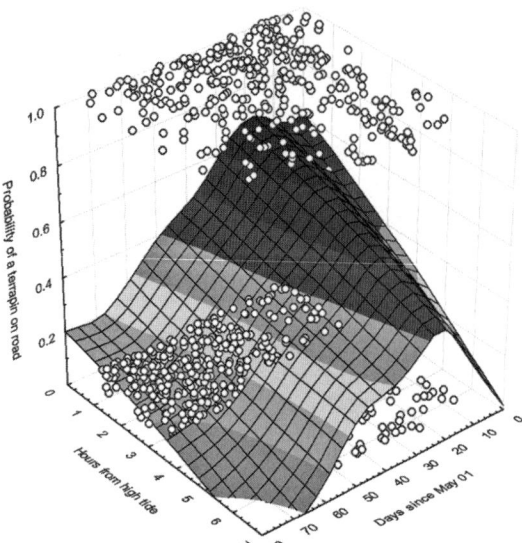

Fig. 14.4. Temporal patterns of adult female terrapin activity over 2 years on the Downing-Musgrove Causeway to Jekyll Island, Georgia. The diagram shows the probability that a female terrapin occurred on the causeway in relation to time within the nesting season and temporal proximity to daytime high tide. Within the first third of the nesting season, the probability that a terrapin was on the causeway within 1 h of the daytime high tide was 0.80. This probability declined as the nesting season progressed and with increasing time from the daytime high tide. From Crawford et al. (2014b)

Integrated Management of Terrestrial Threats

Multiple factors often contribute to declines in wildlife populations, such that management activities that focus on single threats may be insufficient. Ultimately, conservation efforts for species such as the terrapin should identify and integrate management solutions for multiple terrestrial threats as part of a holistic framework (Heppell et al. 1996; Crawford et al. 2014a). Evidence across the species' range indicates that mortality on roads and nest predation are sufficiently high that intervening in one threat without addressing the other is unlikely to stabilize or restore terrapin populations. For example, Crawford et al. (2014a) modeled the joint effects of road mortality and mammalian nest predation on terrapin population growth in Georgia and estimated that complete elimination of road mortality would not stabilize the population without a concurrent reduction in nest predation.

Moreover, the issues of roads, predators, and vegetation changes are interrelated such that their management can and should be coordinated (Crawford et al. 2014a; Grosse et al. 2015). We highlight those interconnections as management opportunities.

Without question, directly reducing predation on turtle nests would have a positive effect on terrapin recruitment and population recovery. Munscher et al. (2012) demonstrated that removal of raccoons from a terrapin nesting beach in northeastern Florida increased terrapin nest survival by 84% to 174%. However, they also showed that 1 year after cessation of raccoon removal, raccoon predation on terrapin nests returned to pre-removal levels—indicating that predator removal can be effective but is likely to require continuous removal efforts. A more efficient approach could employ the removal of mammalian predators only during some terrapin nesting seasons, allowing windows of release from predation pressure. Predators could repopulate removal areas outside the terrapin breeding season or during non-control years. This strategy has been tested and utilized successfully to manage predator effects on threatened and endangered game birds (Tapper et al. 1996). Whether removing mammalian predators will be effective in all areas may depend on the level of compensatory predation, such as by birds or crabs, which may also experience an ecological release after the removal of predatory mammals (Barton and Roth 2008; Roosenburg et al. 2014). We also caution that mammalian predator control can be a publicly sensitive issue; adverse public reaction may counteract support for terrapin conservation (Crawford et al. 2015).

Nonlethal predator controls can be effective for protecting small numbers of known nests, but its effectiveness in conserving terrapin populations is unclear. Mitro (2003) modeled the benefit of protecting a proportion of terrapin nests and estimated that protecting as few as 5% of nests was sufficient to increase hatchling recruitment by 47% and to significantly increase the population growth rate. However, predator excluders do not always work (Grosse et al. 2015), they can require considerable investments of personnel time, and the effect is lost if nest protection interventions are stopped. In some cases, nonlethal predator exclusion can be integrated to improve other management actions such as barriers or anthropogenic nesting habitats that function primarily to prevent road mortality (Fig. 14.5; further discussed below). An area of great potential is the use of vegetation management to nonlethally reduce nest predation while improving nesting habitat, particularly to increase the production of female hatchlings (Grosse et al. 2015). Specifically, restoring coastal nesting habitats to shorter, patchy vegetation types and avoiding the intentional planting of hedgerows along causeways may be effective actions for addressing multiple terrestrial threats to terrapins (Fig. 14.5).

Because coastal development increases road area and traffic volume, road mortality is the other major terrestrial threat for which management solutions will be essential to restore terrapin populations. The identification of hot spots and hot moments of terrapins' movements across roads creates opportunities for targeted management actions to reduce road mortality. Barriers (e.g., Fig. 14.5, A) that prevent animals from crossing roads are effective at reducing wildlife-vehicle collisions when placed at previously observed hot spots (Clevenger et al. 2001; Aresco 2005b). For example, in 2004, researchers at the Wetlands Institute in New Jersey installed barrier fences composed of silt fencing (45 cm high), Tenax mesh material (45 cm), or corrugated plastic pipes (18 cm) along 10 km of the Stone Harbor Boulevard and other roads known to be hot spots of terrapin mortality. In areas where barriers were installed, the number of terrapins killed on the roads was reduced 80% to 100% (R. Wood, Wetlands Institute, unpub. data).

Installing fences along roadsides can reduce terrapin-vehicle collisions, but some limitations warrant consideration when using barriers as a management option. Fencing materials and annual maintenance are costly, which limits their application along extensive causeways. Barriers can also be aesthetically controversial with local stakeholders (Crawford et al. 2015). Strategic use of barriers at local hot spots could effectively reduce costs and conflict, but identifying hot spots presents its own logistical challenges (Crawford et al. 2014b). Determining where hot spots occur along a causeway may require multiple years of intensive monitoring to determine habitat features that predict the location of hot spots. Though Crawford et al. (2014b) found that road mortality hot spots were positively correlated with densely vegetated areas of high marsh, this relationship was not highly predictive because large proportions of roads are adjacent to extensive areas of marsh vegetation. No other measured factors predicted local hot spots.

The difficulty in predicting nesting hot spots may be related to nest philopatry and fidelity. Roosenburg and Place (1995) noted that terrapins' nest site fidelity can interact with spatially variable nest survival to result in increasing nest density. Therefore, hot spots might be areas of historical nesting success that are not easily related to current habitat factors. Crawford et al. (2014b) did find that local hot spots were stable over 3 years, but it is not yet known whether local hot spots are stable over longer periods or whether they occur at sites where terrapins do not exhibit nest site fidelity. Also, we must recognize that while 30% to 60% of terrapin mortality may be concentrated in small areas amenable to using barriers, 40% to 70% of terrapin-vehicle collisions may be spatially diffuse but sufficient to cause population declines (Crawford et al. 2014a, b). Therefore, barriers are potentially part of a multiple approach strategy for reducing terrapin mortality on roads.

If hot spots are identified, it is also important to consider potential secondary, negative effects of barrier fencing on turtles. Fencing can adversely affect adult turtles by increasing the time they spend searching for suitable nest sites, thus exposing them to an elevated risk of thermal stress or predation (Aresco 2005b). Turtles often nest along barriers, providing predators with a means to more efficiently find and destroy nests (Aresco 2005b; Crawford et al. 2014a; Grosse et al. 2015). Fences also serve as barriers to dispersal between habitats that may be important in population dynamics (Andrews et al. 2008).

Anthropogenic nest mounds may be a solution, complementing the use of barriers to reduce terrapin road mortality and improve nesting success at

Fig. 14.5. Management interventions to address terrestrial threats to terrapins on the Downing-Musgrove Causeway to Jekyll Island, Georgia. (A) Tenax barrier fence blocking nesting female terrapins from accessing the causeway. The barrier directs turtles to (B) a manmade nest mound covered with a predator exclusion box. (C) Flashing sign alerting motorists during periods when the probability of turtles emerging on the road is highest. The sign is programmed using tide calendars for the island and the relationship between nesting season and daytime high tide, as described by Crawford et al. (2014) (see Fig. 14.4, B). (D) Removal of overgrown wax myrtle and trees along the marsh line to create more optimal nesting habitat proximate to marsh and farthest from the causeway. Vegetation removal reduces nest depredation rates and increases the production of female hatchlings. (A, C) Images from B. A. Crawford. (B, D) Images from J. C. Maerz. Inset images provided by the Florida Times-Union (Jacksonville)

hot spots. Nest mounds are open, elevated formations with predator excluder boxes that have been used to attract semiaquatic turtles to nest inside the boxes and protect eggs (Buhlmann and Osborn 2011; Paterson et al. 2013). The Georgia Sea Turtle Center (GSTC) is currently testing a hybrid barrier at one hot spot on the causeway to Jekyll Island, using Tenax fencing that leads to a series of nest boxes spanning 21 m of protected roadside habitat (Fig. 14.5, A, B). Grosse et al. (2015) reported high raccoon predation on the nest mounds despite the presence of a predator exclusion cage, as well as high nest temperatures, near levels that can result in deformities or death of embryos. Nonetheless, surviving nests on the mounds produced 100% female hatchlings. Adjustments to more effectively exclude predators (e.g., electrified wires) and the use of sparse vegetation to stabilize mounds and provide some nest shading are currently being evaluated by the GSTC. While more monitoring is needed to measure the viability of nests in nest boxes, a small number of terrapins voluntarily nested within the boxes, and fewer adults have been struck on the road adjacent to the nest box barrier installation (Crawford et al., unpub. data; GSTC, unpub. data).

Because 50% or more of terrapin-vehicle collisions are diffused along long stretches of road outside hot spots, strategies targeted at peak times of terrapins' terrestrial activity are needed to reduce mortality on roads. Studies of terrapins and other turtles have identified general seasonal "hot moments" (May through July) associated with species' nesting activity when frequent overland movements occur (Wood 1997; Szerlag and McRobert 2006; Beaudry et al. 2010; Cureton and Deaton 2012). In some cases, studies have identified more specific seasonal peaks (<20 days) skewed toward the beginning of nesting seasons (Burger and Montevecchi 1975; Feinberg and Burke 2003; Crawford et al. 2014b). These broad activity windows allow management strategies such as the temporary closing of some roads for peak portions of the nesting season or the use of seasonal signage or public awareness campaigns to alert drivers to potential terrapin activity on roads. However, acclimation to static signage can limit the effectiveness of alerting

motorists to wildlife activity along roads (Sullivan et al. 2004). More specific windows of peak activity permit more dynamic interventions to alter drivers' behavior. For example, at many sites, terrapin nesting is highly correlated with tidal amplitude and proximity to the daytime high tide. This creates a situation analogous to school zones: flashing signage can alert motorists during periods when terrapins are highly likely to be on roads. In 2012, the GSTC, University of Georgia, and Georgia Department of Transportation installed two flashing warning signs programmed to the local tide schedule that alerted drivers daily, throughout the nesting season, during the 3 h window when terrapins were most likely to be active on the road (Fig. 14.5, C). The first devices of their kind to target turtles, they have reduced road mortality, improving population persistence (Crawford pers. obs.).

Ultimately, predator, vegetation, and road management are linked under the broader and increasing problem of declining availability of suitable nesting habitat. There have already been significant losses of natural shoreline across the terrapin's range, and efforts to limit coastal erosion are likely to result in increased bulkheading and dredge fill (Roosenburg 1994; Winters 2013; Wnek et al. 2013). There is an urgent need for agencies and local stakeholders to identify, protect, and ensure the quality of remaining terrapin nesting sites. With nesting concentrated into fewer areas, those areas will need active vegetation management to maximize nest survival and maintain high nest temperatures to increase the production of female hatchlings. Vegetation management to prevent the encroachment of shrubs or trees would also complement measures to reduce nest predation by mammalian predators. Roosenburg et al. (2014) have reported on the restoration of Poplar Island in Chesapeake Bay, Maryland, which is free of raccoons and foxes and has had vegetation management that is optimal for terrapin nesting. Female terrapins began nesting on the island within 1 year, and within 3 years, an average of more than 200 terrapin nests were located on the island, with nest survival consistently above 70%.

Conclusions

The terrestrial phases of the terrapin's life are brief but present numerous threats that strongly influence the recovery and sustainability of terrapin populations. New quantitative approaches to evaluating threats and management options show early potential to improve terrapin conservation. Efforts to identify spatial and temporal patterns of threats, such as hot spots and hot moments of terrapin activity on roads, can facilitate targeted management to maximize the efficiency of interventions. Models and experiments are improving the effectiveness of management options, and development of a comprehensive suite of management tools will permit the tailoring of management plans to local situations. However, the quality of models and their ability to inform at scales relevant to management depend on rigorous data on population demographic rates and per capita effects of various threats. Despite widespread attention and significant research on terrapins, our knowledge about the geographic and temporal variation of terrapin demographic rates is limited (Chapter 6). Such information would enable more locally appropriate estimates of threat effects and management effects. In addition, our knowledge of hatchling and juvenile terrapin ecology is still insufficient to identify management needs or opportunities for these life stages. Finally, the majority of research to date addresses threats at relatively local scales. The number of studies that estimate the effects of threats at larger scales is insufficient (e.g., Grosse et al. 2011; Isdell 2014). Such information is important for determining whether threats that have conspicuous local effects are of similar importance across larger spatial scales (e.g., Grosse et al. 2011), for determining whether the importance of threats varies geographically in relation to changes in species ecology or human activities, and for identifying potentially larger-scale factors that threaten the species across its range (e.g., agricultural effects on terrapins; Isdell 2014). Studies examining the regional effects of threats also are important in identifying the role of state and federal agencies and regulations in terrapin management; this will improve prioritization of conservation strategies at broader vs. more localized scales.

ACKNOWLEDGMENTS

Our review of terrestrial threats and conservation in this chapter was made possible by the great work and significant contributions of many conservation and research teams, including the Wetlands Institute, the Georgia Sea Turtle Center, Jekyll Island Foundation and AGL Resources, A. M. Grosse, W. M. Roosenburg, and R. Wood. The chapter was improved greatly by the critique and comments of D. A. Steen. Portions of the research presented here were supported by state wildlife grants through Georgia Department of Natural Resources Coastal Resources Division and the Jekyll Island Foundation and AGL Resources. We are also grateful to the publishers and individuals who permitted the use of figures and photographs.

REFERENCES

Andrews, K.M., J.W. Gibbons, D.M. Jochimsen, J.C. Mitchell and R. Jung Brown. 2008. Ecological effects of roads on amphibians and reptiles: a literature review. Herpetological Conservation 3:121-143.

Aresco, M. 2005a. The effect of sex-specific terrestrial movements and roads on the sex ratio of freshwater turtles. Biological Conservation 123:37-44.

Aresco, M.J. 2005b. Mitigation measures to reduce highway mortality of turtles and other herpetofauna at a north Florida lake. Journal of Wildlife Management 69:549-560.

Auger, P.J. and P. Giovannone. 1979. On the fringe of existence: diamondback terrapins and Sandy Neck Cape. Cape Naturalist 8:44-58.

Avissar, N.G. 2006. Changes in population structure of diamondback terrapins (*Malaclemys terrapin terrapin*) in a previously surveyed creek in southern New Jersey. Chelonian Conservation and Biology 5:154-159.

Baird, R.C. 2009. Coastal urbanization: the challenge of management lag. Management of Environmental Quality: An International Journal 20:371-382.

Barton, B.T. and J.D. Roth. 2008. Implications of intraguild predation for sea turtle nest protection. Biological Conservation 141:2139-2145.

Beaudry, F., P.G. deMaynadier and M.L. Hunter Jr. 2008. Identifying road mortality threat at multiple spatial scales for semi-aquatic turtles. Biological Conservation 141:2550-2563.

Beaudry, F., P.G. deMaynadier and M.L. Hunter Jr. 2010. Identifying hot moments in road-mortality risk for freshwater turtles. Journal of Wildlife Management 74:152-159.

Brennessel, B. 2006. Diamonds in the Marsh: A Natural History of the Diamondback Terrapin. University Press of New England, Hanover, NH, and London. 236 pages.

Buhlmann, K.A. and C.P. Osborn. 2011. Use of an artificial nesting mound by wood turtles (*Glyptemys insculpta*): a tool for turtle conservation. Northeastern Naturalist 18:315-334.

Burger, J. 1976. Behavior of hatchling diamondback terrapins (*Malaclemys terrapin*) in the field. Copeia 1976:742-748.

Burger, J. 1977. Determinants of hatching success in diamondback terrapin, *Malaclemys terrapin*. American Midland Naturalist 97:444-464.

Burger, J. and W.A. Montevecchi. 1975. Nest site selection in the terrapin *Malaclemys terrapin*. Copeia 1975:113-119.

Butler, J.A., C. Broadhurst, M. Green and Z. Mullin. 2004. Nesting, nest predation and hatchling emergence of the Carolina diamondback terrapin, *Malaclemys terrapin centrata*, in northwestern Florida. American Midland Naturalist 152:145-155.

Butler, J.A., R.A. Seigel and B.K. Mealey. 2006. *Malaclemys terrapin*—diamondback terrapin. Pages 279-295 in P.A. Meylan (ed.), Biology and Conservation of Florida Turtles. Chelonian Research Monograph 3. Chelonian Research Foundation, Lunenburg, MA.

Clevenger, A.P., B. Chruszcz and K.E. Gunson. 2001. Highway mitigation fencing reduces wildlife-vehicle collisions. Wildlife Society Bulletin 29:646-653.

Coleman, A.T., T. Wibbels, K. Marion, T. Roberge, D. Nelson and J. Dindo. 2014. Dispersal behavior of diamond-backed terrapin post-hatchlings. Southeastern Naturalist 13:572-586.

Congdon, J.D., A.E. Dunham and R.C. Van Loben Sels. 1993. Delayed sexual maturity and demographics of Blanding's turtles (*Emydoidea blandingii*): implications for conservation and management of long-lived organisms. Conservation Biology 7:826-833.

Crawford, B.A. 2011. Mortality and management: assessing diamondback terrapins (*Malaclemys terrapin*) on the Jekyll Island Causeway. MS thesis, University of Georgia, Athens, GA. 76 pages.

Crawford, B.A., J.C. Maerz, N.P. Nibbelink, K.A. Buhlmann and T.M. Norton. 2014a. Estimating the consequences of multiple threats and management strategies for semi-aquatic turtles. Journal of Applied Ecology 51:359-366.

Crawford, B.A., J.C. Maerz, N.P. Nibbelink, K.A. Buhlmann, T.M. Norton and S.E. Albeke. 2014b. Hot spots and hot moments of diamondback terrapin road-crossing activity. Journal of Applied Ecology 51:367-375.

Crawford, B.A., N.C. Poudyal and J.C. Maerz. 2015. When drivers and terrapins collide: assessing stakeholder attitudes toward wildlife management on the Jekyll Island Causeway. Human Dimensions of Wildlife 20:1-14.

Cureton, J.C. and R. Deaton. 2012. Hot moments and hot spots: identifying factors explaining temporal and spatial variation in turtle road mortality. Journal of Wildlife Management 76:1047-1052.

de Kroon, H., J. van Groenendael and J. Ehrlen. 2000. Elasticities: a review of methods and model limitations. Ecology 81:607-617.

Dorcas, M.E., J.D. Wilson and J.W. Gibbons. 2007. Crab trapping causes population decline and demographic changes in diamondback terrapin over two decades. Biological Conservation 137:334-340.

Draud, M. and S. Zimnavoda. 2004. Predation on hatchling and juvenile diamondback terrapins (*Malaclemys terrapin*) by the Norway rat (*Rattus norvegicus*). Journal of Herpetology 38:467-470.

Dunning, J.B., B.J. Danielson and H.R. Pulliam. 1992. Ecological processes that affect populations in complex landscapes. Oikos 65:165-179.

Dunson, W.A. 1985. Effect of water salinity and food salt content on growth and sodium efflux of hatchling diamondback terrapins (*Malaclemys*). Physiological Zoology 58:736-747.

Enneson, J.J. and J.D. Litzgus. 2008. Using long-term data and a stage-classified matrix to assess conservation strategies for an endangered turtle (*Clemmys guttata*). Biological Conservation 141:1560-1568.

Ernst, C.H. and J.E. Lovich. 2009. Turtles of the United States and Canada. 2nd edition. Johns Hopkins University Press, Baltimore, MD. 840 pages.

Fahrig, L., J.H. Pedlar, S.E. Pope, P.D. Taylor and J.F. Wegner. 1995. Effect of road traffic on amphibian density. Biological Conservation 73:177-182.

Fahrig, L. and T. Rytwinski. 2009. Effects of roads on animal abundance: an empirical review and synthesis. Ecology and Society 14(1):21. http://www.ecologyandsociety.org/vol14/iss1/art21.

Feinberg, J.A. and R.L. Burke. 2003. Nesting ecology and predation of diamondback terrapins, *Malaclemys terrapin*, at Gateway National Recreation Area, New York. Journal of Herpetology 37:517-526.

Forman, R.T.T. and L.E. Alexander. 1998. Roads and their major ecological effects. Annual Review of Ecology and Systematics 29:207-231.

Freedberg, S. and M.J. Wade. 2001. Cultural inheritance as a mechanism for population sex-ratio bias in reptiles. Evolution 55:1049-1055.

Gibbs, J.P. and W.G. Shriver. 2002. Estimating the effects of road mortality on turtle populations. Conservation Biology 16:1647-1652.

Gibbs, J.P. and D.A. Steen. 2005. Trends in sex ratios of turtles in the United States: implications of road mortality. Conservation Biology 19:552-556.

Gilliand, S., R.M. Chambers and M.D. LaMar. 2014. Modeling the effects of crab potting and road traffic on a population of diamondback terrapins. In Proceedings of the Sixth Symposium on Biomathematics and Ecology: Education and Research, 2013. No pagination. http://cas.illinoisstate.edu/ojs/index .php/beer/article/view/808.

Goodwin, C.C. 1994. Aspects of nesting ecology of the diamondback terrapin (*Malaclemys terrapin*) in Rhode Island. MS thesis, University of Rhode Island, Kingston, RI. 84 pages.

Grosse, A.M., B.A. Crawford, J.C. Maerz, K.A. Buhlmann, T.M. Norton, M. Kaylor and T. D. Tuberville. 2015. Effects of vegetation and artificial habitat management on the nesting success and sex ratios of Carolina diamondback terrapins (*Malaclemys terrapin*). Journal of Fish and Wildlife Management 6:19-28.

Grosse, A.M., J.C. Maerz, J. Hepinstall-Cymerman and M.E. Dorcas. 2011. Effects of roads and crabbing pressures on diamondback terrapin populations in coastal Georgia. Journal of Wildlife Management 75:762-770.

Hackney, A.D. 2010. Conservation biology of the diamondback terrapin in North America: policy status, nest predation, and managing island populations. MS thesis, Clemson University, Clemson, SC. 69 pages.

Hart, K.M. 2005. Population biology of diamondback terrapins (*Malaclemys terrapin*): defining and reducing threats across their geographic range. PhD dissertation, Duke University, Durham, NC. 259 pages.

Heppell, S.S. 1998. Application of life-history theory and population model analysis to turtle conservation. Copeia 1998:367-375.

Heppell, S.S., L.B. Crowder and D.T. Crouse. 1996. Models to evaluate headstarting as a management tool for long-lived turtles. Ecological Applications 6:556-565.

Hoden, R. and K. Able. 2003. Habitat use and road mortality of diamondback terrapins (*Malaclemys terrapin*) in the Jacques Cousteau National Estuarine Research Reserve at Mullica River-Great Bay in southern New Jersey. Jacques Cousteau NERR Technical Report 100-24. Tuckerton, NJ. 31 pages.

Holliday, D.K., A.A. Elskus and W.M. Roosenburg. 2009. Impacts of multiple stressors on growth and metabolic rate of *Malaclemys terrapin*. Environmental Toxicology and Chemistry 28:338-345.

Hoyle, M.E. and J.W. Gibbons. 2000. Use of a marked population of diamondback terrapins (*Malaclemys terrapin*) to determine impacts of recreational crab pots. Chelonian Conservation and Biology 3:735-737.

Isdell, R.E. III. 2014. Anthropogenic modifications of connectivity at the aquatic-terrestrial ecotone in the Chesapeake Bay. MS thesis, College of William and Mary, Williamsburg, VA. 76 pages.

Kinneary, J.J. 2008. Observations on terrestrial feeding behavior and growth in diamondback terrapin (*Malaclemys*) and snapping turtle (*Chelydra*) hatchlings. Chelonian Conservation and Biology 7:118-119.

Langen, T.A., K.M. Ogden and L.L. Schwarting. 2009. Predicting hot spots of herpetofauna road mortality along highway networks. Journal of Wildlife Management 73:104-114.

Lazell, J.D.J. and P.J. Auger. 1981. Predation on diamondback terrapin (*Malaclemys terrapin*) eggs by dunegrass (*Ammophila breviligiulata*). Copeia 1981:723-724.

Lima, S.L., B.F. Blackwell, T.L. DeVault and E. Fernández-Juricic. 2015. Animal reactions to oncoming vehicles: a conceptual review. Biological Reviews 90:60-76.

Litvaitis, J.A. and J.P. Tash. 2008. An approach toward understanding wildlife-vehicle collisions. Environmental Management 42:688-697.

Lovich, J.E., A.D. Tucker, D.E. Kling, J.W. Gibbons and T.D. Zimmerman. 1991. Behavior of hatchling diamondback terrapins (*Malaclemys terrapin*) released in a South Carolina salt marsh. Herpetological Review 22:81-83.

Marchand, M.N. and J.A. Litvaitis. 2004. Effects of habitat features and landscape composition on the population structure of a common aquatic turtle in a region undergoing rapid development. Conservation Biology 18:758-767.

Michener, W.K., E.R. Blood, K.L. Bildstein, M.M. Brinson and L.R. Gardner. 1995. Climate change, hurricanes, tropical storms, and rising sea level in coastal wetlands. Ecological Applications 7:770-801.

Mitro, M.G. 2003. Demography and viability analyses of a diamondback terrapin population. Canadian Journal of Zoology 81:716-726.

Moran, N.A. 1994. Adaptation and constraint in the complex life cycles of animals. Annual Review of Ecology and Systematics 25:573-600.

Muldoon, K.A. and R.L. Burke. 2012. Movements, overwintering, and mortality of hatchling diamond-backed terrapins (*Malaclemys terrapin*) at Jamaica Bay, New York. Canadian Journal of Zoology 90:651-662.

Munscher, E.C., E.H. Kuhns, C.A. Cox and J.A. Butler. 2012. Decreased nest mortality for the Carolina diamondback terrapin (*Malaclemys terrapin centrata*) after removal of raccoons (*Procyon lotor*) from a nesting beach in northeastern Florida. Herpetological Conservation and Biology 7:176-184.

Ner, S.E. 2003. Distribution and predation of diamond-back terrapin nests at six upland islands of Jamaica Bay Unit and Sandy Hook Unit, Gateway National Recreation Area. MS thesis, Hofstra University, Hempstead, NY. 104 pages.

Paterson, J.E., B.D. Steinberg and J.D. Litzgus. 2013. Not just any old pile of dirt: evaluating the use of artificial nesting mounds as conservation tools for freshwater turtles. Oryx 47:607-615.

Refsnider, J.M. and F.J. Janzen. 2010. Putting eggs in one basket: ecological and evolutionary hypotheses for variation in oviposition-site choice. Annual Review of Ecology, Evolution, and Systematics 41:39-57.

Reid, G.K. 1955. Reproduction and development in the northern diamondback terrapin, *Malaclemys terrapin terrapin*. Copeia 1955:310-311.

Riitters, K.H. and J.D. Wickham. 2003. How far to the nearest road? Frontiers in Ecology and the Environment 1:125-129.

Riitters, K.H., J.D. Wickham, R.V. O'Neill, K.B. Jones, E.R. Smith, J.W. Coulston, T.G. Wade and J.H. Smith. 2003. Fragmentation of continental United States forests. Ecosystems 5:815-822.

Roosenburg, W.M. 1991. The diamondback terrapin: habitat requirements, population dynamics, and opportunities for conservation. Pages 227-234 in J.A. Mihursky and A. Chaney (eds.), New Perspectives in the Chesapeake System: A Research and Management Partnership. Chesapeake Research Consortium Pub. no. 137. Chesapeake Research Consortium, Solomons, MD.

Roosenburg, W.M. 1994. Nesting habitat requirements of the diamondback terrapin: a geographic comparison. Wetland Journal 6:8-11.

Roosenburg, W.M. 1996. Maternal condition and nest site choice: an alternative for the maintenance of environmental sex determination. American Zoologist 36:157-168.

Roosenburg, W.M., W. Cresko, M. Modesitte and M.B. Robbins. 1997. Diamondback terrapin (*Malaclemys terrapin*) mortality in crab pots. Conservation Biology 11:1166-1172.

Roosenburg, W.M. and K.C. Kelley. 1996. The effect of egg size and incubation temperature on growth in the turtle, *Malaclemys terrapin*. Journal of Herpetology 30:198-204.

Roosenburg, W.M. and A.R. Place. 1995. Nest predation and hatchling sex ratio in the diamondback terrapin: implications for management and conservation. Pages 65-70 in P. Hill and S. Nelson (eds.), Toward a Sustainable Coastal Watershed: The Chesapeake Experiment. Chesapeake Research Consortium Pub. no. 149. Chesapeake Research Consortium, Norfolk, VA.

Roosenburg, W.M., D.M. Spontak, S.P. Sullivan, E.L. Matthews, M.L. Heckman, R.J. Trimbath, R.P. Dunn,

E.A. Dustman, L. Smith and L.J. Graham. 2014. Nesting habitat creation enhances recruitment in a predator-free environment: *Malaclemys* nesting at the Paul S. Sarbanes Ecosystem Restoration Project. Restoration Ecology 22:815-823.

Ruzicka, V.A. 2006. The influence of predation on the nesting ecology of diamondback terrapins (*Malaclemys terrapin*) in the lower Chesapeake Bay. MS thesis, College of William and Mary, Williamsburg, VA. 124 pages.

Saumure, R.A., T.B. Herman and R.D. Titman. 2007. Effects of haying and agricultural practices on a declining species: the North American wood turtle, *Glyptemys insculpta*. Biological Conservation 135:565-575.

Schlaepfer, M.A., M.C. Runge and P.W. Sherman. 2002. Ecological and evolutionary traps. Trends in Ecology and Evolution 17:474-480.

Scholz, A.L. 2007. Impacts of nest site choice and nest characteristics on hatchling success in the diamond-back terrapins of Jamaica Bay, New York. MS thesis, Hofstra University, Hempstead, NY. 97 pages.

Seigel, R.A. 1980. Nesting habits of diamondback terrapins (*Malaclemys terrapin*) on the Atlantic coast of Florida. Transactions of the Kansas Academy of Sciences 83:239-246.

Seigel, R.A. and J.W. Gibbons. 1995. Workshop on the ecology, status, and management of the diamondback terrapin (*Malaclemys terrapin*), Savannah River Ecology Laboratory, 2 August 1994: final results and recommendations. Chelonian Conservation and Biology 1:240-243.

Sheridan, C.M., J.R. Spotila, W.F. Bien and H.W. Avery. 2010. Sex-biased dispersal and natal philopatry in the diamondback terrapin, *Malaclemys terrapin*. Molecular Ecology 19:5497-5510.

Shine, R. and J. B. Iverson. 1995. Patterns of survival, growth, and maturation in turtles. Oikos 72:343-348.

Spencer, R.-J. and F.J. Janzen. 2010. Demographic consequences of adaptive growth and the ramifications for conservation of long-lived organisms. Biological Conservation 143:1951-1959.

Steen, D.A., M.J. Aresco, S.G. Beilke, B.W. Compton, E.P. Congdon, C.K. Dodd Jr., H. Forrester, J.W. Gibbons, J.L. Greene, T.A. Langen, M.J. Oldham, D.N. Oxier, R.A. Saumure, F.W. Schueler, J.M. Sleeman, L.L. Smith, J.K. Tucker and J.P. Gibbs. 2006. Relative vulnerability of female turtles to road mortality. Animal Conservation 9:269-273.

Steen, D.A. and J.P. Gibbs. 2004. Effects of roads on the structure of freshwater turtle populations. Conservation Biology 18:1143-1148.

Steen, D.A., J.P. Gibbs, K.A. Buhlmann, J.L. Carr, B.W. Compton, J.D. Congdon, J.S. Doody, J.C. Godwin, K.L. Holcomb, D.R. Jackson, F.J. Janzen, G. Johnson,

M.T. Jones, J.T. Lamer, T.A. Langen, M.V. Plummer, J.W. Rowe, R.A. Saumure, J.K. Tucker and D.S. Wilson. 2012. Terrestrial habitat requirements of nesting freshwater turtles. Biological Conservation 150:121-128.

Sullivan, T.L., A.F. Williams, T.A. Messmer, L.A. Hellinga and S.Y. Kyrychenko. 2004. Effectiveness of temporary warning signs in reducing deer-vehicle collisions during mule deer migrations. Wildlife Society Bulletin 32:907-915.

Szerlag, S. and S.P. McRobert. 2006. Road occurrence and mortality of the northern diamondback terrapin. Applied Herpetology 3:27-37.

Tapper, S.C., G.R. Potts and M.H. Brockless. 1996. The effect of an experimental reduction in predation pressure on the breeding success and population density of grey partridges *Perdix perdix*. Journal of Applied Ecology 33:965-978.

Trombulak, S.C. and C.A. Frissell. 2000. Review of ecological effects of roads on terrestrial and aquatic communities. Conservation Biology 14:18-30.

van Langevelde, F. and C.F. Jaarsma. 2004. Using traffic flow theory to model traffic mortality in mammals. Landscape Ecology 19:898-907.

Wilbur, H.M. 1980. Complex life cycles. Annual Review of Ecology and Systematics 11:67-93.

Wilson, D.S. 1998. Nest-site selection: microhabitat variation and its effects on the survival of turtle embryos. Ecology 79:1884-1892.

Winters, J. 2013. The effects of bulkheading on diamondback terrapin nesting in Barnegat Bay, New Jersey. PhD dissertation, Drexel University, Philadelphia, PA. 113 pages.

Wnek, J.P., W.F. Bien and H.W. Avery. 2013. Artificial nesting habitats as a conservation strategy for turtle populations experiencing global change. Integrative Zoology 8:209-221.

Wood, R.C. 1997. The impact of commercial crab traps on northern diamondback terrapins, *Malaclemys terrapin terrapin*. Pages 21-27 in J. Van Abbema (ed.), Proceedings: Conservation, Restoration, and Management of Tortoises and Turtles—An International Conference. New York Turtle and Tortoise Society, State University of New York, Purchase, NY.

Wood, R.C. and R. Herlands. 1997. Turtles and tires: the impact of roadkills on northern diamondback terrapin, *Malaclemys terrapin terrapin*, populations on the Cape May Peninsula, southern New Jersey, USA. Pages 46-53 in J. Van Abbema (ed.), Proceedings: Conservation, Restoration, and Management of Tortoises and Turtles—An International Conference. New York Turtle and Tortoise Society, State University of New York, Purchase, NY.

Zimmerman, T.D. 1992. Latitudinal reproductive variation of the salt marsh turtle, the diamondback terrapin (*Malaclemys terrapin*). MS thesis, University of Charleston, Charleston, SC. 54 pages.

15

Interactions with Motorboats

LORI A. LESTER
HAROLD W. AVERY
ANDREW S. HARRISON
EDWARD A. STANDORA

Motorized vehicles such as automobiles and motorboats provide essential connectivity functions for humans, but they pose substantial threats to wildlife populations across temporal and spatial scales (Andrews et al. 2008). Reptiles possess specific physiological, ecological, and behavioral traits that make them vulnerable to vehicles. Roads and surface waters may attract reptiles for thermoregulatory purposes. Furthermore, many reptiles move slowly and immobilize as a predator response when vehicles approach. These behaviors could increase the chances of direct interactions with approaching vehicles. Many studies have focused on the effects of automobiles (i.e., roadkills) on diamond-backed terrapin (*Malaclemys terrapin*) populations (e.g., Wood and Herlands 1997; Szerlag and McRobert 2006; Crawford et al. 2014). In this chapter we discuss the interactions between motorboats and terrapins in coastal ecosystems.

Human alterations to coastal ecosystems are substantial. Approximately 10% of the world population lives in coastal zones below 10 m elevation (McGranahan et al. 2007). Human activities such as recreation and tourism have increased in coastal areas, accelerating development, habitat loss, and interactions with wildlife. In the United States, boating has increased greatly since the end of World War II. In 2010, more than 12.8 million recreational boats were registered in the United States (NMMA 2011). Many of these registered boats are in coastal states throughout the range of the terrapin. Overall, the state with the highest number of registered boats (914,535) is Florida. Many other coastal states also have high numbers of registered boats, including New York (475,689), New Jersey (169,750), Maryland (193,259), and South Carolina (435,491) (NMMA 2011). High levels of boating in coastal ecosystems cause environmental damage, including disturbance to seagrass beds by propellers and anchors (Dawes et al. 1997), accidental and intentional littering by boaters (Backhurst and Cole 2000), changes to shorebird behavior (Bratton 1990; Burger 1998), release of sewage from boats (Strand and Gibson 1990), erosion of shoreline (Schwimmer 2001), and fossil fuel pollution (Voudrias and Smith 1986).

Direct interactions with motorboats threaten marine, estuarine, and freshwater animals. The effects of boat strikes on some aquatic species are well documented, including cetaceans (Nowacek et al. 2001), pinnipeds (Goldstein et al. 1999), manatees (*Trichechus manatus*; Ackerman et al. 1995; Rommel et al. 2007), crocodiles (*Crocodylus niloticus*; Grant and Lewis 2010), and loggerhead sea turtles (*Caretta caretta*; Oros et al. 2005; Work et al. 2010). Aquatic animals are often killed or injured when struck by a boat, and some survive with severe lacerations and/or loss of limbs caused by propellers, which can reduce individual fitness and adult survivorship.

Motorized boating has direct and indirect effects on emydid turtles (Roosenburg 1991; Bulte et al. 2010; Heinrich et al. 2012). Direct effects on terrapins include mutilations or mortality resulting from collisions with boat propellers or hulls (Roosenburg 1991; Gibbons et al. 2001; Cecala et al. 2009); indirect effects include potential behavioral changes due to boat traffic or sounds produced by boat engines (Harrison 2010; Lester et al. 2013).

Terrapin Injuries

Causes of Injuries

Terrapins often sustain major injuries, including missing limbs and substantial shell damage, from natural and anthropogenic sources (Lovich and Gibbons 1990; Roosenburg 1991; Hart and McIvor 2008; Cecala et al. 2009). Natural injuries to terrapins include those resulting from native and non-native predators such as raccoons (*Procyon lotor*; Seigel 1980) and Norway rats (*Rattus norvegicus*; Draud et al. 2004). Anthropogenic injuries of adult terrapins include those caused by automobiles (Wood and Herlands 1997) and motorboats (Cecala et al. 2009). Several studies have assessed terrapin injury rates, but the research methods are not always consistent, particularly in determining the causes of shell injuries. Some boat injuries are distinguished easily by the repetitive pattern, similar to the damage observed in manatees, caused by a rotating propeller. Otherwise, to designate a boat-caused injury, shell damage should involve two or more adjacent vertebral, costal, or plastral scutes, or three or more adjacent marginal scutes (Fig. 15.1). Also, nesting beaches in these study areas should not be located adjacent to roads, so as to eliminate car damage as a cause of injury (Cecala et al. 2009; Lester et al. 2013). Finally, limb loss in adult terrapins is difficult to attribute to boat injury without

Fig. 15.1. Boat injuries to diamond-backed terrapins. Terrapins are typically classified as having an injury from a boat if capture location is not near a road and if damage is present to two or more vertebral or costal scutes, two or more plastral scutes, and/or three or more marginal scutes. (A) An adult female terrapin with anthropogenic injury (some natural regrowth has occurred since injury) to two vertebral and two costal scutes. (B) Injury to two plastral scutes.

accompanying shell damage; predation or developmental defects can cause similar injuries.

Terrapin populations are threatened by boat strikes in many parts of their range. In Chesapeake Bay, motorboats are the primary source of mortality for adult female terrapins (Roosenburg 1991). Terrapins in Barnegat Bay, New Jersey, and Chesapeake Bay, Maryland, exhibit injuries at higher rates than terrapins in other parts of their range. We studied the interactions between motorboats and terrapins (Harrison 2010; Lester et al. 2013) as part of an ongoing terrapin demographic study in Barnegat Bay. Total terrapin injury rate is 20% in Barnegat Bay (Lester 2012), which is similar to that (19.7%) in Chesapeake Bay (Roosenburg 1991). Areas in Texas, Florida, South Carolina, and North Carolina have lower injury rates, from 6% to 16% (Baldwin et al. 2005; Hart and McIvor 2008; Cecala et al. 2009). Frequency of major shell injuries increased from 2006 to 2011 in Barnegat Bay (Lester 2012) and from 1983 to 2006 on Kiawah Island, South Carolina (Cecala et al. 2009), and this increase corresponds with increases in number of motorboats (MTA 2008).

Limb loss rates for terrapins in Barnegat Bay varied from 1% to 5%, depending on location of capture, which is lower than that found in other parts of the terrapin's range (Cecala et al. 2009; Lester 2012). On Kiawah Island, 8% of terrapins were missing one limb and an additional 0.9% were missing more than one limb, and there was no difference between limb loss rates in female and male terrapins (Cecala et al. 2009). Because nesting exposes female terrapins to terrestrial predators, females should be injured more often than males if raccoons are the cause of limb loss. Thus, limb loss found on Kiawah Island was likely to be a result of encounters with motorboats.

Female terrapins have higher rates of major shell injuries than male terrapins at all locations (Roosenburg 1991; Cecala et al. 2009; Lester 2012). Male terrapins inhabit mostly shallow, nearshore waters, whereas adult female terrapins spend more time in open waters (Roosenburg et al. 1999; Sheridan 2010; Tulipani 2013). Female terrapins in open water are probably exposed to higher numbers of longer, faster boats with larger and more powerful engines than in nearshore waters, leading to higher injury rates. Female terrapins are also larger than males and may move more slowly.

Survivorship of Injured Terrapins

Annual survivorship of injured vs. uninjured terrapins has been assessed in Barnegat Bay and on Kiawah Island, using the Cormack-Jolly-Seber open population model in program MARK (Cecala et al. 2009; Lester 2012). Akaike's information criterion was used to assess goodness of fit of the models, and the best fit model assumed constant annual survivorship over time with differences between injured and uninjured terrapins (Lester 2012). In Barnegat Bay, mean uninjured terrapin survivorship was 0.66 for females and 0.72 for males, and mean injured terrapin survivorship was 0.51 for females and 0.11 for males (Lester 2012). On Kiawah Island, uninjured survivorship was 0.8 for females and males, and injured survivorship was 0.67 for females and 0.53 for males (Cecala et al. 2009). Survivorship estimates for uninjured terrapins in Barnegat Bay were similar to estimates for Kiawah Island (Cecala et al. 2009; Lester 2012). However, injured males and females in Barnegat Bay had lower survivorship than those on Kiawah Island.

The decreased survivorship of terrapins with boat injuries will result in decreased population size. We modeled a terrapin population in Vensim 6.2 (based on a model by J. Suss) to simulate the effect of lowered survivorship of injured terrapins due to boat injury in Barnegat Bay. We used the following assumptions, based on our mark-recapture study of this terrapin population: (1) all adult females nest every year; (2) sex ratio is 0.66 females, 0.34 males (Lester 2012); (3) each female lays 1.2 nests of 12 eggs each year (Wnek 2010); (4) hatching success is 0.5 (Wnek 2010); (5) uninjured adult annual survival is 0.8 (Lester 2012); (6) males become sexually mature in 5 years (Roosenburg 1991); (7) females become sexually mature in 10 years (Roosenburg 1991); and (8) hatchlings become juveniles in 1 year. We assigned variables with the following values (Lester 2012): (1) population size was 5,000 adult individuals; (2) 21% of adult females are injured with an annual survivorship of 0.5; and

(3) 15% of adult males are injured with an annual survivorship of 0.1. Our model predicted that, in 100 years, the terrapin population in Barnegat Bay, New Jersey, will decrease by 93% if the boat injury rate remains at current levels.

Anthropogenic injury rates are likely to underestimate the actual injury rates; given that injured terrapins must have survived injury to be captured. Actual terrapin mortality rates due to boat strikes have not been assessed in any part of the range because of difficulties in recovering dead individuals. In Barnegat Bay in 2012, three dead terrapins with boat injuries washed up on Conklin Beach, a major nesting area (Lester 2012). Sea turtle carcasses sink to the bottom of the water column after boat strikes (Crowder et al. 1995), and this probably occurs with terrapins. Additionally, terrapins with missing limbs may be easier to capture, depending on capture technique, possibly leading to higher-than-actual annual survivorship estimates (Cecala et al. 2009).

Terrapin Hearing Capability Overlaps with Motorboat Sounds

Hearing Capability

Hearing is essential to terrapins because visual cues are often limited due to the rapid attenuation of light in the turbid waters of estuaries. Aquatic turtles do not have an external ear, and it was previously incorrectly assumed that they cannot hear (Wever 1978; Lenhardt 1981). Behavioral and physiological research demonstrates that various freshwater turtle (Christensen-Dalsgaard et al. 2012) and sea turtle species (Bartol et al. 1999; Dow Piniak 2012; Martin et al. 2012) can hear sounds of <1,600 Hz. Motorboats produce noise from 0 to 20,000 kHz that may negatively affect turtles. Such sounds can cause behavioral changes, mask other natural sounds, and/or lead to hearing loss or nerve damage (Southall et al. 2000; Popper 2009). For example, desert tortoises (*Gopherus agassizii*) experience physiological responses to anthropogenic sounds, including increased heart rate and stress levels (Bowles et al. 1999). Before assessing the effects of anthropogenic sounds on terrapins, we determined the hearing sensitivity of terrapins and its overlap with motorboat sounds.

We monitored auditory evoked responses in air and underwater to evaluate terrapins' hearing capability (Lester 2012). Terrapins use terrestrial environments for nesting and basking, and aquatic environments for foraging and mating. Thus they must be able to interpret acoustic stimuli in both environments. We measured the auditory evoked potentials of adult terrapins by recording the neural responses to acoustic stimuli of various frequencies and intensities until the threshold was reached and the neural response was no longer apparent. A custom-built Tucker-Davis Technologies system presented acoustic stimuli and recorded evoked neural responses through electrodes inserted subdermally in anesthetized terrapins (Christiansen et al. 2013). Both sexes were tested in air, but we tested only female terrapins underwater because anesthesia required breathing tubes that were too large for males.

Terrapins (females, $N=5$; males, $N=2$) responded to low-frequency sounds in air (50–1,600 Hz; Fig. 15.2) (Lester 2012), as do many other freshwater turtle (Christensen-Dalsgaard et al. 2012) and marine turtle species (Bartol et al. 1999). No differences were found in aerial hearing between adult male and female terrapins (ANCOVA, $F_{1,49}=1.20$; $P=0.28$). Terrapins (females, $N=5$) had a narrower hearing range underwater (50–800 Hz) than in air (Lester 2012); this differs from hearing in red-eared sliders (*Trachemys scripta elegans*), which respond to sounds from 100 to 1,000 Hz both in air and underwater (Christensen-Dalsgaard et al. 2012). Red-eared sliders respond to 1,000 Hz sounds at a mean sound pressure level of 110 dB re 1 µPa. Terrapins did not respond when exposed to 1,000 Hz sounds underwater at a maximum mean sound pressure level of 126 dB re 1 µPa, which should have evoked a response if one existed.

Sound Frequencies of Motorboats

In general, sounds produced by small motorboats consist of a broad range of frequencies up to 20,000 kHz (Richardson et al. 1995). When motorboats travel slowly, the dominant sounds tend to be

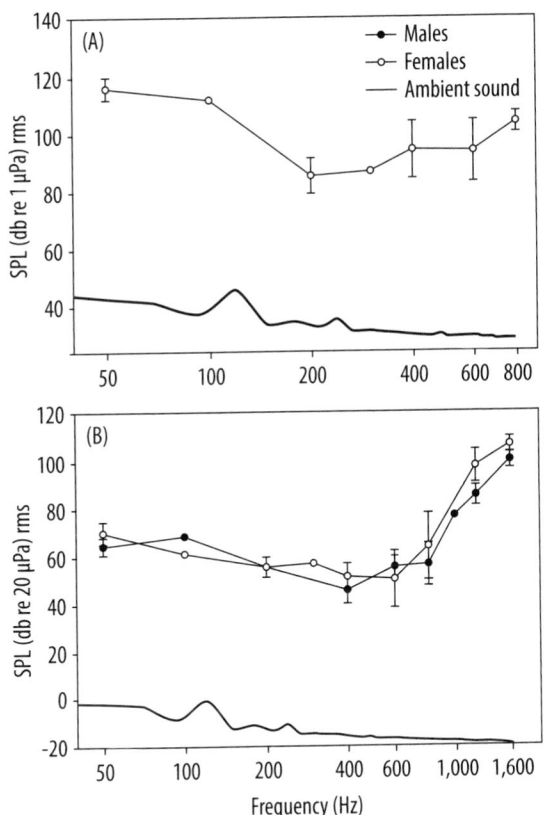

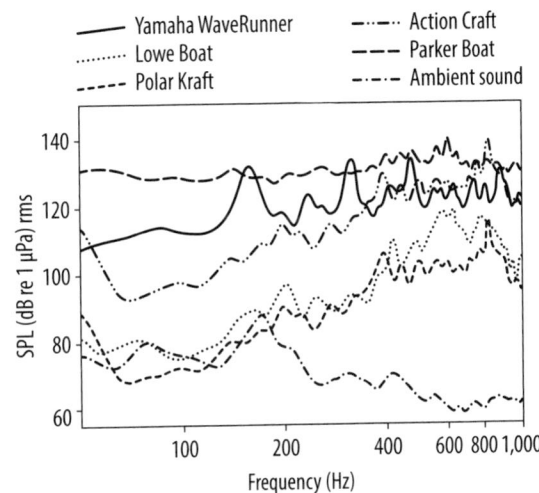

Fig. 15.3. Underwater motorboat sounds in Barnegat Bay, New Jersey. Boats and personal watercraft were found to produce low-frequency sounds in the hearing range of terrapins. The maximum sound pressure level recorded from each boat varied from 90 to 140 dB re 1 μPa, in the range of best hearing for terrapins underwater (400–600 Hz). SPL = sound pressure level.

Fig. 15.2. Amphibious hearing capability of terrapins. Auditory evoked potentials were recorded in terrapins underwater (A) and in air (B). In air, potentials were measured in female (*N* = 6) and male (*N* = 2) terrapins; in water, they were measured only in females (*N* = 5). SPL = sound pressure level; rms = root mean square. Error bars = ±1 SE.

below the frequency of 1,000 Hz, which is within the hearing range of terrapins (Richardson and Wursig 1997). Small outboard motors (defined as those used on boats <6 m in length) have the highest sound pressure level (74 dB re 1 μPa), compared with large outboard motors (68 dB re 1 μPa), medium outboard motors (68 dB re 1 μPa), small inboard motors (68 dB re 1 μPa), and personal watercraft (56 dB re 1 μPa) (Haviland-Howell et al. 2007).

We recorded boat sounds underwater in Barnegat Bay (Lester 2012). Boats and personal watercraft in Barnegat Bay produce frequencies and sound pressure levels in the terrapin's underwater hearing range (compare Figs. 15.2 and 15.3) (Lester 2012). Terrapins can hear sound frequencies from 50 to 800 Hz underwater, and boats and personal watercraft produced sound frequencies in this

range. Mean lowest threshold of terrapin underwater hearing was 86 dB re 1 μPa, and recordings of a personal watercraft (Yamaha WaveRunner, 3 m, 100 hp, 64 km/h) and four boats (Lowe Boat, 4.3 m, 9.9 hp, 23.3 km/h; Polar Kraft, 4.3 m, 25 hp, 41.9 km/h; Action Craft, 5.5 m, 110 hp, 45.4 km/h; Parker, 8.5 m, 250 hp, 53.6 km/h) had sound pressure levels higher than the terrapin's hearing threshold, indicating that boat sounds are within its hearing range.

Anthropogenic sounds may adversely affect terrapins in estuaries by masking essential natural sounds, including conspecific vocalizations and sounds from predators or potential prey items. Although it is unknown whether terrapins vocalize, some aquatic turtle species, including the long-necked turtle (*Chelodina oblonga*; Giles et al. 2009) and the giant South American river turtle (*Podocnemis expansa*; Ferrara et al. 2014), do vocalize.

Behavioral Responses to Sounds of Approaching Motorboats

Marine turtles respond to boat sounds with various behaviors that include increasing submergence time between breaths, spending more time underwater, or swimming to the surface (O'Hara and Wilcox 1990; Samuel 2004). Some of these behaviors may allow sea turtles to avoid being hit by a boat propeller; however, swimming to the surface could increase the chance of being struck by a boat.

We exposed terrapins to playback recordings of approaching boat engines in situ to determine whether and how they responded behaviorally (Lester et al. 2013). Two different size classes—small (400–600 g) and large (1,000–1,200 g)—of nongravid, previously uninjured female terrapins ($N = 80$) were exposed to playback recordings of four different motorboats (as in Fig. 15.3) in Barnegat Bay. We used two different size classes because older (i.e., larger) females may have lower hearing capability than younger (i.e., smaller) females. We also performed control trials in which terrapins swam past the underwater speaker without sound playback to ensure that behavioral responses were not simply due to the presence of the speaker. We measured several behavioral response variables in the water, including swimming speed, swimming depth (using Star-Oddi's Data Storage Tag milli-L temperature and depth data logger), and change in orientation (using Onset Computer's HOBO Pendant G). Only female terrapins were exposed to playback sounds because males were not large enough to ensure data loggers were less than 5% of body mass.

When we played back recordings of approaching boat engines of varying sizes and speeds to terrapins in situ, they did not alter their behavior (Lester et al. 2013). They did not change swimming speed or depth, or make sudden erratic movements, in response to sounds of approaching boats. Behavioral responses did not differ between the two size classes of terrapins, suggesting that hearing capability also did not differ. A lack of behavioral response to boat sounds may explain the high injury rates and mortality of terrapins in Barnegat Bay and other locations and may reduce survivorship within terrapin populations.

Behavioral Responses to Boat Traffic

Aquatic animals may use cues other than sound to alert them to oncoming motorboats, such as seeing the shadow from the vessel or sensing the displace-ment of water from the approaching boat. Green sea turtles (*Chelonia mydas*) are more likely to be struck by a boat when the boat is traveling rapidly, because there is less time for the turtle to respond behaviorally to the approaching boat (Hazel et al. 2007). Freshwater turtles such as yellow-blotched map turtles (*Graptemys flavimaculata*) abandon nesting and basking when approached by motorboats (Moore and Seigel 2006). Given the high numbers of terrapins with boat injuries and the high abundance of motorboats at our study site, we evaluated how terrapins respond to boats in situ.

We studied the behavioral response of terrapins to boat traffic in Barnegat Bay (Harrison 2010). Four terrapins in each of three size classes ($N = 12$; small, 394–443 g; medium, 833–895 g; large, 1,067–1,170 g) were outfitted with data loggers (Data Storage Tag milli-L temperature and depth logger; HOBO Pendant G logger) and sonic and radio transmitters, then released in the bay for 6 h on 2 consecutive days. During the 6 h, we considered the first 2 h to be acclimation time, the next 2 h to be treatment time on one day and control time on the other day, and the final 2 h to be recovery time. To determine whether it was a control or treatment day, we used a two-group simple randomized design and randomly determined whether the first day would be control (even number) or treatment (odd number). During the treatment period, we subjected terrapins to harassment every 15 min as data loggers recorded their swimming depth and motion. We defined harassment as piloting the boat within 3 m of the terrapin, and between harassment intervals the engine was turned off and the motorboat kept more than 20 m from the terrapin. On control days, we did not harass the terrapin during treatment time.

We calculated surface duration and submergence duration for individual events and total surface and submergence time for the entire trial period. Terrapins did not change mean surface duration, mean submergence duration, total time at surface, or total time submerged after harassment by boat traffic (Harrison 2010). Medium and large terrapins increased depth approximately 30 cm from their starting location in the water column when the boat approached. We determined that terrapins were at a depth of ≤20 cm during 50% of the treatment trials, thus medium-sized terrapins increased depth by 17% and large terrapins by 16%. For the terrapins within 20 cm of the surface during treatment, a 30 cm increase in depth would not allow them to avoid a boat (Harrison 2010). We also did not find a change in swimming behaviors such as rates of ascent and descent and duration of sustained banking (i.e., change of swimming direction with incline) when exposed to boat traffic.

Conservation and Management Implications

Terrapins do not change their behavior to avoid boats, and enhanced regulations on boat use need to be implemented and enforced. Boating is likely to continue to increase in estuaries with high terrapin densities. To prevent anthropogenic boat injuries, boating activity should be limited in high-risk areas and in small tidal creeks during certain time periods or through the use of speed limits. High-risk areas include locations where terrapins aggregate, including mating, nesting, hibernating, foraging, and basking areas. In Barnegat Bay, terrapins aggregate in the water off Conklin Island, which is the major nesting beach in the Edwin B. Forsythe National Wildlife Refuge (Winters 2013). This nesting site is located in an area near a public boat ramp, with high boat traffic. During terrapin nesting season, from mid-May to the end of July, boating should be restricted to waterways >100 m from the Conklin Island shoreline to protect terrapins in this high-risk area. Also, boat use should be restricted in the small tidal creeks behind Conklin Beach to protect terrapins traveling from the salt marsh to the nesting beach. Terrapins often traverse shallow tidal creeks in estuarine wetlands (Sheridan 2010) and are prone to boat injuries in these areas, particularly in shallow water. During low tide, many boat propellers drag through the bottom of the creeks, so boating should be limited in these small tidal creeks, especially during low to mid tide when the water is shallow (<0.5 m deep).

Restrictions on boating speed and engine size have been recommended to protect terrapin populations (Roosenburg 1991). Although speed limits have been successful for preventing boat injuries to manatees (*Trichechus manatus latirostris*; Reynolds 1999; Aipanjiguly et al. 2003) and green sea turtles (Hazel et al. 2007), terrapins do not dive deeply enough to consistently avoid the propellers of approaching boats (Harrison 2010). Further research is necessary to determine whether limiting speed and engine size would successfully protect terrapin populations. Until then, the best management strategy would be to seasonally restrict boating in areas of high terrapin density.

Because terrapins do not avoid boats, humans must change their behavior to ensure the sustainability of terrapin populations. Implementing boating regulations in habitats with high terrapin densities and educating boaters about the effects of boats on terrapins and other wildlife will benefit estuarine aquatic wildlife conservation. Educational materials informing the public about the ecology of terrapins and their vulnerability to the effects of motorboats are needed. Many boaters in Barnegat Bay, for example, are unaware that terrapins inhabit the waters. All East and Gulf coast states in the United States require boaters to take a boating safety course before they can legally pilot a boat. Supplementary educational materials describing the presence of terrapins and the effects of boats should be distributed during these courses.

Recommendations for Future Studies

More local evaluations of boat-caused injuries and mortality in terrapins are necessary to allow development of adequate injury prevention and management plans. Terrapin injury rates have been reported for Maryland (Roosenburg 1991), South Carolina (Cecala et al. 2009), and Florida (Hart and McIvor 2008); injury analyses are still needed for other populations along the East and Gulf coasts. Estimates of boat-related mortality are also necessary because terrapin injury analyses are likely to underestimate the effect of boats on terrapin populations. Boat-related mortality has been assessed for manatee populations through a well-publicized carcass recovery and necropsy program (O'Shea et al. 1985). Further studies should evaluate and model how changes in survivorship due to boat injuries might affect terrapin populations. In addition to documenting the effects of boats on terrapin populations, high-risk areas for terrapins must be identified and protected throughout the species' range.

Terrapin hearing audiograms have been produced with electrophysiological methods, but when behavioral methods are used to produce audiograms, the resultant thresholds are sometimes 10 to 30 dB lower than those estimated by electrophysiological methods (Gorga et al. 1988; Brittan-Powell et al. 2002). If terrapins' hearing is 10 to 30 dB lower than predicted by auditory evoked potentials, then they can hear even more of the sounds produced by boats than expected. Further behavioral studies should be conducted to determine the accuracy of the terrapin audiograms created with electrophysiological techniques.

We found that terrapins did not respond behaviorally to playback recordings of boat engines, and they only slightly increased swimming depth in response to physical approaches of boats. However, we did not measure other behavioral and physiological responses. Some fish species exhibit elevated stress levels in response to boat sounds, and this increase in stress causes the fish to be more susceptible to predation (Popper 2009). Future research should measure physiological responses such as stress hormone level and heart rate of terrapins before, during, and after anthropogenic sound exposure. Furthermore, given that terrapins exhibit many other natural behaviors in addition to swimming behavior, future studies should assess whether boat sounds change, for example, mating, basking, and foraging behaviors. Although terrapins do not avoid boats by changing swimming behavior, they may change other behaviors that might have important effects on populations.

ACKNOWLEDGMENTS

Funding and field assistance were provided by the Earthwatch Institute. The Betz Chair of Environmental Science at Drexel University and the Diamondback Terrapin Working Group also provided funding. Amphibious hearing audiograms for terrapins were measured with assistance from W. Dow Piniak, C. Harms, H. Broadhurst, E. Clarke III, and E. Christiansen. The population model was developed by J. Suss.

REFERENCES

Ackerman, B.B., S.D. Wright, R.K. Bonde, D.K. Odell and D.J. Banowetz. 1995. Trends and patterns in mortality of manatees in Florida, 1974-1992. Pages 223-258 in T.J. O'Shea, B.B. Ackerman and H.F. Percival (eds.), Population Biology of the Florida Manatee. National Biological Service Information and Technology Report. National Biological Service, Fort Collins, CO.

Aipanjiguly, S., S.K. Jacobson and R. Flamm. 2003. Conserving manatees: knowledge, attitudes, and intentions of boaters in Tampa Bay, Florida. Conservation Biology 17:1098-1105.

Andrews, K.M., J.W. Gibbons and D.M. Jochimsen. 2008. Ecological effects of roads on amphibians and reptiles: a literature review. Pages 121-143 in J.C. Mitchell, R.E. Jung Brown and B. Bartholomew (eds.), Urban Herpetology. Society for the Study of Amphibians and Reptiles, Salt Lake City, UT.

Backhurst, M.K. and R.G. Cole. 2000. Subtidal benthic marine litter at Kawau Island, north-eastern New Zealand. Journal of Environmental Management 60:227-237.

Baldwin, J.D., L.A. Latino, B.K. Mealey, G.M. Parks and M.R.J. Forstner. 2005. The diamondback terrapin in Florida Bay and the Florida Keys: insights into turtle conservation and ecology. Pages 180-186 in W.E. Meshaka Jr. and K.J. Babbitt (eds.), Amphibians and Reptiles: Status and Conservation in Florida. Krieger, Malabar, FL.

Bartol, S.M., J.A. Musick and M.L. Lenhardt. 1999. Auditory evoked potentials of the loggerhead sea turtle (*Caretta caretta*). Copeia 1999:836-840.

Bowles, A.E., S.A. Eckert, L. Starke, E. Berg, L. Wolski and J. Matesic Jr. 1999. Effects of flight noise from jet aircraft and sonic booms on hearing, behavior, heart rate, and oxygen consumption of desert tortoises (*Gopherus agassizii*). US Air Force Research Laboratory, Final Report AFRL-HE-WP-TR-1999-0170. San Diego, CA. 131 pages.

Bratton, S.P. 1990. Boat disturbance of Ciconiiformes in Georgia estuaries. Colonial Waterbirds 13:124-128.

Brittan-Powell, E.F., R.J. Dooling and O. Gleich. 2002. Auditory brainstem responses in adult budgerigars (*Melopsittacus undulatus*). Journal of the Acoustical Society of America 112:999-1008.

Bulte, G., M.A. Carriere and G. Blouin-Demers. 2010. Impact of recreational power boating on two populations of northern map turtles (*Graptemys geographica*). Aquatic Conservation: Marine and Freshwater Ecosystems 20:31-38.

Burger, J. 1998. Effects of motorboats and personal watercraft on flight behavior over a colony of common terns. Condor 100:528-534.

Cecala, K.K., J.W. Gibbons and M.E. Dorcas. 2009. Ecological effects of major injuries in diamondback terrapins: implications for conservation and management. Aquatic Conservation: Marine and Freshwater Ecosystems 19:421-427.

Christensen-Dalsgaard, J., C. Brandt, K.L. Willis, C.B. Christensen, D. Ketten, P. Edds-Walton, R.R. Fay, P.T. Madsen and C.E. Carr. 2012. Specialization for underwater hearing by the tympanic middle ear of the turtle, *Trachemys scripta elegans*. Proceedings of the Royal Society B: Biological Sciences 279:2816-2824.

Christiansen, E.F., W.E.D. Piniak, L.A. Lester and C.A. Harms. 2013. Underwater anesthesia of diamondback terrapins (*Malaclemys terrapin*) for measurement of auditory evoked potentials. Journal of the American Association for Laboratory Animal Science 52:792-797.

Crawford, B.A., J.C. Maerz, N.P. Nibbelink, K.A. Buhlmann, T.M. Norton and S.E. Albeke. 2014. Hot spots and hot moments of diamondback terrapin road-crossing activity. Journal of Applied Ecology 51:367-375.

Crowder, L.B., S.R. Hopkins-Murphy and J.A. Royle. 1995. Effects of turtle excluder devices (TEDs) on loggerhead sea turtle strandings with implications for conservation. Copeia 1995:773-779.

Dawes, C.J., J. Andorfer, C. Rose, C. Uranowski and N. Ehringer. 1997. Regrowth of the seagrass *Thalassia testudinum* into propeller scars. Aquatic Botany 59:139-155.

Dow Piniak, W.E. 2012. Acoustic ecology of sea turtles: implications for conservation. PhD dissertation, Duke University, Durham, NC. 114 pages.

Draud, M., M. Bossert and S. Zimnavoda. 2004. Predation on hatchling and juvenile diamondback terrapins (*Malaclemys terrapin*) by the Norway rat (*Rattus norvegicus*). Journal of Herpetology 38:467-470.

Ferrara, C.R., R.C. Vogt, R.S. Sousa-Lima, B.M.R. Tardio and V.C.D. Bernardes. 2014. Sound communication and social behavior in an Amazonian river turtle (*Podocnemis expansa*). Herpetologica 70:149-156.

Gibbons, J.W., J.E. Lovich, A.D. Tucker, N.N. FitzSimmons and J.L. Greene. 2001. Demographic and ecological factors affecting conservation and management of the diamondback terrapin (*Malaclemys terrapin*) in South Carolina. Chelonian Conservation and Biology 4:66-74.

Giles, J.C., J.A. Davis, R.D. McCauley and G. Kuchling. 2009. Voice of the turtle: the underwater acoustic repertoire of the long-necked freshwater turtle, *Chelodina oblonga*. Journal of the Acoustical Society of America 126:434-443.

Goldstein, T., S. Johnson, A. Phillips, K. Hanni, D. Fauquier and F. Gulland. 1999. Human-related injuries observed in live stranded pinnipeds along the central California coast 1986-1998. Aquatic Mammals 25:43-51.

Gorga, M.P., J.R. Kaminski, K.A. Beauchaine and W. Jesteadt. 1988. Auditory brainstem responses to tone bursts in normally hearing subjects. Journal of Speech and Hearing Research 31:87-97.

Grant, P.B. and T.R. Lewis. 2010. High speed boat traffic: a risk to crocodilian populations. Herpetological Conservation and Biology 5:456-460.

Harrison, A.S. 2010. Determining behavioral responses of the northern diamondback terrapin (*Malaclemys terrapin terrapin* Schoepff, 1793) to boat traffic: a thesis in biology. MA thesis, State University of New York College, Buffalo, NY. 130 pages.

Hart, K.M. and C.C. McIvor. 2008. Demography and ecology of mangrove diamondback terrapins in a wilderness area of Everglades National Park, Florida, USA. Copeia 2008:200-208.

Haviland-Howell, G., A.S. Frankel, C.M. Powell, A. Bocconcelli, R.L. Herman and L.S. Sayigh. 2007. Recreational boating traffic: a chronic source of anthropogenic noise in the Wilmington, North Carolina, Intracoastal Waterway. Journal of the Acoustical Society of America 122:151-160.

Hazel, J., I.R. Lawler, H. Marsh and S. Robson. 2007. Vessel speed increases collision risk for the green turtle, *Chelonia mydas*. Endangered Species Research 3:105-113.

Heinrich, G.L., T.J. Walsh, D.R. Jackson and B.K. Atkinson. 2012. Boat strikes: a threat to the Suwannee cooter (*Pseudemys concinna suwanniensis*). Herpetological Conservation and Biology 7:349-357.

Lenhardt, M.L. 1981. Evidence for auditory localization ability in the turtle. Journal of Auditory Research 21:255-261.

Lester, L.A. 2012. Direct and indirect effects of recreational boats on diamondback terrapins (*Malaclemys terrapin*). PhD dissertation, Drexel University, Philadelphia, PA. 98 pages.

Lester, L.A., H.W. Avery, A.S. Harrison and E.A. Standora. 2013. Recreational boats and turtles: behavioral mismatches result in high rates of injury. PLoS One 8(12):e82370.

Lovich, J.E. and J.W. Gibbons. 1990. Age at maturity influences adult sex ratio in the turtle, *Malaclemys terrapin*. Oikos 59:126-134.

Martin, K.J., S.C. Alessi, J.C. Gaspard, A.D. Tucker, G.B. Bauer and D.A. Mann. 2012. Underwater hearing in the loggerhead turtle (*Caretta caretta*): a comparison of behavioral and auditory evoked potential audiograms. Journal of Experimental Biology 215:3001-3009.

McGranahan, G., D. Balk and B. Anderson. 2007. The rising tide: assessing the risks of climate change and human settlements in low elevation coastal zones. Environment and Urbanization 19:17-37.

Moore, M.J.C. and R.A. Seigel. 2006. No place to nest or bask: effects of human disturbance on the nesting and basking habits of yellow-blotched map turtles (*Graptemys flavimaculata*). Biological Conservation 130:386-393.

MTA (Marine Trades Association of New Jersey). 2008. Recreational Boating in New Jersey: An Economic Impact Analysis. MTA, Manasquan, NJ. 41 pages.

NMMA (National Marine Manufacturers Association). 2011. 2010 Recreational Boating Statistical Abstract. NMMA, Chicago, IL. 253 pages.

Nowacek, S.M., R.S. Wells and A.R. Solow. 2001. Short-term effects of boat traffic on bottlenose dolphins, *Tursiops truncatus*, in Sarasota Bay, Florida. Marine Mammal Science 17:673-688.

O'Hara, J. and J.R. Wilcox. 1990. Avoidance responses of loggerhead turtles, *Caretta caretta*, to low frequency sound. Copeia 1990:564-567.

Oros, J., A. Torrent, P. Calabuig and S. Deniz. 2005. Diseases and causes of mortality among sea turtles stranded in the Canary Islands, Spain (1998-2001). Diseases of Aquatic Organisms 63:13-24.

O'Shea, T.J., C.A. Beck, R.K. Bonde, H.I. Kochman and D.K. Odell. 1985. An analysis of manatee mortality patterns in Florida, 1976-81. Journal of Wildlife Management 49:1-11.

Popper, A.N. 2009. The effects of anthropogenic sources of sound on fishes. Journal of Fish Biology 75:455-489.

Reynolds, J.E. 1999. Efforts to conserve the manatees. Pages 267-295 in J.R. Twiss and R.R. Reeves (eds.), Conservation and Management of Marine Mammals. Smithsonian Institution Press, Washington, DC.

Richardson, W.J., C.R. Greene Jr., C.I. Malme and D.H. Thomson. 1995. Marine Mammals and Noise. Academic Press, New York, NY. 576 pages.

Richardson, W.J. and B. Wursig. 1997. Influences of man-made noise and other human actions on cetacean behaviour. Marine and Freshwater Behaviour and Physiology 29:183-209.

Rommel, S.A., A.M. Costidis, T.D. Pitchford, J.D. Lightsey, R.H. Snyder and E.M. Haubold. 2007. Forensic methods for characterizing watercraft from watercraft-induced wounds on the Florida manatee (*Trichechus manatus latirostris*). Marine Mammal Science 23:110-132.

Roosenburg, W.M. 1991. The diamondback terrapin: habitat requirements, population dynamics, and opportunities for conservation. Pages 227-234 in J.A. Mihursky and A. Chaney (eds.), New Perspectives in the Chesapeake System: A Research and Management Partnership. Chesapeake Research Consortium, Solomons, MD.

Samuel, Y. 2004. Underwater, low-frequency noise and anthropogenic disturbance in a critical sea turtle habitat. MS thesis, Cornell University, Ithaca, NY. 58 pages.

Schwimmer, R.A. 2001. Rates and processes of marsh shoreline erosion in Rehoboth Bay, Delaware, USA. Journal of Coastal Research 17:672-683.

Seigel, R.A. 1980. Predation by raccoons on diamondback terrapins, *Malaclemys terrapin tequesta*. Journal of Herpetology 14:87-89.

Sheridan, C.M. 2010. Mating system and dispersal patterns in the diamondback terrapin (*Malaclemys terrapin*). PhD dissertation, Drexel University, Philadelphia, PA. 204 pages.

Southall, B.L., R.J. Schusterman and D. Kastak. 2000. Masking in three pinnipeds: underwater, low-frequency critical ratios. Journal of the Acoustical Society of America 108:1322-1326.

Strand, I.E. and G.R. Gibson. 1990. The use of pump-out facilities by recreational boaters in Maryland. Estuaries 13:282-286.

Szerlag, S. and S.P. McRobert. 2006. Road occurrence and mortality of the northern diamondback terrapin. Applied Herpetology 3:27-37.

Tulipani, D.C. 2013. Foraging ecology and habitat use of the northern diamondback terrapin (*Malaclemys terrapin terrapin*) in southern Chesapeake Bay. PhD dissertation, College of William and Mary, Williamsburg, VA. 224 pages.

Voudrias, E.A. and C.L. Smith. 1986. Hydrocarbon pollution from marinas in estuarine sediments. Estuarine, Coastal, and Shelf Science 22:271-284.

Wever, E.G. 1978. The Reptile Ear: Its Structure and Function. Princeton University Press, Princeton, NJ. 1,038 pp.

Winters, J. 2013. The effects of bulkheading on diamondback terrapin nesting in Barnegat Bay, New Jersey. PhD dissertation, Drexel University, Philadelphia, PA. 100 pages.

Wnek, J.P. 2010. Anthropogenic impacts on the reproductive ecology of the diamondback terrapin, *Malaclemys terrapin*. PhD dissertation, Drexel University, Philadelphia, PA. 151 pages.

Wood, R.C. and R. Herlands. 1997. Turtles and tires: the impact of roadkills on northern diamondback terrapins, *Malaclemys terrapin terrapin*, populations on the Cape May Peninsula, southern New Jersey. Pages 46-53 in J. Van Abbema (ed.), Proceedings: Conservation, Restoration, and Management of Tortoises and Turtles—An International Conference. New York Turtle and Tortoise Society, State University of New York, Purchase, NY.

Work, P.A., A.L. Sapp, D.W. Scott and M.G. Dodd. 2010. Influence of small vessel operation and propulsion system on loggerhead sea turtle injuries. Journal of Experimental Marine Biology and Ecology 393:168-175.

16

Bycatch in Blue Crab Fisheries

RANDOLPH M. CHAMBERS

JOHN C. MAERZ

Since the development and use of the commercial crab pot in blue crab (*Callinectes sapidus*) fisheries in the 1940s, captured diamondbacked terrapins (*Malaclemys terrapin*) have drowned as bycatch. Numerous studies have documented the demographic effects of chronic removal of sub-adult and adult terrapins from populations by crab pot mortality. The distribution and intensity of actively fished pots used by both commercial and recreational crabbers vary seasonally and spatially throughout the entire range where crabs and terrapins co-occur. Derelict pots are an additional source of mortality for terrapin populations that can lead to locally important effects, but these effects have not been well quantified at larger regional scales. In the past 10 years, experimental studies on the effectiveness of bycatch reduction devices (BRDs) have corroborated prior work showing up to 100% exclusion of terrapins from commercial pots fitted with BRDs. In some studies, however, crab numbers have been reduced in pots with BRDs, and as a result, BRD regulations have not been enacted in many states. Further, where BRDs are compulsory, the rate of compliance has not been high. Because terrapin habitat and the crab fishery are spread over so many different tidal environments and across so many political jurisdictions, a single, comprehensive regulatory policy to protect terrapins from drowning in crab pots remains elusive. The success of terrapin bycatch reduction and management can be achieved only through collaboration among crabbers, scientists, and resource managers to make commercial and recreational crabbing compatible with terrapin conservation.

Some 14 years after Roosenburg (2004) detailed the effect of crab pot fisheries on terrapin populations and what was needed in terms of future science, policy, and management efforts, we summarize what has happened with respect to the conservation of terrapins and blue crab fisheries. We focus on the effects of commercial hardshell crabbing, recreational crabbing, and peeler crabbing for the softshell industry, all of which use variants of commercial crab pots (for a full description of crabbing gear and methods, see Kennedy et al. 2007). These wire mesh pots are baited, typically with fresh or frozen fish, and crabs move into the pots via narrow funnels and become trapped. The pots are checked and emptied daily during crabbing season, which varies by state. Commercial crabbers may tend hundreds of pots, whereas recreational crabbers in most states are restricted to two pots per person. The peeler crab fishery uses a pot design similar to that in the hardshell crab fishery, with two notable exceptions. First, peeler pots typically are not baited with fish; instead, a live male crab may be placed inside the pot to attract pre-molting females. Some peeler pots contain structure-like artificial seagrass that may provide shelter before shedding, and other pots contain nothing but still attract crabs. Second, peeler pots are constructed of finer mesh wire than regular pots and usually are fished in the relatively shallow waters of tidal creeks where pre-molt crabs tend to congregate and where terrapins tend to live. "Peeler runs"—when crabs are molting in large numbers—occur over limited time periods, so the extent of the peeler crabbing fishery is smaller in space and shorter in time relative to commercial hardshell and recreational crabbing.

Additional fisheries and fishery methods (bank trapping, hardscraping, eelpotting, shrimp trawling, fyke netting, and others) result in some level of terrapin bycatch, but we focus here on terrapin bycatch in pots used for blue crab fisheries and the recent advances in use of BRDs and other efforts to decrease terrapin mortality. As with the Roosenburg (2004) study, we include references to peer-reviewed literature and some non-refereed publications and meeting presentations, as the state of the science, policy, and management of bycatch reduction remains in flux.

Commercial and recreational fisheries' bycatch of turtles is not unique to terrapins. The effect of fisheries' bycatch on marine turtle populations, for example, has been documented for decades and tends to vary globally with respect to the spatial overlap of different fisheries and turtle species (Wallace et al. 2013). In contrast, turtle bycatch in freshwater fisheries remains relatively understudied (Raby et al. 2011) but appears to be a significant threat to population viability of some turtle species (Larocque et al. 2012; Midwood et al. 2014). A number of turtle bycatch reduction strategies have been developed and adapted for different marine and freshwater fisheries. Bycatch in the blue crab fishery, however, is a good model system that demonstrates both the bycatch problem and subsequent efforts to reduce terrapin capture and mortality.

Terrapins occupy estuarine habitats including tidal marshes and mangroves, barrier island lagoons, and hypersaline embayments. This is a testament to the terrapin's unique adaptations among North American turtles to estuarine conditions that vary throughout the species' latitudinal range. However, active commercial and recreational blue crab fisheries occur throughout most of this range. Loss of terrapins as bycatch to crab fisheries is well documented and is recognized as one of the major threats to terrapin populations (Butler et al. 2006; Grosse et al. 2011).

The intensity and effect of this bycatch mortality can be viewed at different spatial scales: at the level of discrete populations within individual tidal creeks, at the level of meta-populations connected by regular gene flow, at the regional level for which populations may or may not be connected by similar regulatory efforts, and at the level of the entire species range from Cape Cod, Massachusetts, to Corpus Christi, Texas. These different scales are important to consider because, for example, the death of 50 terrapins in a single crab pot might represent only a small fraction of the total number of terrapins in an estuary, yet locally, those 50 dead terrapins could comprise the majority of adult terrapins in a section of tidal wetland (e.g., Bishop 1983; Roosenburg et al. 1997; Grosse et al. 2009). For a species like the terrapin that exhibits strong nest and home site fidelity, such a loss could translate into reduced population growth and increased local extinction. We attempt to present the effect of blue crab fisheries with due consideration of these different spatial scales.

Regional Variation in Blue Crab Fisheries

Blue crab fisheries using commercial crab pots extend from Massachusetts to Texas and are most intense in the mid-Atlantic states from New Jersey to North Carolina. In 2011, blue crab landings from 14 coastal states comprised more than 90,000 metric tons, with 81% of that total harvested from four states (Maryland, Virginia, North Carolina, and Louisiana) (NOAA 2013). Most blue crab landings by weight are hardshell and peeler crabs collected by commercial pots. Fewer crabs are collected by other methods such as winter scraping or trot lining. Normalizing blue crab landings per kilometer of tidal shoreline (US Census Bureau 2012) gives a proxy for commercial crabbing intensity. This proxy is based on harvests by state, not by number of pots fished—i.e., some states may have different harvest amounts despite similar efforts. The states with the highest intensity of crabbing are Maryland and Virginia, with more than 4,400 kg and 3,100 kg, respectively, of crabs harvested per kilometer of shoreline in 2011. Crabbing intensity is high because of the extensive, relatively shallow habitat for crabs throughout the open water of Chesapeake Bay. Crabbing intensity translates into high densities of commercial crab pots throughout coastal bays and—in some regions—into tributaries and tidal marsh creeks and rivers.

Estimates of crab harvests by recreational crabbers are difficult to obtain, because reporting of recreational harvests is voluntary in most states. The harvest must be much smaller than commercial landings but is variable by region. The North Carolina Division of Marine Fisheries (2013) estimated crab landings by recreational fishers to be at least 1% of commercial landings in the state, with no breakdown of how many recreational captures were by commercial-style pots. A study along the New Jersey portion of Delaware Bay suggested that commercial potting comprised only 8% of the recreational crab harvest but that recreational landings could be as much as 20% of commercial landings (Muffley et al. 2007). The blue crab stock

assessment in Virginia estimated recreational crabbing at 8% of commercial landings (R. Lipcius, Virginia Institute of Marine Science, pers. comm.). As a first-order approximation, recreational landings of crabs in commercial pots appear to make up a small percentage of the total crab harvest, probably from as low as 1% to as high as 10% (Fogarty and Lipcius 2007).

Of course, recreational crabbers using commercial pots need access to water, by either a dock or a boat launch. To estimate the recreational crabbing potential by state, we normalized the total population of people living in coastal counties (US Census Bureau 2010) by the length of tidal shoreline in those counties (where docks and boat launches are found). We recognize that recreational crabbers make up a small percentage of all people living in coastal counties. Further, the use of commercial-style pots varies geographically, so it is difficult to make state-by-state comparisons. For example, very few crabs in highly populated New York and northern New Jersey are harvested by recreational crabbers using commercial-style pots, and their use to harvest crabs is illegal in Connecticut. Recreational crabbing potential does not necessarily translate into realized recreational crab pot use. Further, the impact of high human populations in coastal counties probably leads to other negative effects on terrapins in addition to recreational crabbing. These effects might include a higher density of coastal roads where nesting female terrapins are killed, greater housing development atop prior terrapin nesting areas, attraction of "subsidized predators" (species that thrive in close association with humans; e.g., raccoons) to terrapin nests, damage to terrapins from boating activities, and the hardening of open shorelines that precludes female terrapins' access to nesting beaches (Chapters 14, 15, and 18).

With these caveats, we plotted commercial crabbing pressure vs. recreational crabbing potential by state (Fig. 16.1; New York, Connecticut, and Massachusetts harvests are <0.25% of the total US commercial harvest and thus were not included). Some states have relatively low commercial crabbing pressure and recreational crabbing potential (Georgia, South Carolina, Mississippi, Alabama).

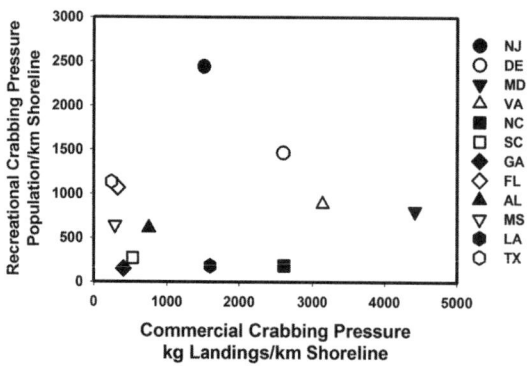

Fig. 16.1. Comparison of commercial crabbing intensity and potential recreational crabbing intensity by state. See text for explanation of axes.

Louisiana and North Carolina have higher commercial crabbing pressure but relatively low recreational crabbing potential. Texas and Florida both have low commercial crabbing pressure but relatively high recreational crabbing potential. Finally, the four mid-Atlantic states (New Jersey, Delaware, Virginia, and Maryland) exhibit relatively high commercial crabbing pressure and high recreational crabbing potential. In fact, the mid-Atlantic states tend to group separately from the southeastern and Gulf Coast states. A high density of both crabs and people along the coast in the mid-Atlantic states is the reason for this separation. Once again, the consideration of spatial scale is important because, for example, recreational crabbing potential may be small in any state, yet at local scales can be intense. In Maryland, for example, legal commercial crabbing with pots is restricted to the mainstem of Chesapeake Bay. More inland waters where terrapins occur are dominated by recreational crabbers, and their bycatch effect on terrapin populations is disproportionate to their effort (Roosenburg et al. 1997; Roosenburg and Green 2000). The movement of terrapins along the margins of tidal creeks populated on either side by scores of docks and piers, each with recreational crab pots in the water, seems akin to running a gauntlet of prospective mortality.

Terrapins spend most of their lives in nearshore habitats (Roosenburg et al. 1999) and are more frequently observed as bycatch in those environments (Hart and Crowder 2011). As stated above, the number of recreational crabbers is unknown but is distributed broadly along developed coastlines. In contrast, commercial crabbing is distributed broadly across open water, although in some southeastern states, commercial crabbing also extends into tidal creeks associated with large river systems that dissect the coastal habitat (M. Dorcas, pers. comm.; Maerz, pers. obs.). The penetration of commercial pots from blue crab fisheries into smaller tidal creek systems is less intense in some mid-Atlantic states. As noted, Maryland prohibits the use of pots for commercial crabbing in most tributaries to Chesapeake Bay, and New Jersey requires the use of BRDs on any pots (commercial or recreational) placed in tidal creeks less than 150 feet wide. No other states impose BRD regulations on commercial crabbing in small tidal creeks where the commercial harvest of peeler crabs is completed seasonally when crabs are molting. Overall, for commercial crabbing, a smaller percentage of a very large total number of pots intersects with terrapin habitat. For recreational crabbing, a larger percentage of a smaller total number of pots intersects with terrapin habitat. Both commercial and recreational pots may end up as derelict—or ghost—pots that intersect with terrapin habitat. Active pots fished in deeper, open water may be lost and carried into terrapin habitat by tides or storms, thereby increasing bycatch risk in shallow water.

Processes Determining Terrapin Bycatch

Commercial crab pots attract and capture terrapins whether the pots are baited (Davis 1942) or unbaited (Bishop 1983). Terrapin bycatch occurs in commercial, recreational, and derelict pots (Roosenburg 1991; Wood and Herlands 1995; Roosenburg et al. 1997; Hoyle and Gibbons 2000; Gibbons et al. 2001; Baldwin et al. 2005; Schaffer et al. 2008; Grosse et al. 2009). Morris et al. (2011) found that crab catch was highest during the first 2 days after fresh baiting but that the daily rate of terrapin bycatch did not change for up to 7 days after baiting; i.e., pots continued to trap terrapins even when the bait was gone. When commercial soak times extend beyond 24 h (as is legal in many states), bycatch mortality increases dramatically (Hart 2005; Grosse et al. 2011). Also, recreational crabbers may not check their pots daily, and derelict pots are not checked at all. Derelict pots are of particular concern because so many are generated from the commercial crab fishery each year. In a study in a Virginia portion of Chesapeake Bay, for instance, some 10,000 derelict pots were removed in each of three consecutive years, with no decrease in pot harvest each year (Bilkovic et al. 2014). Derelict pots are abundant, and every year more commercial pots are added as marine debris to the bottom of shallow estuaries, sometimes directly in terrapin habitat.

The upper size range of individual terrapins captured in pots is restricted by the gape in the pots' entrance funnels. Male terrapins never grow to a size larger than the gape in a crab pot and thus are susceptible to drowning when trapped throughout their entire juvenile and adult lives. Some adult female terrapins eventually may grow shells larger than the gape and thus avoid capture. In other populations, however, adult females may remain small and vulnerable to bycatch, and all female terrapins are vulnerable to bycatch while immature. As a result, size-selective bycatch results in differential demographic effects on male and female terrapins, with male-biased bycatch mortality throughout much of the species' range (Roosenburg et al. 1997). For example, Grosse et al. (2009) reported that 83% of terrapins killed in derelict commercial pots were male, compared with a background sex ratio of 66% male in tidal creeks in the region. Size- and sex-selective mortality tend to be greater at northern latitudes, where female terrapins grow larger than females in more southern populations (Roosenburg 2004). Coleman et al. (2014) found that 85% of adult female terrapins sampled in an Alabama population were susceptible to crab pot mortality. Size-selective mortality imposed by crab pots may be a driver of an evolutionary increase in terrapin growth rate and body size, as has been suggested for populations in Chesapeake Bay (Wolak et al. 2010) and Georgia (Grosse and Maerz 2013).

The estimated effects of terrapin bycatch contrast starkly at state vs. local scales. At the state scale, Bishop (1983) estimated that only 10% of the daily bycatch of some 2,800 terrapins would drown in commercial pots during April and May in South Carolina, and Hart and Lee (2006) suggested that approximately 300 terrapins drowned daily in commercial pots during warm weather in North Carolina. These estimates suggest relatively minor bycatch; however, many direct observations of bycatch serve to demonstrate that effects can be locally extensive. Roosenburg et al. (1997), for example, estimated that 15% to 78% of a Patuxent River, Maryland, population could be lost as bycatch in recreational crab pots during a single season. During five visits over two months, Grosse et al. (2009) documented 133 terrapins drowned in just two derelict commercial pots in a Georgia tidal creek, which they estimated to be 65% of the total terrapin population within that creek (Fig. 16.2). Using 10 unbaited crab pots, Upperman et al. (2014) captured roughly 42% of the terrapins occupying a tidal marsh in Virginia in just 24 days. A mark-recapture study by Tucker et al. (2001) showed that terrapins in a South Carolina creek where bycatch mortality was high rarely survived to reproduce. More recently, the distribution and sizes of terrapin populations in coastal Louisiana appear to be adversely affected by the presence of crab pots (Selman et al. 2014).

The effects of high rates of terrapin bycatch within individual tidal creeks or watersheds can be important. Models developed by Hart (1999) for a Massachusetts terrapin population and Ayers (2010) for a Virginia population indicate that annual removal of just 12% of adult and juvenile terrapins would lead to eventual local population extinction. Consistent with the predictions of those models, Dorcas et al. (2007) documented as much as an 80% decline in terrapin abundance over 20 years among four neighboring tidal creeks in South Carolina, with concurrent increases in female age and size distributions that were consistent with size/sex-selected bycatch mortality. Grosse et al. (2011), in a statewide assessment of crabbing effects on terrapins in Georgia, found an 87% lower mean terrapin abundance among tidal creeks with three or more commercial crab pots present compared with creeks with no recorded or current crabbing activity.

Fig. 16.2. Terrapin entrapment in derelict crab pots. (A) Mortality of more than 100 terrapins in a single crab pot. (B) Live male and female terrapins captured in a crab pot fitted with a chimney to prevent drowning. (A) Reproduced from Grosse et al. (2009); copyright Chelonian Conservation and Biology, Allen Press Publishing Services. (B) Image from Scott Belfit.

Similarly, Isdell (2013) modeled the spatial occurrence of terrapins across the lower Chesapeake Bay and found that predicted terrapin presence was negatively associated with the density of commercial crab pots. Collectively, models using estimated per capita mortality are confirmed by more rigorous regional studies to show conclusively the ongoing, negative effects of commercial crabbing on terrapin populations in different parts of the species range.

At the finest scales, specific risk factors have been linked to the threats a crab pot poses to terrapins (summarized in Grosse et al. 2011). First, crab pots placed in shallow water during the terrapin nesting season present the greatest threat. Davis et al. (1942) demonstrated that pots placed in shallow waters between April and June caught large

numbers of terrapins; Bishop (1983) found that 93% of terrapin bycatch occurred in shallow creeks during the same period; and Hart and Crowder (2011) found that pots placed in shallow waters close to shore in April and May captured more terrapins than pots in deeper water. April through June coincides with the nesting season, when terrapin activity increases in shallow waters near nesting habitats. Harden and Williard (2012), using data from radio-tracked terrapins to model the spatial and temporal overlap of crabbing and terrapins, found that the activities of terrapins and commercial crabbers were separate during cooler months and overlapped extensively during warmer months. Also, the type of bait used in pots may influence bycatch rates, but this has not been tested. In most crab fisheries, menhaden (*Brevoortia tyrannus*) or other oily fish species are used as bait (Kennedy et al. 2007), but in Maryland, some pots are baited with razor clams (*Tagelus plebeius*), which may be a preferred food source for terrapins (W. Roosenburg, Ohio University, pers. comm.). Bishop (1983) found that peeler pots that used live male blue crabs as bait to capture molting female crabs captured more terrapins than pots baited with fish. Finally, Hart and Crowder (2011) suggested that peeler pots, which tend be fished in shallow waters in the spring, may present a particular risk to terrapins. Peeler crabs have been observed in terrapin feces (Roosenburg, pers. comm.) and thus may serve as an attractant for terrapins to crab pots, which may explain the higher catch rates reported in this fishery (Hart and Crowder 2011).

Bycatch Reduction

Bycatch reduction devices are rectangular restrictors placed in the inner gape of crab pot funnels that limit the entry of terrapins based largely on shell height. Many adult and some juvenile terrapin shells are >4.5 cm high, whereas the overwhelming majority of crab shells are <4.5 cm high (Roosenburg and Green 2000). Although many different BRD widths have been used in various studies, in most cases, the height of the BRD limits the entry of terrapins into crab pots; thus, BRDs typically are 4.5 to 5 cm high. The original BRDs were handmade of heavy-gauge wire to minimize possible bending and expansion by terrapins trying to enter pots. More recently, commercially manufactured, plastic BRDs have become available that are more rigid (Fig. 16.3).

Loss of terrapins to bycatch in commercial crab pots has been noted in 13 of the 14 coastal states in which terrapins occur and crab potting is allowed (Seigel and Gibbons 1995). (No bycatch has been reported from Massachusetts; crab pots are not allowed in Connecticut and Rhode Island.) Since the original work on bycatch reduction by Wood (1997), studies on the relative effectiveness of BRDs—also known as turtle excluder devices (TEDs)—have been completed in 12 of those 13 states. New York and Massachusetts are the only states where crab pots are used but no BRD studies

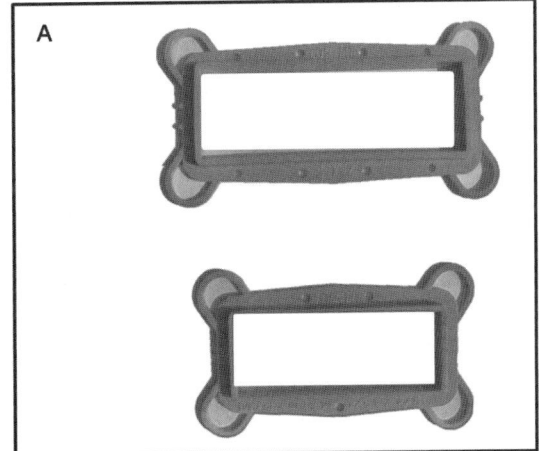

Fig. 16.3. Bycatch reduction devices (BRDs). (A) Plastic BRDs. A size of 4.5 × 12 cm is required for recreational crabbing in Maryland and Delaware; a larger size, 5 × 15 cm, is required in New Jersey and New York. (B) BRDs exclude terrapins primarily by shell height. (C) Crab pot with four BRDs attached to the inner funnels.

have been completed. New York also is one of only four states with BRD regulations (New Jersey, Delaware, and Maryland are the other three). The remaining states in which BRD studies have been completed but do not have BRD regulations are Virginia, North Carolina, South Carolina, Georgia, Florida, Alabama, Mississippi, Louisiana, and Texas.

We present the bycatch results of completed BRD studies in Table 16.1. Most of these studies compared bycatch in pots fitted with BRDs and pots without BRDs. To reduce bycatch mortality, pots were checked frequently or had wire mesh "chimneys" extending above the water surface and attached to the top corners of the pots, thereby giving trapped terrapins access to air so they could breathe. Use of these modified pots is restricted to locations where the tidal range is less than the 2 m height of the pot plus chimney. This limitation explains in part the low number of BRD studies in Georgia and South Carolina, where the tidal range can far exceed 2 m.

The overwhelming result of BRD studies both before and after the review by Roosenburg (2004) is that terrapin bycatch was reduced up to 100% in pots fitted with either wire or plastic BRDs (Table 16.1). The mechanism for terrapin exclusion by the BRD is still unclear since many terrapins small enough to fit through the BRD gape do not enter BRD pots (Upperman et al. 2014), due to physical limitations or behavioral modification, as suggested by Roosenburg (2004). For example, a study by Dominy et al. (2013) demonstrated the tetrachromatic color vision of terrapins that could affect how they "see" crab pots and BRDs. In addition to physical exclusion by size, the orange color might influence how terrapins respond behaviorally to the manufactured plastic BRDs.

From a practical and economic perspective, the clear effectiveness of BRDs in reducing terrapin bycatch is not sufficient to result in widespread implementation if BRDs adversely affect the capture of blue crabs. The effect of BRDs on crab capture is more complicated than the effect on bycatch reduction because the capture of blue crabs must be assessed for both marketable size and number. Further, each study must determine whether experimental methods mimic what is typical of commercial or recreational crabbing techniques in scale, location, baiting, and checking of pots (Roosenburg 2004). Evaluating BRD effects on blue crab capture also requires considering BRD type and dimensions, as well as regional and seasonal differences that could lead to different results both within and among studies.

Despite these caveats, the size of crabs caught generally was unchanged in pots with BRDs relative to controls (Table 16.1). The total number of crabs caught, however, was more variable with the use of BRDs. Wood (1997) and Guillory and Prejean (1998) observed significant increases in crab catch with BRDs in Louisiana and New Jersey, respectively, and a BRD study completed in Virginia found significantly more crabs caught in BRD pots located in seagrass beds (R. Lipcius, pers. comm.). Some studies observed a negative effect of BRDs on total crab catch (Cole and Helser 2001; Lukacovic et al. 2005; Belcher et al. 2007; Powers et al. 2009; Coleman et al. 2011; Hart and Crowder 2011; Morris et al. 2011; Upperman et al. 2014). Other studies documented small or no effects of BRDs on crab catch (Mazzarella 1994; Wood 1997; Cuevas et al. 2000; Roosenburg and Green 2000; Cole and Helser 2001; Butler and Heinrich 2007; Powers et al. 2009; Rook et al. 2010; Graham et al. 2011; Hart and Crowder 2011; Baxter 2014; Chavez 2014). Based on these collective results, a "one-size-fits-all" BRD for both excluding terrapins and maintaining crab catch has not been demonstrated for active pots, nor is it likely given the geographic variation in terrapin size. For derelict pots, BRDs would exclude most terrapins irrespective of crab catch.

Bycatch Management

Despite the clear evidence that bycatch reduction devices are highly effective at excluding terrapins, the variable effect on crab catch has contributed to the creation of different BRD policies among states. Regulations on BRD use in tidal waters of New York are made on an ad hoc basis. In Maryland and Delaware, the use of BRDs by recreational crabbers is compulsory. In New Jersey, BRD regulations do not focus on who is crabbing with commercial pots

Table 16.1. Summary of studies comparing diamond-backed terrapin bycatch and blue crab catch in control pots (standard hardshell or peeler pots) and BRD pots (pots fitted with bycatch reduction devices)

State	BRD type	Dim Dimensions (cm)	Terrapin bycatch, control:BRD	Change in crab size	Change in crab number	Reference
AL	Wire	5×15	22:2	N/A	31% decrease	Coleman et al. 2011
DE	Wire	5×10	97:40	No change	No change	Cole & Helser 2001
DE	Wire	3.8×12	106:0	No change	~26% decrease	Cole & Helser 2001
DE	Wire	4.5×12	106:36	No change	~14% decrease	Cole & Helser 2001
DE	Wire	5×12	106:93	No change	No change	Cole & Helser 2001
FL	Wire	4.5×12	37:4	No change	No change	Butler & Heinrich 2007
GA	Plastic	5×15	136:5	No change	~14% decrease	Belcher et al. 2007
LA	Wire	5×10	0:0	N/A	38% increase	Guillory & Prejean 1998
MD	N/A	4.5×12	1:0	N/A	29% decrease	Lukacovic et al. 2005
MD	Wire	4.5×12	105:19	No change	No change	Roosenburg & Green 2000
MD	Wire	5×10	105:56	No change	No change	Roosenburg & Green 2000
MS	Wire	5×15	0:0	No change	No change	Graham et al. 2011
MS	Wire	5×10	0:0	No change	No change*	Cuevas et al. 2000
NJ	Wire	5×10	40:3	No change	No change	Mazzarella 1994
NJ	Wire	4.5×10	3:0	No change	12% increase	Wood 1997
NJ	Wire	5×10	25:4	No change	10% increase	Wood 1997
NJ	Wire	5×10	46:5	No change	49% increase	Wood 1997
NC	Wire	4×15; 5×15	13:1	No change	No change	Chavez 2014
NC	Wire	5×16	7:0	N/A	No change[†]	Hart & Crowder 2011
NC	Wire	4×16; 4.5×16	1:0	N/A	23% decrease[†]	Hart & Crowder 2011
SC	Plastic	5×15	30:0	N/A	N/A	Powers et al. 2009
SC	Plastic	4.5×12	75:3	N/A	21% decrease	Powers et al. 2009
TX	Plastic	4.5×12	2:0	No change	No change	Baxter 2014
VA	Plastic	4.5×12	2:0	Slight decrease	~25% decrease	R. Lipcius, VIMS, pers. comm.
VA	Plastic	4.5×12	69:2	No change	53% decrease[‡]	Upperman et al. 2014
VA	Plastic	5×15	69:0	No change	No change[‡]	Upperman et al. 2014
VA	Plastic	4.5×12	9:0	N/A	17% decrease	Morris et al. 2011
VA	Plastic	4.5×12	42:0	N/A	47% decrease[‡]	Morris et al. 2011
VA	Plastic	4.5×12	46:2	No change	No change	Rook et al. 2010

*Based on total crab catch data.
[†]Legal male captures in hard crab pots only.
[‡]Unbaited pots.

but rather on where the crabbing is conducted: pots set in tidal creeks less than 150 feet wide must be equipped with BRDs, with the assumption that pots close to shore or in small tidal creeks are more likely to be in terrapin habitat, relative to more open water. Unfortunately, with the lack of general public knowledge regarding regulations and the absence of significant enforcement, compliance typically lags among targeted groups. For example, Radzio et al. (2013) found that <35% of recreational crab pots in Maryland tidal creeks had BRDs, despite a law requiring BRDs on all recreational crab pots. Regulatory agencies in states currently without BRD regulations may be delay-

ing action because of concerns about the lack of enforcement or the suspected absence of economic incentive for crabbers should a BRD regulation be imposed. Especially for fisheries' regulations, clearing the legislative hurdles can be a complex task (Roosenburg et al. 2008).

For some states, the adverse effect of crabbing on terrapin populations has spurred various groups to enact voluntary programs to encourage BRD use. In South Carolina, Georgia, Florida, Mississippi, New Jersey, Virginia, and possibly other states, programs educate citizens about terrapins and provide BRDs for recreational or commercial pots. These programs are run by volunteer citizens' groups and by state departments of natural resources and fish and wildlife agencies that encourage both crab pot manufacturers and buyers to install BRDs. That different stakeholder groups are converging on similar efforts to reduce terrapin bycatch is encouraging and suggests broadening support for terrapin conservation.

With respect to the conservation of terrapin populations, ongoing bycatch mortality in commercial crab pots is thought to hinder population recovery (Tucker et al. 2001). Unfortunately, no data are available that demonstrate a link between BRD use and the subsequent recovery of decimated terrapin populations or the establishment of new populations. One reason for the absence of a link is the time needed for populations to recover once bycatch mortality is reduced. As Hart and Crowder (2011) suggest, use of BRDs on crab pots combined with spatial and temporal restrictions on crabbing could be an effective and economically feasible management strategy for the crab fishery to conserve existing terrapin populations. Management strategies in geographic "hot spots" for conservation where terrapins currently are found (e.g., expansive tidal marsh complexes near accessible nesting beaches), however, would be different from those for population restoration where terrapins are absent (e.g., tidal creeks with hardened shorelines and recreational docks). Finally, superimposed upon BRD implementation strategies are management needs to address other threats to terrapin populations that are associated with human coastal development (Chapter 14). As a species living literally on the edge between upland and open water, the terrapin population's ability to respond to implementation of BRD policies may be constrained by life history characteristics affected by other dynamics at the terrestrial-aquatic interface.

Reducing and managing terrapin bycatch should not be limited to the design of BRDs and the implementation and enforcement of their use. Instead, it should integrate other mechanisms to reduce interactions between terrapins and commercial or recreational crabbing gear. Bilkovic et al. (2012), for example, have developed and tested the use of fully biodegradable panels on crab pots that would allow trapped terrapins and other marine fauna to escape

derelict pots. Both Grosse et al. (2011) and Hart and Crowder (2011) recommended that restrictions on where crab pots may be placed, at least between April and June, could substantially reduce terrapin mortality. Commercial use of crab pots, in fact, is excluded from Maryland tributaries of Chesapeake Bay specifically to reduce the observed bycatch of air-breathing animals, including terrapins (Davis 1942). Grosse et al. (2009) noted that in shallow tidal creeks only accessible by boat at high tide, crab pots were more likely to be neglected, as evidenced by the growth of epibenthos on the pots and their partial burial. Crab pots that are oversoaked, neglected, abandoned, or lost pose a catastrophic risk to terrapin populations (Grosse et al. 2011), although low but persistent background rates of terrapin mortality in active pots may have a greater impact (Roosenburg 2004). Some states have defined soak time limits of 24 hours, some have soak time limits as long as 72 hours, and some states (e.g., Georgia, Maryland) have no defined soak time. Except during midsummer and with exceptions for bad weather, a defined and enforced 24 hour soak time for crab pots, both commercially and recreationally, could in some instances reduce bycatch mortality of terrapins by up to 90%, independent of the effects of BRD implementation.

Management of blue crab fisheries must be integrated with the conservation of both terrapins and nearshore coastal ecosystems. Many tasks will be required to accomplish this objective: (1) development and testing of improved BRDs that exclude terrapins but have little to no effect on crab catch; (2) limits on where or how long crab pots can be soaked; (3) cleanup of derelict pots; (4) enhanced understanding of where crabbing and terrapin habitat intersect, both spatially and temporally; (5) increased regulatory compliance and enforcement with respect to BRDs and soak times; and (6) development of better methods for quantifying annual commercial and recreational crabbing effort within states to help draft regulations and conservation strategies that are reasonable, effective, and adaptable over time. These strategies must include research and monitoring approaches to validate expected management outcomes. So, for example, the demography of terrapin populations along

Maryland tributaries to Chesapeake Bay, where commercial crabbing has been excluded for more than 70 years, might be compared with populations in Virginia tributaries, where commercial crabbing is allowed. Similarly, the outcome of BRD use in New Jersey over the past 15 years might be assessed by comparing terrapin populations in creeks less than 150 feet wide with populations from small tidal creeks in other states where BRDs are not required.

Ultimately, the success of terrapin bycatch reduction and management will not be achieved by pitting crabbing against terrapin conservation. Rather, collaboration among crabbers, scientists, and resource managers is essential to make commercial and recreational crabbing compatible with terrapin conservation. This collaborative effort must be combined with communication, education, and outreach for successful implementation of bycatch reduction strategies (Cox et al. 2007). Because terrapin habitat and crab fisheries are spread over so many different intertidal environments and across so many political jurisdictions, the application of diversified management strategies will facilitate local solutions to terrapin bycatch reduction that can lead to the common goals of sustainable blue crab fisheries and terrapin conservation.

Summary

As was originally concluded more than 20 years ago (Seigel and Gibbons 1995), loss of diamond-backed terrapins as bycatch in commercial crab pots remains a large threat to terrapin populations throughout their range. The past 10 years of research have documented the damaging effects of both active and derelict pots from commercial and recreational crabbing on the demography of terrapin populations in almost every state in which they occur. Ongoing experimentation with bycatch reduction devices continues to demonstrate the clear benefits of BRDs in reducing terrapin mortality, with no to minimal effect on crab harvest. Even in the few states with BRD regulations, however, compliance and enforcement issues hinder the effectiveness of these devices as a conservation strategy. Until clear economic incentives for using BRDs or other technologies are identified, regulation of commercial crabbing will remain elusive, and terrapin populations will continue to decline and remain at risk.

ACKNOWLEDGMENTS

We are grateful to the many dedicated students, managers, conservationists, and crabbers whose contributions to research populate this chapter.

REFERENCES

Ayers, C.A. 2010. A population model for the diamond-back terrapin: implications for conservation management. Honors thesis, College of William and Mary, Williamsburg, VA. 33 pages.

Baldwin, J.D., L.A. Latino, B.K. Mealey, G.M. Parks and M.R.J. Forstner. 2005. The diamondback terrapin in Florida Bay and the Florida Keys: insights into turtle conservation and ecology. Pages 180–186 in W.E. Meshaka Jr. and K.J. Babbitt (eds.), Amphibians and Reptiles: Status and Conservation in Florida. Krieger Publishing Company, Malabar, FL.

Baxter, A.S. 2014. Diamondback terrapin paired crab trap study in the Mission-Aransas Estuary, Texas. Report to Coastal Bend Bays and Estuaries Program, Pub. 92, Project 1335. Corpus Christi, TX. 33 pages.

Belcher, C., L. Liguori, L. Gentit and T. Sheirling. 2007. Evaluation of diamondback terrapin excluders for use in crab traps in Georgia waters. Final Report, University of Georgia Marine Extension Service. Brunswick, GA. 34 pages.

Bilkovic, D.M., K. Havens, D. Stanhope and K. Angstadt. 2012. Use of fully biodegradable panels to reduce derelict pot threats to marine fauna. Conservation Biology 26:957–966.

Bilkovic, D.M., K. Havens, D. Stanhope and K. Angstadt. 2014. Derelict fishing gear in Chesapeake Bay, Virginia: spatial patterns and implications for marine fauna. Marine Pollution Bulletin 80:114–123.

Bishop, J.M. 1983. Incidental capture of diamondback terrapin by crab pots. Estuaries 6:426–430.

Butler, J.A. and G.L. Heinrich. 2007. The effectiveness of bycatch reduction devices on crab pots at reducing capture and mortality of diamondback terrapins (*Malaclemys terrapin*) in Florida. Estuaries and Coasts 30:179–185.

Butler, J.A., G.L. Heinrich and R.A. Seigel. 2006. Third workshop on the ecology, status and conservation of diamondback terrapins (*Malaclemys terrapin*): results and recommendations. Chelonian Conservation and Biology 5:331–334.

Chavez, S. 2014. Reducing diamondback terrapin and blue crab fishery interactions in North Carolina through the implementation of population surveys and bycatch reduction devices. MS thesis, University of North Carolina, Wilmington, NC. 66 pages.

Cole, R.V. and T.E. Helser. 2001. Effect of three bycatch reduction devices on diamondback terrapin, *Malaclemys terrapin*, capture and blue crab, *Callinectes sapidus*, harvest in Delaware Bay. North American Journal of Fisheries Management 21:825–833.

Coleman, A.T., T. Roberge, T. Wibbels, K. Marion, D. Nelson and J. Dindo. 2014. Size-based mortality of adult female diamond-backed terrapins (*Malaclemys terrapin*) in blue crab traps in a Gulf of Mexico population. Chelonian Conservation and Biology 13:140–145.

Coleman, A.T., T. Wibbels, K. Marion, D. Nelson and J. Dindo. 2011. Effect of bycatch reduction devices (BRDs) on the capture of diamondback terrapins (*Malaclemys terrapin*) in crab pots in an Alabama salt marsh. Journal of the Alabama Academy of Sciences 82:145–157.

Cox, T.M., R.L. Lewison, R. Zydelis, L.B. Crowder, C. Safina and A.J. Read. 2007. Comparing effectiveness of experimental and implemented bycatch reduction measures: the ideal and the real. Conservation Biology 21:1155–1164.

Cuevas, K.J., M.J. Buchanan, W.S. Perry and J.T. Warren. 2000. Preliminary study of blue crab catch in traps fitted with and without a diamondback terrapin excluder device. Proceedings of the Annual Southeast Association of Fish and Wildlife Agencies 54:221–226.

Davis, C.C. 1942. A study of the crab pot as a fishing gear. Chesapeake Biological Laboratory Pub. no. 53. Solomons, MD. 20 pages.

Dominy, A.E., E.R. Loew, J.R. Spotila and H.W. Avery. 2013. Turtles in a different light: the underwater visual ecology and mating behavior in the diamondback terrapin. Paper presented at 6th Symposium on the Ecology, Status, and Conservation of Diamondback Terrapins, Seabrook Island, SC, September 13–15.

Dorcas, M.E., J.D. Wilson and J.W. Gibbons. 2007. Crab trapping causes population decline and demographic changes in diamondback terrapin over two decades. Biological Conservation 137:334–340.

Fogarty, M.J. and R.M. Lipcius. 2007. Population dynamics and fisheries. Pages 711–755 in V.S. Kennedy and L.E. Cronin (eds.), The Blue Crab *Callinectes sapidus*. Maryland Sea Grant College Pub. UM-SG-TS-2007-01. University of Maryland, College Park, MD.

Gibbons, J.W., J.E. Lovich, A.D. Tucker, N.N. Fitzsimmons and J.L. Greene. 2001. Demographic and ecological factors affecting conservation and management of diamondback terrapins (*Malaclemys*

terrapin) in South Carolina. Chelonian Conservation and Biology 4:66–74.

Graham, D., H. Perry, D. Gibson, J. Anderson, G. Sanchez, T. Floyd and B. Richardson. 2011. Effect of turtle excluder devices (TEDs) on commercial catch of blue crabs *Callinectes sapidus* in Mississippi. Paper presented at 37th Annual Meeting of the Mississippi Chapter of the American Fisheries Society, Mississippi State, MS, February 16–18.

Grosse, A.M. and J.C. Maerz. 2013. Crabbing alters diamondback terrapin population size structure and growth rates. Paper presented at 6th Symposium on the Ecology, Status, and Conservation of Diamondback Terrapins, Seabrook Island, SC, September 13–15.

Grosse, A.M., J.C. Maerz, J.A. Hepinstall-Cymerman and M.E. Dorcas. 2011. Effects of roads and crabbing pressures on diamondback terrapin populations in coastal Georgia. Journal of Wildlife Management 75:762–770.

Grosse, A.M., J.D. van Dijk, K.L. Holcomb and J.C. Maerz. 2009. Diamondback terrapin mortality in crab pots in a Georgia tidal marsh. Chelonian Conservation and Biology. 8:98–100.

Guillory, V. and P. Prejean. 1998. Effect of a terrapin excluder device on blue crab, *Callinectes sapidus*, trap catches. Marine Fisheries 60:38–40.

Harden, L.A. and A.S. Williard. 2012. Using spatial and behavioral data to evaluate the seasonal bycatch risk of diamondback terrapins *Malaclemys terrapin* in crab pots. Marine Ecology Progress Series 467:207–217.

Hart, K.M. 1999. Declines in diamondbacks: terrapin population modeling and implications for management. MS thesis, Duke University, Durham, NC. 64 pages.

Hart, K.M. 2005. Population biology of diamondback terrapins (*Malaclemys terrapin*): defining and reducing threats across their range. PhD dissertation, Duke University, Durham, NC. 252 pages.

Hart, K.M. and L.B. Crowder. 2011. Mitigating bycatch of diamondback terrapins in crab pots. Journal of Wildlife Management 75:264–272.

Hart, K.M. and D.S. Lee. 2006. The diamondback terrapin: the biology, ecology, cultural history, and conservation status of an obligate estuarine turtle. Studies in Avian Biology 32:206–213.

Hoyle, M.E. and J.W. Gibbons. 2000. Use of a marked population of diamondback terrapins (*Malaclemys terrapin*) to determine impacts of recreational crab pots. Chelonian Conservation and Biology 3:735–737.

Isdell, R. 2013. Geospatial analysis of anthropogenic impacts on diamondback terrapins in the lower Chesapeake Bay. MS thesis, College of William and Mary, Williamsburg, VA. 76 pages.

Kennedy, V.S., M. Oesterling and W. Van Engel. 2007. History of blue crab fisheries on the U.S. Atlantic and

Gulf coasts. Pages 595–649 in V.S. Kennedy and L.E. Cronin (eds.), The Blue Crab *Callinectes sapidus*. Maryland Sea Grant College Pub. UM-SG-TS-2007-01. University of Maryland, College Park, MD.

Larocque, S.M., A.H. Colotelo, S.J. Cooke, G. Blouin-Demers, T. Haxton and K.E. Smokorowski. 2012. Seasonal patterns in bycatch composition and mortality associated with a freshwater hoop net fishery. Animal Conservation 15:53–60.

Lukacovic, R., L.S. Barker and M. Luisi. 2005. Diamond-back terrapin and crab pot interactions and the effect of turtle excluder devices on crab catch in Maryland's coastal bays. Maryland Department of Natural Resources, Fisheries Technical Report 44. Annapolis, MD. 6 pages.

Mazzarella, A.D. 1994. Test of a turtle exclusion device in commercial crab pots. Report of New Jersey Division of Fish, Game, and Wildlife. Port Republic, NJ. 11 pages.

Midwood, J.D., N.A. Cairns, L.J. Stoot, S.J. Cooke and G. Blouin-Demers. 2014. Bycatch mortality can cause extirpation in four freshwater turtle species. Aquatic Conservation: Marine and Freshwater Ecosystems 25:71–80.

Morris, S.A., S.M. Wilson, E.F. Dever and R.M. Chambers. 2011. A test of bycatch reduction devices on commercial crab pots in a tidal marsh in Virginia. Estuaries and Coasts 34:386–390.

Muffley, B., L. Lurig, G.N. Mahnke and H. Driscoll. 2007. Survey of New Jersey's blue crab, *Callinectes sapidus*, recreational fishery, year 1—Delaware Bay. Research Project Summary, New Jersey Department of Environmental Protection, Division of Science, Research and Technology. Trenton, NJ. 4 pages.

NOAA (National Oceanic and Atmospheric Administration) Office of Science and Technology. 2013. Commercial fisheries statistics. www.st.nmfs.noaa.gov/commercial-fisheries/commercial-landings/annual-landings/index.

North Carolina Division of Marine Fisheries. 2013. North Carolina Blue Crab (*Callinectes sapidus*) Fishery Management Plan, Amendment 2. North Carolina Division of Marine Fisheries, Morehead City, NC. 528 pages.

Powers, J., D. Whitaker, B. Gooch, N. West and A. Von Harten. 2009. Cooperative research in South Carolina. National Marine Fisheries Service, NOAA, Final Report, Grant no. NA04NMF4720306. Charleston, SC. 132 pages.

Raby, G.D., A.H. Colotelo, G. Blouin-Demers and S.J. Cooke. 2011. Freshwater commercial bycatch: an understated conservation problem. Bioscience 61:271–280.

Radzio, T.A., J.A. Smolinsksy and W.M. Roosenburg. 2013. Low use of required terrapin bycatch reduction devices in a recreational crab pot fishery. Herpetological Conservation and Biology 8:222–227.

Rook, M.A., R.N. Lipcius, B.M. Bronner and R.M. Chambers. 2010. Bycatch reduction device conserves diamondback terrapins without affecting catch of blue crab. Marine Ecology Progress Series 409:171–179.

Roosenburg, W.M. 1991. The diamondback terrapin: habitat requirements, population dynamics, and opportunities for conservation. Pages 237–244 in J.A. Mihursky and A. Chaney (eds.), New Perspectives in the Chesapeake System: A Research and Management and Partnership. Proceedings of a Conference. Chesapeake Research Consortium Pub. no. 137. Chesapeake Research Consortium, Solomons, MD.

Roosenburg, W. M. 2004. The impact of crab pot fisheries on terrapin (*Malaclemys terrapin*) populations: where are we and where do we need to go? Pages 23–30 in C. Swarth, W.M. Roosenburg and E. Kiviat (eds.), Conservation and Ecology of Turtles of the Mid-Atlantic Region: A Symposium. Bibliomania, Salt Lake City, UT.

Roosenburg, W.M., J. Cover and P.P. van Dijk. 2008. Legal issues: legislative closure of the Maryland terrapin fishery: perspectives on a historical accomplishment. Turtle and Tortoise Newsletter 12:27–30.

Roosenburg, W.M., W. Cresko, M. Modesitte and M.B. Robbins. 1997. Diamondback terrapin (*Malaclemys terrapin*) mortality in crab pots. Conservation Biology 5:1166–1172.

Roosenburg, W.M. and J.P. Green. 2000. Impact of a bycatch reduction device on diamondback terrapin and blue crab capture in crab pots. Ecological Applications 10:882–889.

Roosenburg, W.M., K.L. Haley and S. McGuire. 1999. Habitat selection and movements of diamondback terrapins, *Malaclemys terrapin*, in a Maryland estuary. Chelonian Conservation and Biology 3:425–429.

Schaffer, C., R.C. Wood, T.M. Norton and R. Schaffer. 2008. Terrapins in the stew. Iguana 15:78–85.

Seigel, R.A. and J.W. Gibbons. 1995. Workshop on the ecology, status, and management of the diamondback terrapin (*Malaclemys terrapin*), Savannah River Ecology Laboratory, 2 August 1994: final results and recommendations. Chelonian Conservation and Biology 1:241–243.

Selman, W., B. Baccigalopi and C. Baccigalopi. 2014. Distribution and abundance of diamondback terrapins (*Malaclemys terrapin*) in southwestern Louisiana. Chelonian Conservation and Biology 13:131–139.

Tucker, A.D., J.W. Gibbons and J.L. Greene. 2001. Estimates of adult survival and migration for diamondback terrapins: conservation insight from local extirpation within a metapopulation. Canadian Journal of Zoology 79:2199–2209.

Upperman, A.J., T.M. Russell and R.M. Chambers. 2014. Diamondback terrapins and the influence of recreational crabbing regulations. Northeastern Naturalist 211:12-22.

US Census Bureau. 2010. Coastline population trends in the United States: 1960 to 2008. www.census.gov/prod/2010pubs/p25-1139.pdf.

US Census Bureau. 2012. Statistical abstract of the United States. www.census.gov/compendia/statab/2012/tables/12s0364.pdf.

Wallace, B.P., C.Y. Kot, A.D. DiMatteo, T. Lee, L.B. Crowder and R.L. Lewison. 2013. Impacts of fisheries bycatch on marine turtle populations worldwide: toward conservation and research priorities. Ecosphere 4(3):1-49.

Wolak, M.E., G.W. Gilchrist, V.A. Ruzicka, D.M. Nally and R.M. Chambers. 2010. A contemporary, sex-limited change in body size of an estuarine turtle in response to commercial fishing. Conservation Biology 24:1268-1277.

Wood, R.C. 1997. The impact of commercial crab traps on northern diamondback terrapins, *Malaclemys terrapin terrapin*. Pages 21-27 in J. Van Abbema (ed.), Proceedings: Conservation, Restoration, and Management of Tortoises and Turtles—An International Conference. New York Turtle and Tortoise Society, State University of New York, Purchase, NY.

Wood, R.C. and R. Herlands. 1995. Terrapins, tires, and traps: conservation of the northern diamondback terrapin (*Malaclemys terrapin*) on the Cape May peninsula, New Jersey, USA. Pages 254-256 in J. Van Abbema (ed.), Proceedings of an International Congress of Chelonian Conservation. SOPTOM, Gonfaron, France.

17

Conservation through Environmental Education

GEORGE L. HEINRICH
TIMOTHY J. WALSH
WILL WILLIAMS

Turtles are ancient creatures that shared the earth with the dinosaurs and today are important and visible elements in many ecosystems. Some species serve as indicators of environmental health, while others are classified as keystone species (playing a vital ecological role in a given habitat), umbrella species (for which conservation efforts on their behalf benefit the larger ecological community), or flagship species (as iconic symbols of habitat conservation efforts) (Mills et al. 1993; Roberge and Angelstam 2004; Eckert and Hemphill 2005). Thus, conservation of turtles benefits the ecosystems in which they are found. Certainly, the threats to this highly endangered vertebrate group and its associated habitats present broad and immediate conservation challenges (Turtle Taxonomy Working Group 2014). Despite the urgency of turtle conservation, opportunities for conservation are still abundant, and the charismatic attraction of turtles makes them an excellent group on which to base education and outreach efforts to enhance ecological stewardship, conservation, and environmental awareness.

Historically, environmental education has played a significant role in successful conservation efforts related to a broad range of taxa. The National Association of Audubon Societies' public awareness campaign led to the end of the legal bird plume trade in the 1920s (Doughty 1975). Also in the early twentieth century, public awareness efforts led by the American Bison Society contributed to the recovery of a species that was on a path to almost certain extinction in the wild (Coder 1975; Danz 1997). Another excellent example concerning the value of environmental education to conserving wildlife is Rachel Carson's (1962) book *Silent Spring*, which alerted a broad audience to the environmental threat of DDT, a dangerous organochlorine insecticide (Black 2006). *Silent Spring* led to the US ban on

DDT that contributed to the recovery of the endangered American bald eagle (*Haliaeetus leucocephalus*). These examples primarily involved public awareness campaigns, whereas modern wildlife conservation education is a multifaceted discipline targeting diverse audiences. These efforts may include issue-specific to broad-focused educational materials for a wide range of groups and stakeholders, teaching curricula for educators, web-based resources, site-based and outreach programming, and field programs. All of these methods have been successfully integrated into various local, regional, and worldwide chelonian conservation programs.

Global sea turtle conservation efforts strongly incorporate environmental education. The work of the Sea Turtle Conservancy (www.conserveturtles.org), a nongovernmental organization (NGO) focusing on marine turtle conservation within the Caribbean, Atlantic, and Pacific regions, highlights environmental education. This group offers resources for children, educators, and other stakeholders, as well as opportunities for volunteer and fee-based fieldwork. The conservancy's success demonstrates the importance of including an education component when developing program objectives and goals. Education programs coupled with sound management practices that employ both ecologically and economically sustainable strategies are the hallmark of sound conservation programs.

Working on an international scale, the Turtle Survival Alliance (www.turtlesurvival.org) is a conservation NGO that developed in response to the unsustainable harvest of Asian turtles. Since its inception in 2001, it has grown into a global leader in addressing threats to highly endangered tortoise and freshwater turtle species on several continents. Its work incorporates capacity-building initiatives that include community education programs as an integral component of in situ efforts within range countries. The Turtle Survival Alliance's diverse efforts in this regard include training teachers in India, supporting an annual graduate-level turtle research course in Brazil, and using grassroots community outreach and education to address the poaching of endangered tortoises in Madagascar.

An example of a successful regional education program for tortoise conservation is that conducted by the Gopher Tortoise Council (www.gophertortoisecouncil.org). The group focuses on the imperiled gopher tortoise (*Gopherus polyphemus*) in the southeastern coastal plain of the United States. Its programs include educational material for diverse audiences, grants for educators, and workshops/symposia for diverse stakeholder groups. Its work is often accomplished and strengthened by working in partnership with federal and state agencies, as well as other conservation NGOs. The council led a 10-year educational campaign that resulted in the Florida Fish and Wildlife Conservation Commission prohibiting the harvest of gopher tortoises statewide in 1988 (Berish 2001). That state wildlife agency now facilitates its own extensive education program related to the tortoise's ecological importance and conservation.

Environmental education attempts to show the interconnectedness, beauty, benefits, and importance of natural communities for both the individual and society, often from both ecological and economic perspectives. In doing so, the effects on the environment may become clearer and solutions brought into focus. Unfortunately, scientific literacy has taken a downturn over the past few generations, despite increased environmental education efforts. Louv (2005, 2012) sees this decline as a response to society's broad disconnect with nature. Knowledge about natural systems, ecology, biodiversity, and environmental literacy aids in creating empathy for nature and wildlife, which can be a potent public motivator of management practices that benefit conservation initiatives (Myers et al. 2009).

Benefits of Education Efforts in Terrapin Conservation

Increased awareness of the natural history, anthropogenic threats, and conservation of the imperiled coastal diamond-backed terrapin (*Malaclemys terrapin*) is required to improve its conservation status. In a 2004 survey of 54 researchers, agency biologists, and other individuals with knowledge of diamond-backed terrapins range wide, 63% of respondents ranked conservation education programs as a needed action (Butler et al. 2006). Due to the secretive nature of terrapins, the greater public is mostly unfamiliar with the species over substantial portions of its range. An understanding of the threats facing terrapins and the conservation actions needed would better position individuals and stakeholder groups to offer support.

Positioning the diamond-backed terrapin as a flagship species can generate support for coastal habitat conservation; this may be most beneficial in local areas where it could garner support for addressing specific conservation issues (Bowen-Jones and Entwistle 2002), in addition to broader environmental issues. The terrapin's narrow linear distribution along the Atlantic and Gulf coasts of the United States and its being the only turtle in the world restricted to brackish water ecosystems make it a prime candidate for such a designation.

Terrapin researchers must increase their efforts to disseminate their findings (applicable to the species' conservation) beyond academic publications. Researchers should serve as environmental educators (Brewer 2001) and participate in education efforts, either by incorporating a conservation education component into their own programs or by assisting others who are undertaking such activities. This can be accomplished with minimal effort, such as offering an interpretive hike at a coastal natural area or presenting relevant information to a conservation NGO or coastal habitat management agency. Although reaching out to the latter two organizations may be "preaching to the choir," it is important for potential partners to clearly understand conservation issues and messages.

Multifaceted conservation education programs that target a wide range of audiences can have the greatest effect. Although smaller issue-specific programs are important, they reach only a narrow audience and often do not address the greater conservation needs of a species. Targeting a broader audience through diverse environmental education efforts can generate comprehensive support, even to the point of developing new and unexpected conservation partnerships. Incorporating information on actions needed to conserve terrapins into existing programs of other NGOs and government agencies can extend the reach of conservation efforts. Long-term education efforts are required to address persistent conservation threats. Audience demographics continually change, requiring a committed and sustained effort.

Integrating Research and Environmental Education

The Wetlands Institute (www.wetlandsinstitute.org), in southern New Jersey, facilitates a long-term terrapin environmental education and conservation project. This comprehensive program blends research, conservation education, and community involvement. The research program focuses on several issues, including long-term investigations of specific threats such as mortality in crab pots and on coastal roadways. Since 1989, this work has been facilitated by the participation of undergraduate college/university students and incorporates fieldwork, laboratory work, and community outreach. During the course of an annual 10 week summer program, students learn field research techniques and the importance of environmental education to conserving diamond-backed terrapins. Participation provides valuable experience toward becoming effective environmental educators and researchers. Outreach programming plays a significant role in increasing community awareness and involvement in the institute's conservation efforts. The Wetlands Institute also includes live terrapin exhibits along with signage carrying effective natural history and conservation messaging in its interpretive center's "Terrapin Station."

The relatively young (established in 2007) Georgia Sea Turtle Center (www.georgiaseaturtlecenter.org), on Jekyll Island, has adopted a similar approach toward developing a comprehensive conservation program, including elements of research, rehabilitation, education, and community involvement. Fieldwork addressing vehicle-related terrapin mortality on a local causeway is supported by community education efforts that focus on reducing the effects of this threat. The center's site-based education efforts feature a diamond-backed terrapin exhibit that teaches about life history, threats, and management efforts and includes live animal displays. A window into the treatment room from the exhibit gallery allows visitors to observe terrapins and other turtles undergoing emergency care, diagnostic workups, and various treatments, including surgery. A broad range of outreach strategies—including standards-based classroom curricula, homeschool programs, distance learning opportunities, and on-site visitor programming—also support the Georgia Sea Turtle Center's education mission.

Signage along with exhibits is an effective communication mechanism to educate the general public about terrapin conservation issues. Signage associated with terrapin exhibits must be factually correct and include locally relevant messages on conservation issues. Messages should encourage specific actions that individuals can take to assist with conservation efforts, such as helping terrapins cross the road safely during the nesting season and using bycatch reduction devices on crab pots (Chapters 14 and 16). Content should be designed specifically for the targeted audience. Visitors are more likely to view and read signage that attracts and holds their attention (Bitgood 2000). Exhibit signage is the primary route of information delivery, so detailed attention to its creation is warranted. Creating signage that engages visitors can be a challenging task; incorporating storylines and narratives can present the information in a more interesting and readable format. The inclusion of large, subject-rich photographs can help draw attention to signage and encourage thoughtful review of the associated text. Selected images should specifically illustrate key topics. Video footage, interactive components, and touchable objects can increase dwell time at an exhibit and assist visitors' understanding of the information. Serrell (1996) provided a thorough review of exhibit signage presentation and format.

Use of nonreleasable and head-started terrapins for on-site exhibition and outreach provides an opportunity for the public to understand the ecological roles of the species, as well as the threats and conservation actions needed. Live animals serve as wildlife ambassadors that can create emotional responses, aid in storytelling, focus attention, and encourage inquiry when used in exhibits and pro-

grams (Ballantyne et al. 2007; Genovesi 2011). Most states require permits for the collection, possession, and exhibition of terrapins, and care must be given to the ethical procurement of live specimens. Acquisition of ambassador terrapins should be limited to captive-bred individuals, or those that are rehabilitated but nonreleasable, or juveniles being head-started for eventual release. Keeping wild-caught animals, including rehabilitated and releasable individuals, must be avoided because this delivers the mixed message that although the species is in need of conservation, it is acceptable to remove individuals from the wild. Illegally trafficked terrapins are occasionally confiscated by state and federal wildlife agencies and put up for adoption by educational institutions around the country. Additionally, many local herpetological societies and animal rescue organizations offer discarded pet terrapins for adoption. Confiscated or adoptable turtles are an excellent source for educational animals that might otherwise be euthanized.

Seigel and Dodd (2000) provided a detailed overview of the manipulation of turtle populations for conservation, including the use of head-started turtles in conservation programs. Although there is a need to evaluate the value of head-starting of diamond-backed terrapins (Chapter 18), head-started animals can be used by conservation NGOs to raise community awareness and generate support. Use of head-started terrapins is a significant component of the Wetlands Institute's conservation education efforts. Eggs salvaged from road-killed terrapins are incubated, and the resulting hatchlings are head-started for eventual release. These terrapins are used in educational programming that includes organized releases with school groups and other members of the community (Herlands et al. 2004).

One environmental education program that integrates scientific research with education and engages students in head-starting is the Paul S. Sarbanes Ecosystem Restoration Project at Poplar Island in Maryland. Terrapins began nesting on Poplar Island shortly after nesting habitat was constructed (Chapter 18). Since 2005, hatchling terrapins have been collected from nests on the island

and placed into Maryland schools, through environmental outreach programs at Arlington Echo Outdoor Education Center (www.arlingtonecho.org), Maryland Environmental Services (www.menv.com), and the National Aquarium in Baltimore (www.aqua.org). The collection of hatchlings is conducted by terrapin researchers who document nesting success on Poplar Island and distribute 180 to 250 hatchlings to the project partners annually. As part of Maryland's environmental literacy efforts, teachers use classroom terrapins in an integrated curriculum for learning on a variety of environmental topics and use an issue investigation model (Hungerford and Volk 1990) to explore pertinent issues facing terrapins (e.g., habitat loss and mortality in crab pots). The curriculum creates awareness of the species and related environmental problems, develops understanding of the biology and threats, and implements grassroots action. Students raise the terrapins throughout the school year, record growth data at regular intervals, and then engage in grade-specific activities that teach not only terrapin perils but a broad spectrum of environmental issues, ranging from development and habitat loss to nutrient dynamics and dead zones in Chesapeake Bay. At the culmination of the school year, the turtles are fitted with PIT (passive integrative transponder) tags, and the students return them to Poplar Island for release.

Several components of the Poplar Island head-start program are unique. First, although the terrapins serve as ambassadors, the goal is not terrapin conservation per se. Rather, the terrapins are a tool for environmental and STEM (science, technology, engineering, and math) education. The curricula include a variety of environmental issues and science skills, such as data graphing and interpretation. At the Arlington Echo Outdoor Education Center, in particular, the Terrapin Connection program is part of a larger Chesapeake Connection program in which teachers and classrooms receiving terrapins must also participate in other environmental and restoration projects.

Second, the students contribute to ongoing terrapin research. Many of the terrapins are large enough at the time of release to allow sex determination based on external morphology, without the need to sacrifice the animal. Furthermore, all head-started turtles enter a mark-recapture program that compares growth, age of maturity, and reproductive output in head-started turtles and in non-head-started hatchlings from the island (Chapter 18). High school students learn data management skills and evaluate the growth and changes in the population by exploring a digital copy of the mark-recapture database. Educators and scientists meet during annual workshops that review the most recent scientific findings from Poplar Island and provide detailed instructions on husbandry and veterinary concerns. Third, during the students' trip to release the head-started terrapins on Poplar Island, they learn about wetland ecology, habitat restoration, and the socioeconomic benefits of such projects.

Finally, the program's reach extends beyond the classroom when students later share their experiences with others. Perhaps the biggest hurdle to terrapin conservation is making common knowledge of the presence and role of terrapins in the salt marsh ecosystem. Their cryptic behavior and the difficulty in seeing them in the wild minimize the human-turtle interaction that generates awareness and concern. Educational programs such as that at Poplar Island enhance awareness about the terrapin, and that program clearly contributed to generating public approval for closure of the terrapin fishery in Maryland in 2007. More importantly, the program illustrates how appropriately blending education and research can result in greater conservation gains than regulatory management alone.

The incorporation of educational resources is critical to the success of any comprehensive conservation education program. Use of issue-specific to broad-focused educational material (e.g., brochures, educator's guides, and web-based information) for a wide range of groups and stakeholders can support efforts to effect positive change (Fig. 17.1). The Diamondback Terrapin Working Group (www.dtwg.org) offers an excellent listserve and website, including an extensive bibliography with links to digital (PDF) files. Many members of this working group incorporate educational components in their research and conservation

Fig. 17.1. These educational products created by the Florida Turtle Conservation Trust, as part of a regional diamond-backed terrapin awareness campaign, illustrate material intended for use by a wide range of groups and stakeholders.

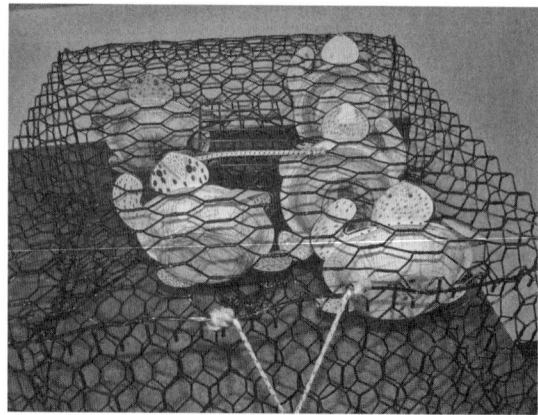

Fig. 17.2. Example of an art project created after a lesson on diamond-backed terrapin mortality in crab pots given during a summer nature camp for children ages 7 to 11.

programs, some of which are posted on the website. Educators searching for specific information or data are encouraged to get in touch with researchers, as they are often willing to provide assistance.

Educators, both formal and nonformal, can play critical roles in conserving wildlife and natural areas. Providing educators with a solid introduction to the ecology and conservation of diamond-backed terrapins through classroom presentations and field experiences can help them become more effective at teaching about conservation issues. Educators should be introduced to natural history, causes of decline, and conservation measures, and should be provided with educational activities and resources. The goal should be to provide a lifetime experience that will allow educators to return to their formal and nonformal settings to teach and motivate others to join and support conservation efforts.

Educational material developed for use by formal educators should be presented in a manner that makes it easy to use. Classroom educators have multiple demands on their time, so providing information in condensed or summarized formats increases the likelihood of it being used. Cross-curricular integration across multiple subjects such as art, history, math, and science furthers the value of educational materials to a broad range of

educators (Fig. 17.2). Including sources for additional information will provide access to more extensive material. Heinrich et al.'s (2010) publication is an example of a regionally focused educator's guide developed with these concerns in mind. Correlating educational material with Common Core State Standards, as well as individual state and local school district standards, is imperative if one expects the material to be used in both public and private schools. Limited resources such as funding and time restrict what classroom teachers can include in their curriculum and offer as field trip opportunities. Careful development and production of educational material is essential to successfully address all of these needs. As with correlating material to the standards noted above, one should consider the content's relevancy to STEM subjects and their increasing importance in classroom settings. These subjects serve critical roles in developing STEM literacy (leading to a more informed society), providing individuals with the knowledge and skill competencies required for future careers, and furthering research and development of new technologies (Bybee 2013).

All conservation education programs require sustained funding for programming and staffing. Conservation funding fluctuates with the economic climate, and during downturns can become limited and highly competitive. Consequently, organizations must consider long-term funding needs and diverse sources to sustain programs and ensure success. Partnering with public/private schools and other nonprofit organizations can offer increased opportunities for funding to sustain long-term educational programs. The Poplar Island program has multiple funding sources (Arlington Echo Outdoor Education Center, Maryland Environmental Services, National Aquarium in Baltimore, US Army Corps of Engineers, Ohio University, and the Port of Baltimore) that contribute to the program's success, with either fiscal resources or in-kind services.

For many programs, the definition of "successful" may be subjective without some type of formal evaluation. When critically reviewing any program, one should investigate effectiveness, impact, and potential improvements, then make necessary adjustments (Jacobson 2009). Without program evaluation, several consequences may remain undetected: not reaching the intended audience, misinterpreting information, reinforcing wrong behaviors, and not reaching the program's full potential (Jacobson et al. 2006).

Guidelines for Developing a Conservation Education Program

1. Identify program objectives and goals before initiating education efforts (think big, start small). Strive to build programs that deliver greater environmental and conservation messages, beyond terrapins.
2. Develop education, research, and conservation partnerships that will increase the likelihood of funding, success, and sustainability.
3. Identify and make use of opportunities to incorporate your message into existing programs of other conservation NGOs and government agencies.
4. Develop long-term, multifaceted programming that reaches out to a variety of audiences and will support the program's objectives and goals. Consider nontraditional audiences such as government agency staff, regulators, and politicians.
5. Identify and secure varied sources of funding or enter into diverse partnerships that can share the burden of cost.
6. Create educational material (e.g., brochures, educator's guides, and websites) targeting a wide range of groups and stakeholders. Provide material in both printed and web-based formats.
7. Involve researchers with education efforts, including in-service training and a variety of programming (e.g., site-based, outreach, and field-based).
8. Offer visits to coastal natural areas that include guided interpretive programs for multiple audiences.
9. Provide educators with field experiences highlighting the ecology and conservation needs of diamond-backed terrapins.
10. Use only captive-bred terrapins, or those that are rehabilitated but nonreleasable, or juveniles being head-started as exhibit animals and for outreach. Secure all necessary permits for possession and exhibition of animals.
11. Present accurate, consistent information and messaging throughout all programs, products, and exhibits.
12. Incorporate program evaluation methods, then rework programs and update products as needed.

Summary

Environmental education is essential to conserving diamond-backed terrapins and their coastal habitat. Environmental education components must be incorporated into research and conservation programs. We believe that partnerships between researchers, coastal habitat managers, conservationists, and educators are vital to the success of these programs and to the future of diamond-backed terrapins.

ACKNOWLEDGMENTS

We thank T. Marcinkowski for assistance with locating a difficult-to-find publication. Thoughtful reviews by B. K. Atkinson, J. G. Byrd, V. S. Kennedy, C. B. Manis, B. K. Mealey, T. Norton, W. M. Roosenburg, and B. Sullivan significantly improved earlier drafts of this chapter.

REFERENCES

Ballantyne, R., J. Packer, K. Hughes and L. Dierking. 2007. Conservation learning in wildlife tourism settings: lessons from research in zoos and aquariums. Environmental Education Research 13:367-383.

Berish, J.E. 2001. Management considerations for the gopher tortoise in Florida. Final Report, Florida Fish and Wildlife Conservation Commission. Tallahassee, FL. 44 pages.

Bitgood, S. 2000. The role of attention in designing effective interpretive labels. Journal of Interpretation Research 5(2):31-45.

Black, B. 2006. Nature and the Environment in Twentieth-Century American Life. Greenwood Press, Westport, CT. 264 pages.

Bowen-Jones, E. and A. Entwistle. 2002. Identifying appropriate flagship species: the importance of culture and local contexts. Oryx 36:189-195.

Brewer, C. 2001. Cultivating conservation literacy: "trickle-down" education is not enough. Conservation Biology 15:1203-1205.

Butler, J.A., G.L. Heinrich and R.A. Seigel. 2006. Third workshop on the ecology, status, and conservation of diamondback terrapins (*Malaclemys terrapin*): results and recommendations. Chelonian Conservation and Biology 5:331-334.

Bybee, R.W. 2013. The Case for STEM Education: Challenges and Opportunities. National Science Teachers Association Press, Arlington, VA. 116 pages.

Carson, R.L. 1962. Silent Spring. Houghton Mifflin Company, Boston, MA. 368 pages.

Coder, G.D. 1975. The national movement to preserve the American buffalo in the United States and Canada between 1880 and 1920. PhD dissertation, Ohio State University, Columbus, OH. 348 pages.

Danz, H.P. 1997. Of Bison and Man. University Press of Colorado, Niwot, CO. 248 pages.

Doughty, R.W. 1975. Feather Fashions and Bird Preservation: A Study in Nature Protection. University of California Press, Berkeley, CA. 184 pages.

Eckert, K.L. and A.H. Hemphill. 2005. Sea turtles as flagships for protection of the wider Caribbean region. Maritime Studies 3(2) and 4(1):119-143.

Genovesi, J.S. 2011. An exploratory study of a new educational method using live animals and visual thinking strategies for natural science teaching in museums. PhD dissertation, Drexel University, Philadelphia, PA. 156 pages.

Heinrich, G.L., T.J. Walsh and J.A. Butler. 2010. Diamondback Terrapins of Tampa Bay: An Educator's Guide. Florida Turtle Conservation Trust, St. Petersburg, FL. 29 pages.

Herlands, R., R. Wood, J. Pritchard, H. Clapp and N. Le Furge. 2004. Diamondback terrapin (*Malaclemys terrapin*) head-starting project in southern New Jersey. Pages 13-21 in C. Swarth, W.M. Roosenburg and E. Kiviat (eds.), Conservation and Ecology of Turtles of the Mid-Atlantic Region: A Symposium. Bibliomania, Salt Lake City, UT.

Hungerford, H.R. and T.L. Volk. 1990. Changing learner behavior through environmental education. Journal of Environmental Education 21(3):8-21.

Jacobson, S.K. 2009. Communication Skills for Conservation Professionals. Island Press, Washington, DC. 480 pages.

Jacobson, S.K., M.D. McDuff and M.C. Monroe. 2006. Conservation Education and Outreach Techniques. Oxford University Press, New York, NY. 496 pages.

Louv, R. 2005. Last Child in the Woods: Saving Our Children from Nature-Deficit Disorder. Algonquin Books, Chapel Hill, NC. 324 pages.

Louv, R. 2012. The Nature Principle: Reconnecting with Life in a Virtual Age. Algonquin Books, Chapel Hill, NC. 332 pages.

Mills, L.S., M.E. Soulé and D.F. Doak. 1993. The keystone-species concept in ecology and conservation: management and policy must explicitly consider the complexity of interactions in natural systems. BioScience 43:219-224.

Myers, O.E. Jr., C.D. Saunders and S.M. Bexell. 2009. Fostering empathy with wildlife: factors affecting free-choice learning for conservation concern and behavior. Pages 39-55 in J.H. Falk, J.E. Heimlich and S. Foutz (eds.), Free-Choice Learning and the Environment. AltaMira Press, Lanham, MD.

Roberge, J.-M. and P. Angelstam. 2004. Usefulness of the umbrella species concept as a conservation tool. Conservation Biology 18:76-85.

Seigel, R.A. and C.K. Dodd Jr. 2000. Manipulation of turtle populations for conservation: halfway technologies or viable options? Pages 218-238 in M.W. Klemens (ed.), Turtle Conservation. Smithsonian Institution Press, Washington, DC.

Serrell, B. 1996. Exhibit Labels: Interpretive Approach. AltaMira Press, Walnut Creek, CA. 261 pages.

Turtle Taxonomy Working Group [P.P. van Dijk, J.B. Iverson, A.G.J. Rhodin, H.B. Shaffer and R. Bour]. 2014. Turtles of the world, 7th edition: annotated checklist of taxonomy, synonymy, distribution with maps, and conservation status. Pages 329-479 in A.G.J. Rhodin, P.C.H. Pritchard, P.P. van Dijk, R.A. Saumure, K.A. Buhlmann, J.B. Iverson and R.A. Mittermeier (eds.), Conservation Biology of Freshwater Turtles and Tortoises: A Compilation Project of the IUCN/SSC Tortoise and Freshwater Turtle Specialist Group. Chelonian Research Monographs 5(7). Chelonian Research Foundation, Lunenburg, MA. DOI 10.3854/crm.5.000.checklist. v7.2014.

18

Habitat Restoration and Head-starting

WILLEM M. ROOSENBURG

Conservation efforts to protect and restore turtle populations have included a variety of techniques and regulatory efforts, including prohibitions and restrictions on harvest (Roosenburg et al. 2008), sustainable management (reviewed in Thorbjarnarson et al. 2000), habitat restoration (Mitchell and Klemens 2000), head-starting (reviewed in Burke 2015), and predator control on nesting beaches (Ratnaswamy et al. 1997; Ratnaswamy and Warren 1998; Munscher et al. 2012). A population growth rate ($\lambda > 1$) is the hallmark goal that suggests a sustainable population, but we know relatively little about the actual population dynamics or demographic parameters required to achieve and maintain this benchmark for freshwater turtles. Further, we have no historical data other than anecdotal accounts that hint at the population size or density to set as targets to identify a recovered or restored population.

The rehabilitation and restoration of turtle populations is constrained by their slow demographic response to mitigation strategies to reestablish numbers or reverse declining numbers. The turtle's life history phenotype of late and relatively low reproductive output that selected for longevity to ensure survival also makes turtles susceptible to population decline (Congdon et al. 1993, 1994; Heppell 1998). Decades may pass before a critical test can evaluate the effectiveness of a management strategy, and as generation time increases due to increasing age at maturity, the population response may become even further delayed.

Most turtle restoration projects target two potential mechanisms to rehabilitate populations: protection and rebuilding of critical habitat (particularly for reproduction) and supplementation of populations by increasing survivorship of young cohorts, a process also known as head-starting (reviewed in Burke 2015). However, the first and most important step in the conservation and restoration of any population remains elimination or reduction of the extrinsic source of mortality or the environmental stressors responsible for reducing population size (Frazer 1992, 1997; Seigel and Dodd 2000).

The diamond-backed terrapin, *Malaclemys terrapin*, is no exception to these principles, and throughout its range many initiatives have the goal of maintaining or restoring terrapin populations. Because only anecdotal records exist of the abundance of terrapins before the extensive harvesting that occurred during the late nineteenth and early twentieth centuries (Chapter 13), meaningful targets for terrapin recovery are lacking. Natural resource management agencies of many states have adopted regulations to protect terrapins, but most active programs to supplement or restore terrapin populations are undertaken by nongovernmental organizations, private individuals, or educational initiatives. Resource management agencies are often involved in habitat restoration through programs that promote living shorelines, a hybrid of stabilization techniques that emphasize natural habitat features (Duhring 2006), or other, broader ecosystem restoration initiatives that rarely identify terrapins as a potential benefactor. Often, the discovery that terrapins are positively affected by restoration is used as an *a posteriori* benefit of the project. To date, there are no government-funded efforts directly targeted to rebuilding terrapin populations through the use of head-starting, habitat restoration, or other means. In this chapter, I present data from two restoration initiatives in Chesapeake Bay that have created habitat with potential benefit to terrapin populations. These restoration projects did not identify terrapin population restoration as a goal, but they have affected terrapin behavior and population dynamics as a consequence. I also present preliminary data from a head-start program associated with one of these restoration projects and evaluate its potential demographic effects.

Nesting Habitat Restoration

As with other aquatic turtles, terrestrial and aquatic environments are integral parts of the terrapin's habitat, necessary to complete its life cycle. As an estuarine turtle, the diamond-backed terrapin lives in salt marshes and mangroves and frequently nests on low-lying beaches that fringe the shorelines. The salt marsh provides important habitat where terrapins forage and, in many populations, live most of their lives (Tucker et al. 1995). In the Chesapeake Bay region, terrapins rely on fringe and salt marsh habitat in the earlier parts of their life cycle, but adults, particularly females, move offshore to more open water (Roosenburg et al. 1999). Nesting habitat in many parts of the terrapin's range is frequently the narrow sandy strip that lies on the windward shore of a salt marsh or between the water and upland areas, and can often be agricultural fields (Roosenburg and Place 1995). Erosion and accretion of sand, particularly during storms, maintain natural nesting areas by depositing sand and creating open areas above mean high water. Hardening of these shorelines creates a barrier at the aquatic-terrestrial interface that disrupts the dynamic processes that maintain open areas on nesting beaches and disrupts access for nesting females (Winters et al. 2015). However, terrapins are able to quickly locate and use restored or newly created habitats (Wnek et al. 2013; Crawford et al. 2014; Roosenburg et al. 2014).

Development and climate change contribute to physical loss and decline of both salt marsh and nesting habitat throughout the terrapin's range. Sea level rise associated with climate change is likely to result in more frequent nest site inundation and salt marsh loss (Najjar et al. 2000; Simas et al. 2001; Hartig et al. 2002). Also, the increase in human densities along coastal areas adversely affects terrapin habitat through the destruction of wetlands and nesting areas by shoreline hardening (by bulkheads, sea walls made of concrete, metal, or wood; and by riprap, continuous stone revetment), which builds barriers between the aquatic and terrestrial habitats (Roosenburg 1991; Roosenburg et al. 2014; Winters et al. 2015).

The aquatic-terrestrial interface is critical to two components of the terrapin's life cycle. First, female terrapins must be able to cross the interface to reach nesting sites above mean high water, where eggs can incubate without risk of inundation

(Roosenburg and Place 1995). With the reduction of nearshore nesting habitat, terrapin nesting in the remaining habitat becomes more concentrated and potentially more vulnerable to nest predators (Chapters 7 and 14). Mesopredators such as raccoons and foxes, whose populations have grown in the absence of apex predators (Ritchie and Johnson 2009), quickly identify and repeatedly depredate nests in high-density nesting areas. In terrapin conservation, restoring nesting habitat ranks second to reducing adult and crab pot mortality (Seigel and Gibbons 1995; Chapter 14). Second, newly emerged hatchlings frequently retreat to the wrack line that accumulates in the upper intertidal zone and overwinter there terrestrially (Muldoon and Burke 2012; Chapter 8). The wrack line provides sheltered areas where hatchling terrapins can feed, avoid predation (Draud et al. 2004), and possibly overwinter. Adult terrapins noticeably use nesting areas for the 2 to 3 months of nesting season (Chapter 7), but overwintering hatchlings can result in a terrapin presence on nesting beaches almost year-round (Baker et al. 2006; Chapter 8).

Living shorelines maintain a permeable aquatic-terrestrial interface and allow estuarine organisms to traverse the interface (Davis et al. 2006). Offshore barriers diffuse wave energy, while nearshore supplementation of vegetation holds the shoreline in place. Terrapin populations benefit from living shoreline in two ways. First, a living shoreline can build or create access to nesting habitat that attracts females to newly created, open sandy areas visible from offshore (Wnek et al. 2013; Roosenburg et al. 2014). Second, it provides habitat for younger terrapins and smaller size classes that remain closer to shore because of the greater availability of their preferred food resources in these habitats (Roosenburg et al. 1999). These observations suggest that the aquatic-terrestrial interface is critical terrapin habitat, and its loss can dramatically affect terrapins and other organisms that traverse or require the interface to complete their life cycle. Further, alternative shoreline protection (such as living shorelines) or restoration may greatly benefit terrapin populations in parts of their range that have extensive habitat loss due to development, sea level rise, or shoreline contamination.

I have evaluated two shoreline restoration projects and their effects on terrapin nesting. The first is Jefferson Patterson Park (the Park), Calvert County, Maryland. Here, a living shoreline was constructed using offshore U-shaped riprap to deflect wave energy and using vegetation (*Spartina alterniflora* and *S. patens*) to stabilize the sand placed behind the riprap (Fig. 18.1). The project was one of the first to combine the use of offshore riprap with a living shoreline to protect beach habitat and reduce erosion. One unique component was that the riprap was constructed with openings that provided both access to the beach and beach visibility from offshore.

The second project is the Paul S. Sarbanes Ecosystem Restoration Project at Poplar Island, Talbot County, Maryland. In 2000, the US Army Corps of Engineers began reconstructing Poplar Island using dredge material, recreating tidal wetland habitat

Fig. 18.1. Riprap structures at Jefferson Patterson Park. These U-shaped structures face in the direction of the prevailing winds. Notice the areas behind the riprap that have been backfilled with sand and vegetation to prevent erosion.

and upland coastal plain forest lost due to erosion and subsidence. Poplar Island is in the middle of Chesapeake Bay. Its area exceeded 400 ha in the mid-1800s, but by 1998, when the restoration began, the island had eroded to less than 3 ha. The last section of the perimeter dike was completed in 2001. Weather-sheltered areas of the dike were made of sand harvested from the site, while exposed areas were reinforced with continuous riprap that created a shoreline impassable to terrapins. We discovered the first terrapins nesting on the sandy portions of the perimeter dike in 2002, the first summer after the dike was completed. During 2002 and 2003, we monitored nesting and hatching success on the island. In 2004, we began daily detailed surveys of potential nesting sites on the island by recording all discovered nests and documenting nesting success. In 2005, the first wetland cell on Poplar Island was completed, and since then more than 115 ha of wetlands have been restored. We also developed an education-based head-start program in 2005 (described in Chapter 17), and in 2008, we initiated mark-recapture to evaluate population dynamics on the island and document the fate of head-starts.

At both Poplar Island and Jefferson Patterson Park, I used GPS to geo-locate terrapin nests in reconstructed areas and then mapped their locations (Roosenburg et al. 2014). I recorded intact and depredated nests and monitored intact nests for predation. Terrapins readily discovered and used the nesting habitat at both sites (Fig. 18.2). However, at both the Park and Poplar Island, nesting was precluded where the shoreline was hardened by continuous rock revetment, creating an impediment that resulted in distinct boundaries to nesting areas and concentrating nesting in accessible areas (Fig. 18.2). Similarly, from 2002 to 2008, an embayment on the southwest corner of Poplar Island provided access to the island interior, but when the perimeter dike was completed (in 2008), closing the embayment, terrapins lost access and nesting was eliminated (Fig. 18.2, C, D). However, when wetland cells were completed, access to new nesting areas in the island interior was created by openings in the perimeter dike. Note the cluster of nests in the central portion of the island (Fig. 18.2, D) in an area where no nesting occurred before 2008

(Fig. 18.2, C). These observations demonstrate the importance of sheltered areas with accessibility across the aquatic-terrestrial interface for terrapin nesting. Further, high-density nesting areas are open sandy areas easily visible from offshore, which possibly enhances detection. The data from the Park and Poplar Island show that terrapins can locate and will use newly available nesting habitat. However, there were existing terrapin populations at both sites before the restoration projects (D. Brinker, Maryland Department of Natural Resources, and P. McGowan, US Fish and Wildlife Service, pers. comm.).

One major qualitative difference between the two sites is the presence of mesopredators (raccoons and foxes) only at the Park. Because Poplar Island lacks established populations of raccoons and foxes, the primary nest predators in the Chesapeake Bay area (Roosenburg 1991), nest success (measured as hatchling production) is unusually high (58.6%–85.3% across years; Roosenburg et al. 2014). By contrast, at the Park, none of the nests identified in Fig. 18.2 (A, B), produced hatchlings during either year. The proportion of surviving nests at the Park was probably similar to that in nearby mainland sites, where nest survival varied between 0% and 30% across years (Roosenburg 1991). The Park is a mainland site with easy predator access to nesting areas and does not allow hunting or trapping to control fox and raccoon populations. Raccoons have not immigrated to Poplar Island, and foxes are controlled as part of the management plan.

This contrast between Poplar Island and the Park identifies the dramatic effect nest predators can have on terrapin populations, suggesting that effective terrapin restoration must monitor and reduce nest predation (Chapter 14). Because shoreline stabilization can increase nesting density, predator removal may be even more important as the more concentrated nests are easier for predators to locate, and predators can quickly learn the nest areas. Munscher et al. (2012) demonstrated that the removal of raccoons in terrapin nesting areas increased nest survival. Similarly, predator removal on sea turtle nesting beaches with high nest predation increased nest survival (Ratnaswamy et al.

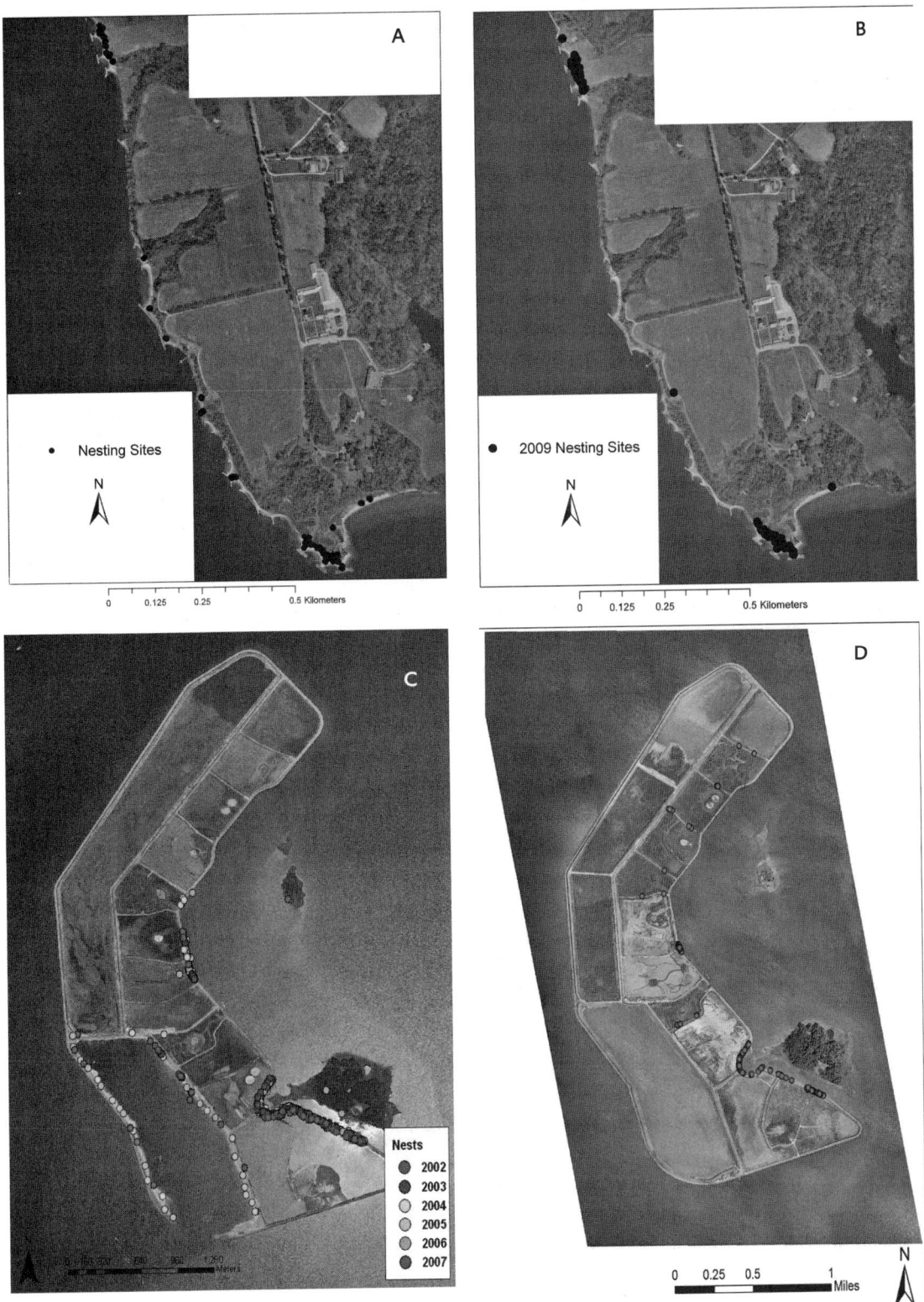

Fig. 18.2. (A, B) Terrapin nesting sites at Jefferson Patterson Park, in 2008 and in 2009, after shoreline reconstruction/restoration. Black dots represent terrapin nests. Note the offshore positions and the opening between riprap that gives terrapins access to sandy nesting areas behind and between groins. (C, D) Nesting sites at Poplar Island before closure of Cell 6 in 2008 (C) and after closure (D). Areas where nesting occurs are not obstructed by stone riprap on the perimeter dike. (C, D) Printed with permission from John Wiley & Sons, Inc.

1997; Ratnaswamy and Warren 1998). This management tool, however, requires annual implementation, as raccoon populations in both studies recovered quickly once removal was stopped.

Management of nest predator populations, although frequently controversial due to opposition to hunting and trapping, should be part of any long-term restoration project. Trapping of mammalian predators may offer opportunity for public-private partnerships (e.g., allowing commercial fur trappers to reduce raccoon and fox populations). Alternatively, nests can be protected with anti-predator cages. However, caging requires not only finding nests before the predators do but also building and installing cages that effectively prevent nest predation. Such methods have been used in Cape Cod (Brennessel 2002) and South Carolina (Chapter 14) with some success. These techniques are costly and labor intensive, but once it is protected there is no further need to revisit the nest. The Poplar Island project revealed that the combination of habitat restoration and predator removal on isolated islands holds great promise for terrapin populations. Eliminating nest predation could transform population sinks into sources that could help restore terrapin populations on a wider regional scale through dispersal of high numbers of juveniles. On the other hand, terrapin-friendly shoreline restoration that concentrates nesting in areas with high predator density or without predator control might produce ecological traps that could accelerate population decline (Schlaepfer et al. 2002).

Both Jefferson Patterson Park and Poplar Island reveal how vegetation can affect nesting behavior. When first planted, vegetation was sparse and low in height, but over a period of 5 to 6 years it increased substantially in height and density. Much of what once was a continuous open nesting habitat at both sites has now become an area of dense vegetation, primarily *S. patens* and *Panicum virgatum*, eliminating many open sandy areas that previously were attractive nesting sites. Most of the nesting in the Park occurs in areas where open beach is maintained, at the northern and southern ends of the site (Fig. 18.2, A, B). Nesting in areas behind the riprap occurred in years immediately after completion, but the vegetation spread quickly

and occluded open nesting sites. On Poplar Island, nesting activity has shifted from areas that were open in 2008 but are now heavily vegetated to open areas that persisted due to windblown movement of sand (Roosenburg, pers. obs.). Clowes (2013) conducted a vegetation removal experiment on Poplar Island, revealing higher nesting activity in open areas. Her observations suggest that vegetation influences terrapin nesting behavior and that the restoration of shoreline habitat should create and maintain open areas. Shoreline stabilization eliminates the natural erosion and accretion that maintain the open sandy areas that both attract terrapins to nesting sites and are where they prefer to nest. Therefore, shoreline restoration with terrapins as a target species should maintain open nesting areas. We created open areas using mowing and rototilling (Clowes 2013), but this is a labor-intensive process that requires critical timing—after all overwintering hatchlings have emerged (Kitson 2016) and before the subsequent nesting season begins. At Poplar Island, this window is less than 2 weeks in mid-May and thus requires careful monitoring of terrapin behavior.

Head-starting

Head-starting involves collecting eggs or hatchlings from the wild and raising the young to a larger body size in captivity to increase juvenile survival. Many see it as a panacea for turtle conservation (see Burke 2015). It creates a "hands on" satisfaction for conservationists, but frequently its effectiveness remains untested. Millions of dollars have been invested in sea turtle head-starting programs that remain underevaluated in terms of their effect on population growth or an economic cost-benefit analysis weighing effectiveness relative to conservation achievements. Early criticisms of head-starting as "halfway technology" (Frazer 1992) warned of treating the symptom of population decline without treating the cause. Additional criticisms include genetic contamination resulting from translocation, disease introduction, and behavioral modifications during captivity (Seigel and Dodd 2000). Long-term, multicohort, demographic studies of turtle head-start survival are limited.

Radio-telemetry documented high post-release survivorship in wood turtles (Michell and Michell 2015) and western pond turtles (Vander Haegen et al. 2009). However, post-release survival varied dramatically among sites and between years for gopher tortoises (Tuberville et al. 2015). Furthermore, chelonian biologists have yet to establish a consensus on an appropriate metric by which to evaluate success vs. failure of a head-starting program. For example, the western pond turtle program that accelerated growth also may have contributed to a 29% to 49% prevalence of ulcerative shell disease (WDFW 2016).

Head-starting, when implemented with a well-delivered education program (Chapter 17), can have a significant conservation impact by raising public awareness of the threats facing turtle populations and by promoting alternative management and conservation strategies. Most turtle species are "charismatic macrofauna," which helps solicit public sympathy for their cause; this has helped mitigate the plight of the terrapin. For example, in Maryland, closing the commercial fishery for terrapins was easily accomplished (Roosenburg et al. 2008), whereas the snapping turtle harvest remains open despite a similar life history and management concerns. Still, head-start programs where the public is broadly involved can generate social awareness and thus political attention to the fate and conservation needs of turtles. Successful legislation closing Maryland's commercial terrapin harvest greatly benefited from the large-scale education programs that head-started terrapins in local schools. Similarly, closing New Jersey's commercial terrapin fishery was probably facilitated by a long-term education program with terrapin head-starting (Wood and Herlands 1997; Herlands et al. 2004). The indirect benefit of head-starting with broad public involvement cannot be overstated. However, many pitfalls for head-starting remain (Seigel and Dodd 2000). Here, I compare head-started vs. released hatchlings to evaluate head-starting as a conservation tool and provide preliminary findings on life history traits key to population restoration.

Beginning in 2005, I collected 160 to 250 hatchlings annually from natural nests on Poplar Island, and, with instruction, these hatchlings were raised in classrooms (kindergarten–12th grade) during the academic year as part of an environmental education curriculum (Chapter 17). Before distribution to the classrooms, all hatchlings were uniquely notched in their marginal scutes (Cagle 1939) and internally injected with a binary coded wire tag (BCWT; Chapter 2). All hatchlings not inducted into the head-start program were marked similarly and released on Poplar Island to establish a control (wild-released) group. The following spring, all head-starts were PIT tagged (Biomark tags) and measured before release on Poplar Island in the late spring and early summer (8–9 months in captivity). In 2008, mark-recapture began on the island to establish a long-term study of head-started vs. released hatchlings. Because of the high nest survival rates and the large numbers of hatchlings produced, we have large sample sizes with which to compare demographic metrics of wild-released vs. head-started individuals. Thus, we can evaluate the merits of head-starting as a turtle conservation strategy in a population context to evaluate how demographic parameters are affected.

By the end of 2015, we had released 2,171 head-started and 12,426 wild-released hatchlings. We have recaptured 232 head-started and 415 released hatchlings. Our objectives are to compare the population of head-starts with the population of released hatchlings, including (1) evaluating the optimal size of a head-started individual, (2) determining whether age of first reproduction in females is affected by the accelerated growth of head-starting, and (3) determining the survival rate of head-started vs. released individuals. Here, I address questions 1 and 2 and briefly discuss question 3, based solely on recapture statistics. A more thorough analysis of this data set is forthcoming.

Head-starts when released were larger (plastron length [PL] mean = 72.3 mm, SD = 14.7) than individuals released at time of hatching (PL mean = 28.2 mm, SD = 2.0). Head-starts' size at time of release did not differ between all head-starts and those recaptured in subsequent years (Fig. 18.3; t-test, $t = 1.085$; $P > 0.05$; variance ratio = 1.058). Thus, maximizing the size of the head-start to a minimal or larger size confers no greater survival advantage.

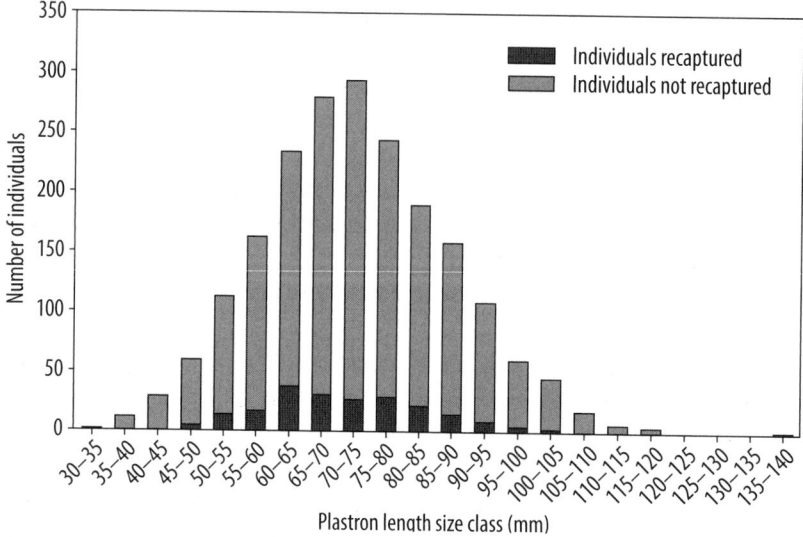

Fig. 18.3. Distribution of terrapins by size class at the time of completing the head-start period, for individuals recaptured and those not recaptured.

The head-starts' recapture rate after release (0.107, combined catch and survival probability) was greater than the wild-released hatchlings' recapture rate (0.033), indicating that head-starts have a threefold higher combined catch and survival probability to age > 3 years and suggesting a benefit of head-starting. Increased survival may result from potential size-related predator immunity, although we have no data or observations to support this claim. We believe that the primary predators of hatchling and juvenile terrapins on Poplar Island are birds. The island supports nesting colonies of a variety of gulls, herons, and egrets and a large double-crested cormorant rookery. Many of the juvenile terrapins that we recover dead have carapace punctures suggesting avian predation. We suggest that the head-starting process increases juvenile survival via an increase in body size that provides immunity to many of these gape-limited avian predators, and this size advantage results in an increase in survival relative to the wild-released hatchlings.

Our data suggest, then, that head-starts have higher survival than released hatchlings. However, one of the pitfalls of head-starting is demographic stochasticity (Morris and Doak 2002), which estimates the probability that supplementation of the population will be successful. Because we have a preliminary estimate of head-starts' survival rate, we estimate the survival probability of two individuals as $(0.107)^2 = 0.011$ that both individuals will survive, while the probability that both will die is $(1 - 0.107)^2 = 0.797$. Similarly, the probability that one of the two individuals will survive is 0.191. Because our metric of survival is based on a combined capture and survival probability, actual values for survival are likely to be greater; nonetheless, the consequences of demographic stochasticity remain comparable. These calculations indicate that if head-starting were to be used as a tool to reestablish extirpated terrapin populations, large numbers would need to be released to guarantee successful restoration with an effective or sustainable population size. The combined survival and mortality probabilities quickly reveal subtle demographic nuances that are frequently overlooked when initiating head-starting programs. We are unaware of studies of demographic stochasticity before evaluating the need for head-starting or considering the magnitude of supplementation needed to bolster existing populations or establish new ones. The goal would be to determine *a priori* what the minimum number of individuals released into a population should be given demographic stochasticity.

Accelerating growth can also affect the timing of maturity or first reproduction, which can dramatically affect population growth (Cole 1954). The earliest finding on first reproduction for released hatchlings on Poplar Island was a single individual aged 7 years and several gravid 8-year-olds, whereas we now have data on numerous head-starts gravid at 6 years. However, numerous head-starts recaptured beyond these ages were not yet reproductive. Maturation age in terrapins from other areas in Chesapeake Bay ranges from 8 to 13 years (Roosenburg 1994). Still, our data suggest that head-starting can reduce the time to maturity in some individuals and thereby reduce generation time. The data are not yet thorough enough to evaluate what proportion of the head-starts have accelerated maturity, but given the range in maturity among wild individuals, we may never be able to identify such an effect. However, if on average we are accelerating the age of first reproduction for head-starts, this could have a substantial effect on the population rate of growth that is potentially greater than the increase in survivorship (Cole 1954).

Head-starts' behavior is yet another trait that can be affected through growth acceleration, particularly if behavioral development is age and not size dependent. For example, we have recaptured wild hatchlings and 2-year-old terrapins under the wrack and in salt marshes. We also have captured head-starts within their first year of release in these habitats instead of in aquatic habitats where their wild size-class counterparts are captured (Roosenburg et al. 1999). We frequently observe that, on release, head-starts move into the semi-terrestrial salt marsh habitat instead of remaining in the open waters where they are placed. Although in other areas of the terrapin's range, larger size classes are found in salt marshes (Tucker et al. 1995), in Chesapeake Bay they are more aquatic (Roosenburg et al. 1999). Clearly, additional work to evaluate ontogenetic behavioral and habitat shifts is needed. We have attempted to use both radio and sonic telemetry to track head-started terrapins to understand their behavior after release. Unfortunately, this endeavor has met with limited success because most individuals could not be relocated after their first post-release night. Our meager observations suggest, however, that there may be age- or experience-related effects in the behavioral development of terrapins. These findings raise interesting questions that require additional research and novel tracking technologies that can provide precise location data in both terrestrial and aquatic estuarine environments.

In summary, our preliminary results imply that head-starting can have a positive effect on population growth rate. The increased recapture rate and acceleration in age of first reproduction of head-starts suggest a positive demographic effect, but this requires further evaluation. Based on our recapture rates, calculation of demographic stochasticity effects indicates that the founding of populations using head-starts requires large numbers of individuals. Further, the cost and time required to head-start >200 individuals per year demands that the practice be carefully evaluated. We estimate expenses for setting up and administering the program between $150,000 and $200,000 per year, which is based mostly on the time of those involved. Time spent raising and caring for turtles and running the education component is not included in this accounting, and teachers contribute significant amounts of time to maintaining turtles in their classrooms. Fortunately, the costs are distributed across and contributed by a number of agencies and volunteers, which minimizes the fiscal burden on a single entity or program participant. But targeted head-start programs not linked to an education program should be prepared to sustain costs of this magnitude for several years to provide the number of animals necessary to overcome demographic stochasticity.

Conclusions and Recommendations for the Future

The rebuilding of Poplar Island's terrapin population exemplifies a model conservation/restoration program combining several factors that simultaneously benefited growth of the population, including high-quality raccoon-free nesting areas, increased wetland habitat for juveniles, and a head-start program. High hatchling recruitment fueled by the absence of nest predators is the single most effective measure that has resulted in population growth on Poplar Island (Roosenburg et al. 2014). Further, on Poplar Island the recruits have available 115 ha of newly created wetlands that provide high-quality habitat. Because of high nest survivorship, the island's terrapin population did not need to be supplemented to achieve positive growth, given the high recruitment. The island provided the ideal system to experimentally evaluate the potential merits of head-starting, establishing that, in conjunction with other protective measures, this may be an effective tool for resource managers in areas where terrapins have been extirpated. The Jefferson Patterson Park example further demonstrates that newly constructed habitat can be located and used by terrapins, but nest predators have an overwhelming effect on terrapin recruitment.

How can we use the example of Poplar Island as a model for terrapin restoration? Poplar Island is a successful terrapin restoration effort and confirms the value of the holistic ecosystem approach strategy to turtle conservation advocated by Klemens (2000). Our observations also reinforce the conclusions (Frazer 1992; Seigel and Dodd 2000) that a comprehensive conservation and restoration strategy should address first the problems and then the symptoms to ensure successful recovery of turtle populations. In fact, Poplar Island illustrates that solving the problems eliminates the need to address the symptoms. One interesting prospect that emerges from the Poplar Island example is that high recruitment can contribute directly to population growth, while most models of turtle conservation focus on high adult survival (Congdon et al. 1993, 1994; Heppell 1998). Adult survival remains paramount, but high recruitment in small populations can substantially contribute to population recovery. Our findings suggest that small numbers of adults can replenish a population when high-quality juvenile habitat and predator-free nesting areas are available. Populations such as these could serve as sources for the recolonization of other areas via natural dispersal from high-recruitment areas. Offshore islands with high-quality habitat may be ideal for terrapins because predator control is facilitated by long-range dispersal barriers; islands could serve as excellent estuarine sanctuaries or marine reserves (Halpern and Warner 2002; Halpern 2003) where terrapin populations can increase, more so than in mainland areas where nest predation is overwhelming.

Such reserves could potentially function as source populations from which terrapins could disperse and repopulate depleted areas. This strategy would require removal of potential terrapin threats such as crab pots to promote dispersal, as well as the patience to allow restoration to move forward at a turtle's pace. Although we have not trapped the nearby mainland to determine dispersal rates from Poplar Island, this certainly will be an interesting project for the future.

Restoring and preserving turtle populations remains an emotionally charged topic with often rancorous debate on the myriad opinions about the most appropriate ways to proceed and how to measure success. I suggest that the implementation of terrapin restoration should follow two simple guidelines. First, establish a clear and quantifiable metric that can be used to evaluate the success or failure of the program with a concrete *a priori* goal. Second, implement the restoration project so that its success can be evaluated in an objective, scientific study. A meaningful evaluation requires adequate control groups, sample sizes large enough to provide rigorous scientific comparisons, and long-term data to overcome the effects of demographic and environmental stochasticity. This level of scrutiny and self-evaluation should not be reserved for academics but should also be expected from all parties designing and implementing restoration until best management practices are identified. Such an approach would pinpoint the most cost-effective restoration strategies and prevent the duplication of ineffective techniques. In the midst of a biodiversity crisis, it is imperative that conservationists and resource managers be critical of restoration techniques so that limited funds can be appropriately allocated to effective solutions.

REFERENCES

Baker, J.P., J.P. Costanzo, R. Herlands, R.C. Woods and R.E. Lee Jr. 2006. Inoculative freezing promotes winter survival in the diamondback terrapin, *Malaclemys terrapin*. Canadian Journal of Zoology 84:116–124.

Brennessel, B.A. 2006. Diamonds in the Marsh: A Natural History of the Diamondback Terrapin. University Press of New England, Hanover, NH, and London. 236 pages.

Burke, R.L. 2015. Head-starting turtles: learning from experience. Herpetological Conservation and Biology 10:299-308.

Cagle, F.R. 1939. A system for marking turtles for future identification. Copeia 1939:170-173.

Clowes, E.L. 2013. Influence of vegetation on northern diamondback terrapin (*Malaclemys terrapin terrapin*) nesting. Honors tutorial thesis, Ohio University, Athens, OH. 86 pages.

Cole, L.C. 1954. The population consequences of life history phenomena. Quarterly Review of Biology 2:103-107.

Congdon, J.D., A.E. Dunham and R.C. van Loben Sels. 1993. Delayed sexual maturity and demographics of Blanding's turtles (*Emydoidea blandingii*): implications for conservation and management of long-lived organisms. Conservation Biology 7:826-833.

Congdon, J.D., A.E. Dunham and R.C. van Loben Sels. 1994. Demographics of common snapping turtles (*Chelydra serpentina*): implications for conservation and management of long-lived organisms. American Zoologist 34:397-408.

Crawford, B.A., J.C. Maerz, N.P. Nibblelink, K.A. Buhlmann, T.M. Norton and S.E. Albeke. 2014. Hot spots and hot moments of diamondback terrapin road-crossing activity. Journal of Applied Ecology 51:367-375. DOI 10.1111/1365-2664.12195.

Davis, J.L., R.L. Takacs and R. Schnabel. 2006. Evaluating ecological impacts of living shorelines and shoreline habitat elements: an example from the upper western Chesapeake Bay. Pages 55-61 in S.Y. Erdle, J.L.D. Davis and K.G. Sellner (eds.), Management, Policy, Science, and Engineering of Nonstructural Erosion Control in the Chesapeake Bay. Chesapeake Research Consortium Pub. no. 08-164. Chesapeake Research Consortium, Solomons, MD.

Draud, M., M. Bossert and S. Zimnavoda. 2004. Predation on hatchling and juvenile diamondback terrapins (*Malaclemys terrapin*) by the Norway rat (*Rattus norvegicus*). Journal of Herpetology 38:467-470.

Duhring, K.A. 2006. Overview of living shoreline options. Pages 13-18 in S.Y. Erdle, J.L.D. Davis and K.G. Sellner (eds.), Management, Policy, Science, and Engineering of Nonstructural Erosion Control in the Chesapeake Bay. Chesapeake Research Consortium Pub. no. 08-164. Chesapeake Research Consortium, Solomons, MD.

Frazer, N.B. 1992. Sea turtle conservation and halfway technology. Conservation Biology 6:179-184.

Frazer, N.B. 1997. Turtle conservation and halfway technology: what is the problem? Pages 422-425 in J. Van Abbema (ed.), Proceedings: Conservation, Restoration, and Management of Tortoises and Turtles—An International Conference. New York Turtle and Tortoise Society, State University of New York, Purchase, NY.

Halpern, B.S. 2003. The impact of marine reserves: do reserves work and does reserve size matter? Ecological Applications 13:S117-S137.

Halpern, B.S. and R.W. Warner. 2002. Marine reserves have rapid and lasting effects. Ecology Letters 3:361-366.

Hartig, E.K., V. Gornitz, A. Kolker, F. Mushacke and D. Fallon. 2002. Anthropogenic and climate-change impacts on salt marshes of Jamaica Bay, New York City. Wetlands 22:71-89.

Heppell, S.S. 1998. Application of life-history theory and population model analysis to turtle conservation. Copeia 1998:367-375.

Herlands, R., R. Wood, J. Pritchard, H. Clapp and N. Le Furge. 2004. Diamondback terrapins (*Malaclemys terrapin*) head-starting project in southern New Jersey. Pages 13-18 in C.W. Swarth, W.M. Roosenburg and E. Kiviat (eds.), Conservation and Ecology of Turtles of the Mid-Atlantic Region: A Symposium. Bibliomania, Salt Lake City, UT.

Kitson, S.R. 2016. *Malaclemys terrapin* hatchlings: variation in seasonal emergence. MS thesis, Ohio University, Athens, OH. 80 pages.

Klemens, M.W. 2000. From information to action: developing more effective strategies to conserve turtles. Pages 239-258 in M.W. Klemens (ed.), Turtle Conservation. Smithsonian Institution Press, Washington, DC.

Michell, K. and R.G. Michell. 2015. Use of radio-telemetry and recapture to determine the success of head-started wood turtles (*Glyptemys insculpta*) in New York. Herpetological Conservation and Biology 10:525-534.

Mitchell, J.C. and M.W. Klemens. 2000. Primary and secondary effects of habitat alteration. Pages 5-32 in M.W. Klemens (ed.), Turtle Conservation. Smithsonian Institution Press, Washington, DC.

Morris, W.F. and D.F. Doak. 2002. Quantitative Conservation Biology: Theory and Practice of Population Viability Analysis. Sinauer Associates, Sunderland, MA. 480 pages.

Muldoon, K.A. and R.A. Burke. 2012. Movements, overwintering, and mortality of hatchling diamond-backed terrapins (*Malaclemys terrapin*) at Jamaica Bay, New York. Canadian Journal of Zoology 90:651-662.

Munscher, E.C., E.H. Kuhns, C.C. Cox and J.A. Butler. 2012. Decreased nest mortality for the Carolina diamondback terrapin (*Malaclemys terrapin centrata*) following the removal of raccoons (*Procyon lotor*) from a nesting beach in northeastern Florida. Herpetological Conservation and Biology 7:167-184.

Najjar, R.G., H.A. Walker, P.J. Anderson, E.J. Barron, R.J. Bord, J.R. Gibson, V.S. Kennedy, C.G. Knight, J.P. Megonigal, R.E. O'Connor, C.D. Polsky, N.P. Psuty,

B.A. Richards, L.G. Sorenson, E.M. Steele and R.S. Swanson. 2000. The potential impacts of climate change on the mid-Atlantic coastal region. Climate Research 14:219-233.

Ratnaswamy, M.J. and R.J. Warren. 1998. Removing raccoons to protect sea turtle nests: are there implications for ecosystem management? Wildlife Society Bulletin 26:846-850.

Ratnaswamy, M.J., R.J. Warren, M.T. Kramer and M.D. Adam. 1997. Comparisons of lethal and nonlethal techniques to reduce raccoon depredation of sea turtle nests. Journal of Wildlife Management 61:368-376.

Ritchie, E.G. and C.N. Johnson. 2009. Predator interactions, mesopredator release and biodiversity conservation. Ecology Letters 9:982-998.

Roosenburg, W.M. 1991. The diamondback terrapin: habitat requirements, population dynamics, and opportunities for conservation. Pages 237-234 in J.A. Mihursky and A. Chaney (eds.), New Perspectives in the Chesapeake System: A Research and Management and Partnership. Proceedings of a conference. Chesapeake Research Consortium Pub. no. 137. Chesapeake Research Consortium, Solomons, MD.

Roosenburg, W.M. 1994. Nesting habitat requirements of the diamondback terrapin: a geographic comparison. Wetlands Journal 6(2):9-12.

Roosenburg, W.M., J. Cover and P.P. van Dijk. 2008. Legal issues: legislative closure of the Maryland terrapin fishery: perspectives on a historical accomplishment. Turtle and Tortoise Newsletter 12:27-30.

Roosenburg, W.M., K.L. Haley and S. McGuire 1999. Habitat selection and movements of diamondback terrapins, Malaclemys terrapin, in a Maryland estuary. Chelonian Conservation and Biology 3:425-429.

Roosenburg, W.M. and A.R. Place. 1995. Nest predation and hatchling sex ratio in the diamondback terrapin: implications for management and conservation. Pages 65-70 in P. Hill and S. Nelson (eds.), Toward a Sustainable Coastal Watershed: The Chesapeake Experiment. Proceedings of a conference. Chesapeake Research Consortium Pub. no 149. Chesapeake Research Consortium, Solomons, MD.

Roosenburg, W.M., D.M. Spontak, S.P. Sullivan, E.L. Matthews, M.L. Heckman, R.J. Trimbath, R.P. Dunn, E.A. Dustman, L. Smith and L.J. Graham. 2014. Nesting habitat creation enhances recruitment in a predator free environment: Malaclemys nesting at the Paul S. Sarbanes ecosystem restoration project. Restoration Ecology 22:815-823.

Schlaepfer, M.A., M.C. Runge and P.W. Sherman. 2002. Ecological and evolutionary traps. Trends in Ecology and Evolution 17:474-480.

Seigel, R.A. and C.K. Dodd. 2000. Manipulation of turtle populations for conservation: halfway technologies or viable options. Pages 218-238 in M.W. Klemens (ed.), Turtle Conservation. Smithsonian Institution Press, Washington, DC.

Seigel, R.A. and J.W. Gibbons. 1995. Workshop on the ecology, status, and management of the diamondback terrapin (Malaclemys terrapin) Savannah River Ecology Laboratory, 2 August 1994: final results and recommendations. Chelonian Conservation and Biology 1:241-243.

Simas, T., J.P. Nunes and J.G. Ferreira. 2001. Effects of global climate change on coastal salt marshes. Ecological Modelling 139:1-15.

Thorbjarnarson, J., C.J. Lagueux, D. Bolze, M.W. Klemens and A.B. Meylan. 2000. Human use of turtles: a worldwide perspective. Pages 33-84 in M.W. Klemens (ed.), Turtle Conservation. Smithsonian Institution Press, Washington, DC.

Tuberville, T.D., T.M. Horton, K.A. Buhlman and V. Greco. 2015. Head-starting as a management component for gopher tortoises (Gopherus polyphemus). Herpetological Conservation and Biology 10:455-471.

Tucker, A D., N.N. Fitzsimmons and J.W. Gibbons. 1995. Resource partitioning by the estuarine turtle Malaclemys terrapin: trophic, spatial and temporal foraging constraints. Herpetologica 51:167-181

Vander Haegen, W.M., S.L. Clark, K.M. Perillo, D.A. Anderson and H.L. Allen. 2009. Survival and causes of mortality of head-started western pond turtles on Pierce National Wildlife Refuge, Washington. Journal of Wildlife Management 73:1402-1406.

WDFW (Washington Department of Fish and Wildlife). 2016. Western pond turtle shell disease in Washington. http://wdfw.wa.gov/conservation/western_pond_turtle/western_pond_turtle_shell_disease_2pager_9august2016.pdf.

Winters, J.M., H.W. Avery, E.A. Standora and J.R. Spotila. 2015. Between the bay and a hard place: altered diamondback terrapin nesting movements demonstrate the effects of coastal barriers upon estuarine wildlife. Journal of Wildlife Management 79:682-688.

Wnek, J.P., W.F. Bien and H.W. Avery. 2013. Artificial nesting habitats as a conservation strategy for turtle populations experiencing global change. Integrative Zoology 8:209-213.

Wood, R.C. and R. Herlands. 1997. Turtles and tires: the impact of roadkills on northern diamondback terrapin, Malaclemys terrapin terrapin, populations on the Cape May Peninsula, southern New Jersey, USA. Pages 46-53 in J. V. Abbema (ed.), Proceedings: Conservation, Restoration, and Management of Tortoises and Turtles—An International Conference. New York Turtle and Tortoise Society, State University of New York, Purchase, NY.

19

The Future for Diamond-backed Terrapins

JOSEPH A. BUTLER
WILLEM M. ROOSENBURG

Despite a growing wealth of scientific knowledge on diamond-backed terrapins, public knowledge and understanding of this turtle remain limited. Nonetheless, the anthropogenic threats to terrapins' existence continue, and their future survival is tenuous in some locales. Our ability to implement appropriate conservation and management strategies is frequently hampered by political wrangling associated with shoreline development and commercial fishing. However, as is clear from the discussions and the cited references in this book, there are and have been many scientists, researchers, government agencies, resource managers, naturalists, and members of the general public who have discovered these turtles and their habits and have become fascinated by them. Many of these individuals have dedicated themselves to preserving and improving terrapin populations both locally and throughout their range.

In 1994, a group of about 20 turtle biologists who studied or had an interest in diamond-backed terrapins met at the Savannah River Ecology Laboratory, University of Georgia. Assembled by Whit Gibbons and Rich Seigel, we discussed the future and fate of diamond-backed terrapins. Since that initial meeting, dubbed Workshop on the Ecology, Status, and Conservation of Diamond-back Terrapins (Seigel and Gibbons 1995), this core group has met regularly and expanded to more than 80 active terrapin researchers and managers. At the third workshop, in 2004, a group of nearly 80 attendees voted to establish the Diamondback Terrapin Working Group (DTWG; http://www.dtwg.org) with the purpose of promoting terrapin conservation and the preservation of intact, wild terrapin populations and their associated ecosystems throughout their range. We started the nonprofit group with the goal of maintaining communication and interactions among researchers and state wildlife personnel. Our objectives were to enhance terrapin populations and their habitats while increasing our knowledge of terrapin ecology and promoting sound, scientifically based conservation strategies.

These objectives are accomplished partly through triennial symposia, where projects from throughout the terrapins' range are presented and members can become more acquainted, establish collaborations, and learn from one another about the issues that face terrapins. Because the terrapin has an extensive geographic range spanning 16 states, five regional groups (Gulf Coast, Florida, Southeast, mid-Atlantic, and Northeast) were established to discuss local conservation issues and promote management targeted to the specific regional needs. This is accomplished through regular meetings organized and spearheaded by regional representatives elected through the group.

A major accomplishment of the DTWG since its inception includes support of legislation banning the commercial harvest of diamond-backed terrapins in Maryland, New Jersey, Virginia, North Carolina, South Carolina, and, during the writing of this book, New Jersey and New York. Louisiana remains the only state that allows a regulated commercial terrapin harvest. The working group will continue to pursue "no harvest" regulations. The successful proposals to place diamond-backed terrapins on CITES (Convention on International Trade in Endangered Species of Wild Fauna and Flora) Appendix II and the IUCN (International Union for Conservation of Nature) revision of the Red List Assessment of terrapins were organized and written as collaborative efforts by members of the group. Furthermore, each year the working group distributes all monies collected from membership dues and gifts to fund projects on terrapin research and conservation. The working group provides modest research grants to working group members, primarily students, and over the years has awarded more than $25,000. We have supported four to six such studies each year since 2004, with topics as varied as genetic diversity, head-starting, feeding ecology, mark-recapture, crab pot mortality, and skeletochronology, many of

which are described in this book. Thanks to the attention and funding derived through the working group, terrapin populations are now being studied or monitored in all 16 states throughout their range, many funded through the group, which was not the case before 2004.

Terrapin mortality in crab pots continues to be one of the major threats to terrapins and has been studied in nearly every state in the range (Chapter 16). The DTWG has supported research and advocated for the implementation of bycatch reduction devices (BRDs), which provide an inexpensive and effective solution to terrapin mortality in crab pots. Four states have adopted regulations requiring BRDs, but unfortunately, compliance remains low (Chapter 16). Furthermore, the politically powerful crab fishery vehemently opposes BRDs and presents a major obstacle to adopting regulations or legislation requiring limited BRD use in states where crab pots continue to be a conservation threat to terrapins.

Perhaps another approach is necessary to protect terrapins from crab pots. In some regions, terrapins live in narrow (30 m wide) tidal creeks, and regulations prohibiting crabbing within some distance, e.g. 50 m, from shore may protect those populations. Maryland's spatial restrictions on commercial use of crab pots provide a working example. Depth restrictions on the use of crab pots might also improve terrapin protection by eliminating pots from water less than 4 m deep. Given the ubiquitous problem of crab pot mortality and the widespread overlap between terrapins and the blue crab fishery, perhaps it is time for federal regulations requiring BRDs on crab pots, instead of progressing one state at a time. The variation in successful BRD dimensions throughout the terrapins' range warrants additional work to develop a universal BRD, and Mike Arendt of South Carolina Department of Natural Resources is spearheading range-wide efforts to develop and test alternative sizes and dimensions. The DTWG continues to direct funds and energy to the problems of terrapins and crab pot mortality and also encourages collection of abandoned crab pots (ghost pots), which has helped justify many states' establishment of active ghost pot removal programs.

There are ample opportunities for research and discovery in our understanding of diamond-backed terrapins, and many biologists are drawn to this charismatic species. Conflict remains in the evidence from molecular and morphological data on the seven subspecies designations of terrapins (Chapters 4 and 5). Recently, researchers at the University of Maryland (M. Pop, S. Mount, and M. Cummings, unpub. data) have sequenced the terrapin genome and developed a transcriptomics library for the species.

All of the current molecular work on terrapins has been based on two sets of primers, one specific to the terrapin and the other developed for the bog turtle. These new sequence data can identify a sorely needed set of new markers and polymorphic loci that can be used to understand both the population genetics and gene flow in terrapins. Additionally, this information can generate a character set that can identify and generate a well-supported subspecific phylogeny. Philopatry and its effect on phylogeography of terrapins, particularly since the last glaciation, also remains an interesting question, although the movement of turtles during the early years of the terrapin trade may have obscured the genetic signatures left by evolutionary history (Chapter 5). The use of transcriptomics provides a potentially powerful tool to continue the exploration of interactions between genetic and environmental mechanisms of gender determination to further develop the terrapin as a model for temperature-dependent sex determination. Furthermore, the development of new molecular markers could be used to explore and validate genetic techniques vs. traditional mark-recapture to understand mating dynamics, effective population size, and population structure. Such validated quantitative genetic tools potentially could allow researchers to evaluate population dynamics cost effectively relative to the funding of long-term demographic studies.

Significant gaps remain in the basic natural history of terrapin populations throughout their range. Although our knowledge of terrapins and the number of researchers studying them have increased exponentially in the past 20 years, parts of the terrapin's range are still lacking in basic data

necessary for the development of sound local conservation strategies. For example, eastern Louisiana, the Florida Gulf Coast, and Pamlico Sound in North Carolina are regions where limited life history data are available. One of the goals of this book is to provide standardized techniques (Chapter 2) so that data will be comparable among populations. Although geographic variation has been a focus in some comparative studies (Chapter 6), teasing out the effects of salinity and its interaction with latitude has never been attempted. Another interesting topic is geographic comparisons of head morphology and size in relation to prey availability and hardness, which as yet has received little detailed study. The opportunities for compelling terrapin research are readily available.

Finally, conservation remains a paramount objective throughout the terrapin's range. Terrapins exemplify the turtle life history phenotype of delayed reproduction, low nest/juvenile survival, and longevity that is indicative of species in which increased adult mortality results in rapid population decline. Unfortunately, the waterfront habitat of terrapins is immensely popular for development, and interactions with commercial fisheries often impose political and economic challenges that complicate terrapin management and conservation. Perhaps estuarine reserves and managed restoration sites such as Poplar Island (Chapter 18), in which resources and management strategies use a multispecies rather than a single species approach focusing on the terrapin alone, would offer a better opportunity for sustaining populations. Combining forces with those interested in shorebirds, horseshoe crabs, and other sensitive estuarine organisms could result in a pooling of resources and expertise in an ecosystem approach, as opposed to tackling each species independently. Terrapins, in this capacity, may be an excellent indicator of general ecosystem decline.

Despite continued concern over the well-being of terrapins, we are uplifted by the increasing awareness and efforts of many to protect and conserve these animals. Education, research, and science-based management have identified problems and solutions that have greatly improved conservation since our original meeting in 1994. We feel confident that the charismatic nature of the diamond-backed terrapin will attract top-flight educators, researchers, and conservationists who will continue to improve the outlook for terrapins so that they remain natural treasures for future generations.

REFERENCES

Seigel, R.A. and J.W. Gibbons. 1995. Workshop on the ecology, status, and management of the diamondback terrapin (*Malaclemys terrapin*), Savannah River Ecology Laboratory, 2 August 1994: final results and recommendations. Chelonian Conservation and Biology 1:240-243.

Contributors

Benjamin K. Atkinson, PhD
Department of Natural Sciences
Flagler College
St. Augustine, FL 32084

Harold W. Avery, PhD
Department of Biology
The College of New Jersey
Ewing, NJ 08628-0718

Patrick J. Baker, PhD
Department of Geography and
    Environmental Engineering
United States Military Academy
West Point, NY 10996

Ralph E. J. Boerner, PhD
3304 Bybrook Drive
Cape May Beach, NJ 08251

Russell L. Burke, PhD
Biology Department
Hofstra University
Hempstead, NY 11549

Joseph A. Butler, PhD
Department of Biology
University of North Florida
Jacksonville, FL 32224

Randolph M. Chambers, PhD
Keck Environmental Lab College
    of William and Mary
Williamsburg, VA 23187

Paul E. Converse, PhD
Department of Biological Sciences
Ohio University
Athens, OH 45701

Brian A. Crawford, PhD
D. B. Warnell School of Forestry
    and Natural Resources
University of Georgia
Athens, GA 30602

Rusty D. Day, PhD
Marine Science and Nautical
    Training Academy (MANTA)
417 Planters Trace Drive
Charleston, SC 29412

Dana J. Ehret, PhD
New Jersey State Museum
205 West State Street
Trenton, NJ 08625-0530

J. Whitfield Gibbons, PhD
Savannah River Ecology
    Laboratory
University of Georgia,
    Drawer E
Aiken, SC 29802

Kathryn M. Greene, BS
Department of Biology
Davidson College
Davidson, NC 28035-7118

Leigh Anne Harden, PhD
College of Science
Benedictine University
5700 College Road
Lisle, IL 60532

Andrew S. Harrison, MS
Williamsville Central School
    District
105 Casey Road
East Amherst, NY 14051

Kristen M. Hart, PhD
USGS Wetland and Aquatic
    Research Center
3321 College Avenue
Davie, FL 33314-7799

George L. Heinrich, BPS
Heinrich Ecological Services
1213 Alhambra Way S.
St. Petersburg, FL 33705-4620

Dawn K. Holliday, PhD
Department of Biology and
    Environmental Science
Westminster College
Fulton, MO 65251

Victor S. Kennedy, PhD
UMCES Chesapeake Biological
    Laboratory
146 Williams Street
Solomons, MD 20688

Shawn R. Kuchta, PhD
Department of Biological Sciences
Ohio University
Athens, OH 45701

Lori A. Lester, PhD
Division of Science, Research &
    Environmental Health
New Jersey Department of
    Environmental Protection
Trenton, NJ 08608

Jeffrey E. Lovich, PhD
USGS Southwest Biological
    Science Center
2255 North Gemini Drive, MS-9394
Flagstaff, AZ 86001

John C. Maerz, PhD
D. B. Warnell School of Forestry
    and Natural Resources
University of Georgia
Athens, GA 30602

David Owens, PhD
Grice Marine Laboratory
University of Charleston, South
    Carolina, at the College of
    Charleston
Charleston, SC 29412

Allen R. Place, PhD
Institute of Marine and
    Environmental Technology
Columbus Center
701 East Pratt Street
Baltimore, MD 21202

Taylor Roberge, PhD
Department of Biology
University of Alabama at
    Birmingham
Birmingham, AL 35242-1170

Willem M. Roosenburg, PhD
Center for Ecology and Evolution-
    ary Studies
Department of Biological Sciences
Ohio University
Athens, OH 45701

Richard A. Seigel, PhD
Department of Biological Sciences
Towson University
Towson, MD 21252

Edward A. Standora, PhD
Department of Biology
SUNY Buffalo State
Buffalo, NY 14222

Anton D. Tucker, PhD
Department of Biodiversity,
    Conservation and Attractions
Marine Science Program
17 Dick Perry Avenue
Kensington, WA 6151, Australia

Diane C. Tulipani, PhD
Rappahannock Community
    College
12745 College Drive

Glenns, VA 23149

Timothy J. Walsh, MA
Bruce Museum
1 Museum Drive
Greenwich, CT 06830-7157

Thane Wibbels, PhD
Department of Biology
University of Alabama at
    Birmingham
1300 University Blvd
Birmingham, AL 35294-1170

Will Williams
Arlington Echo Outdoor
    Education Center
975 Indian Landing Road
Millersville, MD 21108

Amanda Southwood Williard,
    PhD
Department of Biology and Marine
    Biology
University of North Carolina
    Wilmington
601 South College Rd
Wilmington, NC 28403

Roger C. Wood, PhD
20 E. Mechanic Street
Cape May Court House,
    NJ 08210

Index

Pages in *italics* refer to figures or tables.